Serono Symposia USA
Norwell, Massachusetts

Springer

New York
Berlin
Heidelberg
Barcelona
Budapest
Hong Kong
London
Milan
Paris
Santa Clara
Singapore
Tokyo

PROCEEDINGS IN THE SERONO SYMPOSIA USA SERIES

Continued after Index

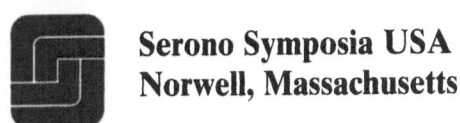

Serono Symposia USA
Norwell, Massachusetts

Clarence J. Gibbs, Jr.
Editor

Bovine Spongiform Encephalopathy

The BSE Dilemma

With 58 Figures

Springer

Clarence J. Gibbs, Jr., Ph.D.
Laboratory of Central Nervous System Studies
Basic Neurosciences Program
Division of Intramural Research
National Institute of Neurological Disorders and Stroke
National Institutes of Health
Bethesda, MD 20892-4122
USA

Proceedings of the Sixth International Workshop on Bovine Spongiform Encephalopathy: The BSE Dilemma, sponsored by Serono Symposia USA, Inc., and the National Institute of Neurological Disorders and Stroke of the National Institutes of Health, held February 26 to March 1, 1995, in Williamsburg, Virginia.

For information on previous volumes, please contact Serono Symposia USA, Inc.

Library of Congress Cataloging-in-Publication Data
Bovine spongiform encephalopathy: the BSE dilemma/[edited by]
  Clarence J. Gibbs, Jr.
     p. cm.—(Proceedings in the Serono Symposia)
  Proceedings of a seminar held at the National Institutes of Health
in December, 1994.
  Includes bibliographical references and index.

    1. Bovine spongiform encephalopathy—Congresses. I. Gibbs,
Clarence J., Jr., 1924– . II. Series.
QR201.B74B68   1996
616′.0194—dc20   96-13586

Printed on acid-free paper.

Production coordinated by Chernow Editorial Services, Inc., and managed by Francine McNeill; manufacturing supervised by Jacqui Ashri.
Typeset by Best-set Typesetter Ltd., Hong Kong.

9 8 7 6 5 4 3 2 1

ISBN-13: 978-1-4612-7527-5   e-ISBN-13: 978-1-4612-2406-8
DOI: 10.1007/ 978-1-4612-2406-8

SIXTH INTERNATIONAL WORKSHOP ON BOVINE SPONGIFORM ENCEPHALOPATHY: THE BSE DILEMMA

Scientific Committee

Clarence J. Gibbs, Jr., Ph.D.
Laboratory of Central Nervous System Studies
Basic Neurosciences Program
Division of Intramural Research
National Institute of Neurological Disorders and Stroke
National Institutes of Health
Bethesda, Maryland

Organizing Secretary

Leslie Nies
Serono Symposia USA, Inc.
100 Longwater Circle
Norwell, Massachusetts

This book is dedicated to Richard Masland, M.D., and the late Joseph E. Smadel, M.D., both of whom had the vision and wisdom to recognize in kuru the opening of vast new fields of knowledge and advances into the etiology and pathogenesis of degenerative diseases of the central nervous system.

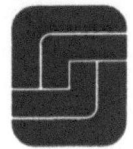

Foreword

The very first international working discussion on slow infections of the nervous system was entitled "Slow, Latent, and Temperate Virus Infections" and was held at the National Institutes of Health (NIH) in December 1964. The primary impetus was the discovery and investigation of kuru in New Guinea by D. Carleton Gajdusek, M.D. This working discussion brought together investigators in human and veterinary medicine, virologists, microbiologists, and neuropathologists actively engaged in laboratory work with viruses that illustrated properties of latency, masking, slowness, or temperateness, with emphasis on subacute and chronic neurologic diseases of unknown etiology.

In the Preface to the monograph of published papers presented at the working discussion, Gajdusek and Gibbs wrote the following:

After microbiology had given solution to the etiology of most acute infections of the central nervous system and after fungi and bacteria had been incriminated in important chronic disorders of the nervous system such as torula and tuberculosis meningitis, we have been left, in neurology, with a wide range of subacute and chronic affections of the central nervous systems of unknown etiology. Some of these diseases, still listed as idiopathic, are among the most prevalent afflictions of the central nervous system. Many others with familial patterns of occurrence do not yet have their basic pathogenesis or underlying metabolic defect elucidated, although we tend to think of them as genetically mediated. If any of these diseases fall into the category of slow, latent or temperate virus infections—if even a single one of these many syndromes of unknown etiology can be traced to a virus or virus-gene interaction we shall win significant advance in our understanding of diseases of the human brain.

Our subsequent demonstration of infection as the etiology of kuru, Creutzfeldt-Jakob disease (CJD), and Gerstmann-Sträussler-Scheinker syndrome (GSS), and most recently fatal familial insomnia (FFI), and the identification of genetic mutations in familial forms of CJD, GSS, and FFI, which are expressed in autosomal dominant forms in the three diseases, have substantiated the prophetic statement quoted above. Moreover, there is now an enormous compendium on the pathogenesis of what we have classified as the subacute spongiform encephalopathies as well as the biologic, physical, biochemical, and molecular biologic properties of the transmissible agent.

The sudden recognition of the epidemic of bovine spongiform encephalopathy (BSE) in the United Kingdom mandated concern for public health not only in the U.K. but in other industrial and agricultural nations of the world. It was this concern that led to convening the six international workshops on BSE and related animal and human diseases. The first, third, fourth, and fifth workshops were held at NIH, whereas the second met at the Central Veterinary Laboratory, Weybridge, U.K. Workshops I through V were cosponsored and financially supported by Professor C. Liana Bolis, M.D., Ph.D., Association Internationale pour le Recherche et l'Enseignement en Neurosciences (AIREN) and the National Institute of Neurological Disorders and Stroke (NINDS). The proceedings of the first and fifth workshops were published in the *Journal of the American Veterinary Medical Association* (Bolis, L. and Gibbs, C.J., Jr., 1990;196(10):1673; Gibbs, C.J., Jr., Bolis, C.L., Asher, D.M. et al., 1992;200(2):164–97).

This, the sixth international workshop, was cosponsored by Serono Symposia USA and the National Institute of Neurological Disorders and Stroke of the NIH. The participants are deeply grateful to Ms. Leslie Nies, Dr. Gerry Stiles, and the staff of Serono Symposia USA for their interest, encouragement, and financial support in making this workshop a successful scientific undertaking. Our appreciation is also extended to Ms. Devera Schoenberg, M.S., NINDS, who served as the technical editor of this publication; to my secretary, Ms. N. LaDonna Tavel, for working out all the administrative details that accompany an international workshop; and especially to my colleagues who made the workshop a resounding scientific success by their presentations, discussion, and manuscript submissions in a timely manner.

CLARENCE J. GIBBS, JR.

Preface

The outbreak of bovine spongiform encephalopathy (BSE) that started in 1985 in the United Kingdom has all the characteristics of an epidemic of infectious disease. Whether or not it is contagious, however, has remained a critical question. Whether or not it presents any danger to humans has been debated in countless meetings, newspapers, and correspondences. How effectively it has been confined to the U.K. or disseminated elsewhere remains uncertain. This volume addresses all of these questions and many more.

Since we are dealing with a problem of nucleation of conformational change in a normal host precursor by a "heteronucleant," namely, a different infectious protein from that produced in the infected cattle, presumably derived from sheep, the amino acid sequence of the infectious amyloid protein from the infected cattle will be that of the cow, not that of the sheep. The crucial question is this: What properties of the infectious nucleant have changed as a result of this breach of the species barrier? Have the properties of host range, target organs, brain regions, and route of infectability changed as occurs when sheep scrapie is passed into mink and ferrets? If this is the case, is the bovine agent now capable of causing disease in humans that the sheep agent presumably cannot do? Can the cattle—or humans—now be infected by the oral route? Is the agent now present in significant titer in muscle, milk, or even urine or saliva? Has it become communicable? Can it be passed vertically from an infected mother to her offspring? Will we be seeing early in the twenty-first century a sharp rise in Creutzfeldt-Jakob disease–Gerstmann-Sträussler-Scheinker syndrome–fatal familial insomnia (CJD-GSS-FFI) cases in the U.K. resulting from exposure to the BSE agent? These are the Andromeda strain–like hypotheses of countless journalists, and scientists cannot give a certain, definitive answer to these legitimate questions. This, then, is largely the reason for worldwide concern about the BSE outbreak in the U.K.

Recently, similar concerns have caused hysterical reactions in journalistic discussions of the long-known fact that many CJD patients have been blood donors during the silent incubation period of their disease. Does the agent ever appear in the peripheral blood leukocytes as it has in experimentally scrapie-infected mice? This is now a paramount concern of the blood banks and the pharmaceutical industry using bovine tissues and blood for the production of biologics.

The basic quantum dynamics of conformational change of the normal host precursors when nucleated to fold into the infectious cross-beta-pleated conformation of lower free energy—the critical conversion of the host precursor into the infectious amyloid protein—is now the focus of concerted attention in several laboratories. Can the nucleated conversion to infectious protein be accomplished in vitro even from synthetic polypeptides? What conformation differences determine host range, route of infectability, incubation period, targeting, and other aspects of virulence pathogenesis of different strains? The answers to these questions may be a long way off as they are for all other microbes, but they are of critical importance. We may find that infectious amyloid proteins might afford a simpler system for solving these problems than the polioviruses, herpes viruses, influenza viruses, or retroviruses, or for even the relatively simple small circular RNAs of the viroids wherein these problems have still proved elusive.

This Sixth International Workshop on BSE and related animal and human diseases brought together 75 scientists in medicine, veterinary medicine, microbiology, genetics, protein conformational chemistry, and pharmacology and representatives of industry and regulatory agencies to present and discuss the latest data on all of these matters and to discuss their implications for the control of possible hazards to humans and other animals and the directions of future research. The meeting took place February 26 to March 1, 1995, in the pleasant setting of the Kingsmill Resort in Williamsburg, Virginia, under the sponsorship of Serono Symposia USA and the National Institute of Neurological Disorders and Stroke of the National Institutes of Health. Clarence J. Gibbs, Jr., planned and organized the program, selecting the participants as he had done for the previous five international workshops on BSE. There was ample time for long discussions after the presentations and between sessions during the three days of relaxed conversations over good food. Many new collaborations were established, which was surely one of the most valuable aspects of the meeting. It was, in the opinion of all who attended, an unusually successful symposium in this often polemical field known for its vituperative exchanges. The conference demonstrates clearly that there is a large body of data on which there is full agreement and consensus about these infectious diseases and their etiological agents, and only a few areas of legitimate debate and disagreement wherein new concentrated research is now required.

D. Carleton Gajdusek

Contents

Part V. Public Health Considerations of Human and Animal Spongiform Encephalopathies

Contributors

ANTHONY R. AUSTIN, Central Veterinary Laboratory, New Haw, Addlestone, Surrey, UK.

ALFRED E. BACOTE, Laboratory of Central Nervous System Studies, Basic Neurosciences Program, Division of Intramural Research, National Institute of Neurological Disorders and Stroke, National Institutes of Health, Bethesda, Maryland, USA.

ELIZABETH BALDAUF, Department of Virology, Robert Koch-Institute, Berlin, Germany.

MICHAEL BEEKES, Department of Virology, Robert Koch-Institute, Berlin, Germany.

PETER B.G.M. BELT, Department of Bacteriology, Institute for Animal Science and Health (ID-DLO), Lelystad, The Netherlands.

JAVIER BLANCO VIERA, Institute of Pathobiology, Center for Research in Veterinary Science, National Institute of Agricultural Technology, Moron, Buenos Aires, Argentina.

ALEX BOSSERS, Department of Bacteriology, Institute for Animal Science and Health (ID-DLO), Lelystad, The Netherlands.

RAY BRADLEY, Central Veterinary Laboratory, New Haw, Addlestone, Surrey, UK.

BERNARDO G. CANÉ, Animal Health Service, Secretary of Agriculture and Fisheries, Buenos Aires, Argentina.

GEORGE A. CARLSON, McLaughlin Research Institute, Great Falls, Montana, USA.

RICHARD I. CARP, Department of Virology, New York State Institute for Basic Research in Developmental Disabilities, Staten Island, New York, USA.

BERNARDO J. CARRILLO, Center for Research in Veterinary Science, National Institute of Agricultural Technology, Moron, Buenos Aires, Argentina.

SVEN CASSENS, Department of Virology, Robert Koch-Institute, Berlin, Germany.

BYRON CAUGHEY, Laboratory of Persistent Viral Diseases, Rocky Mountain Laboratories, National Institute of Allergy and Infectious Diseases, National Institutes of Health, Hamilton, Montana, USA.

MAURO CERONI, Neurological Institute, Pavia University, IRCCS, C. Mondino Foundation, Pavia, Italy.

BRUCE CHESEBRO, Laboratory of Persistent Viral Diseases, Rocky Mountain Laboratories, National Institute of Allergy and Infectious Diseases, National Institutes of Health, Hamilton, Montana, USA.

JON COME, Department of Chemistry, Massachusetts Institute of Technology, Cambridge, Massachusetts, and Laboratory of Persistent Viral Diseases, Rocky Mountain Laboratories, National Institute of Allergy and Infectious Diseases, National Institutes of Health, Hamilton, Montana, USA.

RANDALL C. CUTLIP, United States Department of Agriculture, Agricultural Research Service, National Animal Disease Center, Ames, Iowa, USA.

MICHAEL DAWSON, Central Veterinary Laboratory, New Haw, Addlestone, Surrey, UK.

IAN DEXTER, Central Veterinary Laboratory, New Haw, Addlestone, Surrey, UK.

ALESSANDRO DI MARTINO, Fidia Research Laboratories, Abano Terme, Italy.

HEINO DIRINGER, Department of Virology, Robert Koch-Institute, Berlin, Germany.

DARWIN ERNST, Laboratory of Persistent Viral Diseases, National Institute of Allergy and Infectious Diseases, National Institutes of Health, Rocky Mountain Laboratories, Hamilton, Montana, USA.

REGINA FERSKO, Department of Virology, New York State Institute for Basic Research in Developmental Disabilities, Staten Island, New York, USA.

D. CARLETON GAJDUSEK, Laboratory of Central Nervous System Studies, National Institute of Neurological Disorders and Stroke, National Institutes of Health, Bethesda, Maryland, USA.

CLARENCE J. GIBBS, JR., Laboratory of Central Nervous System Studies, Basic Neurosciences Program, Division of Intramural Research, National Institute of Neurological Disorders and Stroke, National Institutes of Health, Bethesda, Maryland, USA.

EMILIO J. GIMENO, School of Veterinary Sciences, National University of La Plata, La Plata, Buenos Aires, Argentina.

LEV G. GOLDFARB, Clinical Neurogenetics Unit, National Institute of Neurological Disorders and Stroke, National Institutes of Health, Bethesda, Maryland, USA.

ROBERT B. GREEN, Central Veterinary Laboratory, New Haw, Addlestone, Surrey, UK.

CHRISTIAN GRIOT, Institute of Virology and Immunoprophylaxis, Mittelhäusern, Switzerland.

WILLIAM J. HADLOW, Veterinary Pathologist, Hamilton, Montana, USA (retired).

STEPHEN A.C. HAWKINS, Central Veterinary Laboratory, New Haw, Addlestone, Surrey, UK.

DAGMAR HEIM, Institute of Virology and Immunoprophylaxis, Mittelhäusern, Switzerland.

MARK W. HORIGAN, Central Veterinary Laboratory, New Haw, Addlestone, Surrey, UK.

BEAT HORNLIMANN, Institute of Virology and Immunoprophylaxis, Mittelhäusern, Switzerland.

JAMES L. HOURRIGAN, United States Department of Agriculture, Animal and Plant Health Inspection Service, Veterinary Services, Hyattsville, Maryland, USA (retired).

LOREDANA INGROSSO, Section of Persistent and Slow Virus Infections, Laboratory of Virology, Istituto Superiore di Sanità, Rome, Italy.

ALLEN L. JENNY, United States Department of Agriculture, Animal and Plant Health Inspection Service, National Veterinary Services Laboratories, Ames, Iowa, USA.

RICHARD T. JOHNSON, Department of Neurology, The Johns Hopkins University School of Medicine, Baltimore, Maryland, USA.

RICHARD J. KASCSAK, Department of Virology, New York State Institute for Basic Research in Developmental Disabilities, Staten Island, New York, USA.

RICHARD H. KIMBERLIN, Scrapie and Related Diseases Advisory Service (SARDAS), Edinburgh, UK.

ALBERT L. KLINGSPORN, United States Department of Agriculture, Animal and Plant Health Inspection Service, Veterinary Services, Hyattsville, Maryland, USA (retired).

DAVID A. KOCISKO, Department of Chemistry, Massachusetts Institute of Technology, Cambridge, Massachusetts, and Laboratory of Persistent Viral Diseases, Rocky Mountain Laboratories, National Institute of Allergy and Infectious Diseases, National Institutes of Health, Hamilton, Montana, USA.

ANNA LADOGANA, Section of Persistent and Slow Virus Infections, Laboratory of Virology, Istituto Superiore di Sanità, Rome, Italy.

PETER T. LANSBURY, JR., Department of Chemistry, Massachusetts Institute of Technology, Cambridge, Massachusetts, USA.

HOWARD D. LEHMKUHL, United States Department of Agriculture, Agricultural Research Service, National Animal Disease Center, Ames, Iowa, USA.

PAUL P. LIBERSKI, Department of Oncology, Medical Academy, Lodz, Poland.

RICHARD F. MARSH, Department of Animal Health and Biomedical Sciences, University of Wisconsin, Madison, Wisconsin, USA.

JANICE M. MILLER, United States Department of Agriculture, Agricultural Research Service, National Animal Disease Center, Ames, Iowa, USA.

ALFREDO NADER, Animal Health Service, Secretary of Agriculture and Fisheries, Buenos Aires, Argentina.

SANTOSH NANDAN, Department of Chemistry, Massachusetts Institute of Technology, Cambridge, Massachusetts, and Laboratory of Persistent Viral Diseases, Rocky Mountain Laboratories, National Institute of Allergy and Infectious Diseases, National Institutes of Health, Hamilton, Montana, USA.

MUHSIN ÖZEL, Department of Virology, Robert Koch-Institute, Berlin, Germany.

MICHAEL C. PAPINI, Department of Virology, New York State Institute for Basic Research in Developmental Disabilities, Staten Island, New York, USA.

PAOLA PERGAMI, Laboratory of Central Nervous System Studies, Basic Neurosciences Program, Division of Intramural Research, National Institute of Neurological Disorders and Stroke, National Institutes of Health, Bethesda, Maryland, USA.

PEDRO PICCARDO, Department of Pathology, Indiana University School of Medicine, Indianapolis, Indiana, USA.

MAURIZIO POCCHIARI, Section of Persistent and Slow Virus Infections, Laboratory of Virology, Istituto Superiore di Sanità, Rome, Italy.

SUZETTE A. PRIOLA, Laboratory of Persistent Viral Diseases, Rocky Mountain Laboratories, National Institute of Allergy and Infectious Diseases, National Institutes of Health, Hamilton, Montana, USA.

STANLEY B. PRUSINER, Departments of Neurology and of Biochemistry and Biophysics, University of California, San Francisco, California, USA.

RICHARD E. RACE, Laboratory of Persistent Viral Diseases, National Institute of Allergy and Infectious Diseases, National Institutes of Health, Rocky Mountain Laboratories, Hamilton, Montana, USA.

GREGORY RAYMOND, Laboratory of Persistent Viral Diseases, National Institute of Allergy and Infectious Diseases, National Institutes of Health, Rocky Mountain Laboratories, Hamilton, Montana, USA.

MARK M. ROBINSON, United States Department of Agriculture, Agricultural Research Service, Animal Disease Research Unit, Pullman, Washington, USA.

ADRIENNE A. RUBENSTEIN, Department of Virology, New York State Institute for Basic Research in Developmental Disabilities, Staten Island, New York, USA.

RICHARD RUBENSTEIN, Department of Virology, New York State Institute for Basic Research in Developmental Disabilities, Staten Island, New York, USA.

JIRI SAFAR, Laboratory of Central Nervous System Studies, Basic Neurosciences Program, Division of Intramural Research, National Institute of Neurological Disorders and Stroke, National Institutes of Health, Bethesda, Maryland, USA.

RAIMUNDO ANDRES SAN MARTIN, Laboratory of Central Nervous System Studies, Basic Neurosciences Program, Division of Intramural Research, National Institute of Neurological Disorders and Stroke, National Institutes of Health, Bethesda, Maryland, USA.

CAROL L. SCALICI, Department of Virology, New York State Institute for Basic Research in Developmental Disabilities, Staten Island, New York, USA.

BRAM E.C. SCHREUDER, Department of Pathophysiology and Epidemiology, Institute for Animal Science and Health (ID-DLO), Lelystad, The Netherlands.

ALEJANDRO A. SCHUDEL, Institute of Virology, Center for Research in Veterinary Science, National Institute of Agricultural Technology, Moron, Buenos Aires, Argentina.

MICHAEL SCOTT, Department of Neurology, University of California, San Francisco, California, USA.

MARION M. SIMMONS, Central Veterinary Laboratory, New Haw, Addlestone, Surrey, UK.

MARI A. SMITS, Department of Bacteriology, Institute for Animal Science and Health (ID-DLO), Lelystad, The Netherlands.

MICHAEL P. SULIMA, Laboratory of Central Nervous System Studies, Basic Neurosciences Program, Division of Intramural Research, National Institute of Neurological Disorders and Stroke, National Institutes of Health, Bethesda, Maryland, USA.

GLENN C. TELLING, Department of Neurology, University of California, San Francisco, California, USA.

EDGARDO ULLOA, Serono Argentina S.A., Martinez, Buenos Aires, Argentina.

CARLOS VAN GELDEREN, Serono Argentina S.A., Martinez, Buenos Aires, Argentina.

ELBA L. WEBER, Institute of Virology, Center for Research in Veterinary Science, National Institute of Agricultural Technology, Moron, Buenos Aires, Argentina.

GERALD A.H. WELLS, Central Veterinary Laboratory, New Haw, Addlestone, Surrey, UK.

JOHN W. WILESMITH, Central Veterinary Laboratory, New Haw, Addlestone, Surrey, UK.

R.G. WILL, National Creutzfeldt-Jakob Disease Surveillance Unit, Western General Hospital, Edinburgh, UK.

1

The Potential Risk to Humans of Amyloids in Animals

D. Carleton Gajdusek

Nucleating Induction of Infectious Amyloids from Host Precursors

The subacute spongiform virus encephalopathies (transmissible virus dementias or unconventional virus infections) are amyloidoses of brain caused by infectious amyloid proteins derived from normal host precursors by nucleated conformational transition to a more stable cross-β-pleated lower energy state. We now call these diseases—scrapie, kuru, Creutzfeldt-Jakob disease (CJD), Gerstmann-Sträussler-Scheinker syndrome (GSS), fatal familial insomnia (FFI), transmissible mink encephalopathy (TME), chronic wasting disease (CWD), bovine spongiform encephalopathy (BSE)—transmissible cerebral amyloidosis. Normal aging of the brain, Alzheimer's disease, and dementias of Down syndrome are nontransmissible amyloidoses based on a different precursor protein.

When a cross-β-pleated protein molecule induces the same conformation it possesses in its own precursor, it is an *infectious amyloid*. An infectious amyloid molecule may even induce this conformational transition to cross-β-pleated structure in a different protein. This essential nucleation induces conformational change by tight epitaxic (geometrical) matching to the normal precursor monomer by hydrogen bonding, and perhaps stronger bond formation, with or without the aid of a ligand.

This nucleated autopatterning of the conversion of the normal precursor monomer to the conformation of the cross-β-pleated nucleating form is infectious behavior. It surely occurs by raising the energy of the normal precursor over the thermodynamic barrier for transition into a lower energy β-pleated state. The tight geometric association of the infectious amyloid molecule (or oligomer) with the monomeric precursor results in generation of another β-pleated molecule—a copy of the nucleant—that polymerizes into fibrils. Alternatively, the infectious amyloid may initiate the process by first nucleating the patterning of polymerization or ordered aggregation of the normal precursor, which then undergoes the transition to β-pleated infectivity after crystalloid packing (ordered aggregation) in a semisolid, noncrystalline, or vitrous state. Transient, intermediate conformational states such as molten globular forms of the infectious nucleant may be more active nucleants for this induction of patterned transition to β-pleated form.

The nontransmissible amyloidoses, both of brain and systematic, will probably also be found to be infectious amyloidoses that do not evoke the full clinical picture of the human disease because of different organ targeting, less pronounced amyloid formation, heteronucleation in a different species, or imperfect geometric matching of nucleant with precursor in passing the host barrier. Thus, it is probably naive to believe that we can read host range and biologic processes of targeting, routes of infectivity, and incubation periods from sequence data and homology.

The presence of both mutated and unmutated amyloid polypeptides in the amyloid fibillary deposits in familial transthyretin amyloidoses, and in the 39 amino acid peptide deposits in hereditary cerebral hemorrhage with amyloidosis in Dutch families from the brain β-amyloid precursor (APP) of normal aging and Alzheimer's disease, are both examples of scrapie-like induction of conformation change. So are all amyloid enhancing factors. Likewise, copolymerization of insulin-associated protein with insulin amyloid, β-A4 amyloid with CJD amyloid deposits, and glycosaminoglycan and proteoglycan associations with amyloid deposits are also scrapie-like inductions of conformational change.

Potential Risk to Humans of BSE

Andromeda strain–like predictions that sheep scrapie may have been modified by rendering plant procedures and/or its passage through cattle, so that it is now virulent for humans and can infect humans by the oral route, and that bovine tissues such as muscle, milk, and blood, not usually infectious in sheep, may now contain more infectivity in cattle, have led to apocalyptic prophecies that a high percentage of the population of the United Kingdom will come down after long incubation periods with fatal CJD-like disease. These prophecies have been the cause of hysterical responses among the media, the public, government advisors, and legislators. Scientific certainty that such a scenario is impossible cannot be produced. An honest scientist cannot state that we are certain that such prophecies are false. We can only present the indirect evidence, which indicates that such dire predictions are probably wrong.

We can devise no experiment that will answer the demand of the media, the public, and the government for certainty. The British are now titrating muscle, milk, and blood from BSE-affected cattle intracerebrally (i.c.) into mice. We could feed the BSE agent to different susceptible species in different suspensions to determine the generality of its virulence by the oral route—a very costly and time-consuming experiment requiring decades of observation. Only if we fed previously titrated quantities of BSE agent to hundreds of "disposable" children and allowed several decades for the result would we have the basis for more certain predictions. Obviously we shall never do this. To do this crude feeding experiment using monkeys or chimpanzees would not provide the desired answers for human health. The cynic might say that the experiment is already on. By careful i.c. titration of bovine tissues in cattle and mice, we could state the possible dose of the BSE agent that the United Kingdom population has ingested, if any, but this would not allow for the loss of infectivity in cooking. By the first decade of the next century, we may have our answer.

Clearly, the infectious β-pleated protein from BSE has several amino acid differences from that of sheep scrapie, the number of amino acid residues by which the scrapie amyloid precursor protein of sheep differs from its homologue in cattle. No analysis of the genomic sequence of any microorganism yet permits us to assess either its host range or its virulence and pathogenicity. Nothing in the genomic sequencing of a host permits us to predict its susceptibility or resistance to a given microbe, other than the mutations causing hereditary immunodeficiency syndromes. Only the neurotropic loop in polioviruses and some other enteroviruses (encephalomyocarditis [EMC], Theiler's GDVII virus [GDVII]) has given us insight into specific neurotropism or other tropism. Even full amino acid sequencing and conformational study of receptor proteins have still failed to give us predictive knowledge for host range, virulence, pathogenicity, or route of susceptibility. For some retroviruses, such as HIV, the polioviruses, and influenza, we may be approaching that goal. For the infectious amyloids of the transmissible spongiform encephalopathies, we are a long way off. This means that we cannot, by any molecular genetic study of the infectious amyloid precursor gene, on chromosome 20 in humans and chromosome 2 in mice, predict any of these properties for the host or for the infectious amyloid derived from its precursor. It has not been possible to extrapolate from experiments in any host the host range, virulence, tropism, or pathogenicity (production of amyloid plaques, incubation period, localization of lesions) for any strain of scrapie (or TME, CWD, BSE) to a different host. Transgenic mice with many copies of the human/mouse genes for the infectious amyloid precursor protein (PrPc) (on chromosome 20 in humans and 2 in mice) and knockout mice with many copies of only the human CJD amyloid precursor gene are now available to make more sophisticated inferences. Mice with chimeric chromosomes containing the human gene are now available. Extrapolation of these systems to the human host will require caution.

On what, then, do we base our rejections of the doomsday prophecies of the media for the threat of BSE to human health? The total failure in 25 years of epidemiologic study in the United States, France, Japan, Slovakia, Israel, Chile, New Zealand, and Australia to correlate the eating and handling of scrapie-infected sheep with human CJD (or its variants, GSS and FFI). Even sheep brain handlers in abattoirs, where a significant number of scrapied sheep were known to have been handled, have failed to demonstrate a higher risk for developing CJD than the worldwide incidence of 0.5 to 1.0/million population/year. Scrapie-free countries like New Zealand and Australia have the same worldwide 1.0/million population incidence rates of CJD as countries heavily contaminated with scrapie, such as the U.K. and France.

We studied the oral transmission of kuru, CJD, and scrapie to susceptible nonhuman primates in the 1960s and 1970s and reported our experience in the *Journal of Infectious Diseases* in 1980 (1). Of many species, including the chimpanzee, only the squirrel monkey ever developed disease. Doses given by gavage feeding were 10^9 to 10^{10} times the i.c. minimum effective dose (MID) and yet no infection occurred, except in the squirrel monkeys fed infected whole tissues over several days, where incubation periods were longer than for i.c. infection. Thus, an astronomical dosage is required to produce infection even irregularly in the one species susceptible, and the long incubation periods suggest the entry via the gut of very low dosage infection. Thus, even if BSE-infected bovine tissues were pathogenic for humans, it is unlikely that they would ever cause disease, unless consumed in extremely large quantities.

That two different strains of scrapie in infected sheep and mouse brain suspensions given i.c. to five chimpanzees failed to cause disease, although such inoculations regularly produced infection in susceptible monkeys, indicates how unpredictable the species barrier may be.

Disappearance of Kuru

In recent years, with diminishing numbers of kuru patients yearly, those patients who still get kuru are of higher age each year. Now only persons over 40 years of age develop kuru; there were only six cases diagnosed last year. No person born after the opening of the Pandora's box of the cranium and spreading of liquefied brain tissues over the bodies of infants and small toddlers by their mother's hands—a practice that disappeared with the cessation of cannibalism—has ever developed kuru. There is no source of infection either vertically or horizontally or from any environmental contamination or reservoir species. Infants and small children were massively contaminated by the never-washed hands of their caregivers in scratching their scabies, impetigo, and insect bite sores, and cleaning their eyes and nose, but none ingested kuru-infected tissues. Most such children of the dozens present at each cannibalistic mourning ritual who have not died of other causes have died of kuru, often with incubation periods of several decades. The high incidence of infection must indicate widespread high-level contamination and uniform susceptibility.

Genetic Determination of De Novo Spontaneous Creation of Infectious Proteins Without an Infectious Chain

We have hypothesized that the low incidence and prevalence rates of CJD of 1 per 10^6 per annum worldwide, the same everywhere from tropical to arctic regions in both hemispheres, indicate a spontaneous transition of the host precursor protein to an infectious β-pleated conformation as a rare stochastic event, accounting for most of the sporadic cases without any infectious chain. We have never established a CJD contact or source of infection in 95% of CJD cases. In familial CJD (and its GSS and FFI variants), we have now found eight point mutations causing an amino acid change, one causing a stop codon, and seven different octapeptide insertions into a region where we normally have four or five such tandem repeats. Each increases the likelihood of the transition to infectious conformation about 10^6-fold, the worldwide incidence rate of CJD. Thus, each member carrying the mutation in an affected family generates his own infectious protein from the mutation-bearing precursor protein.

At codon 129, there is a silent polymorphism replacing methionine by valine, common (17%) in the normal population. If the codon 178 pathogenic mutation has the less common valine at codon 129, rapidly fatal classic CJD results. If there is, instead, methionine on the same chromosome as the codon 178 mutation, a clinically and pathologically vastly different syndrome develops, namely fatal familial insominia (FFI). This same rare and normally silent valine polymorphism has appeared homozygously much more commonly in iatrogenic cases of CJD from contaminated human growth hormone, and also in i.c. inoculated iatrogenic cases (dura mater, corneal transplant, stereotactic electrodes), suggesting a genetic factor of enhanced susceptibility to low-dose peripheral infection. This indicates the complexity of interaction of host and infectious protein genetics in host range, virulence, and pathogenesis.

Changes in Host Range, Virulence, Cell Targeting, Routes of Susceptibility, and Incubation Period in Crossing the Species Barrier

When sheep scrapie was transmitted to mink on mink farms or to mink or ferrets in the laboratory, the clinical picture of disease produced was markedly altered. The infectious amyloid subunits with the mink or ferret amino acid sequence would no longer infect mice, even if the original inoculum had been mouse-adapted scrapie. Furthermore, transmission mink-to-mink occurred by biting during their usual fighting. The clinical picture of disease was markedly altered from that of sheep scrapie or mouse scrapie.

Infected Leukocytes in Peripheral Blood During the Presymptomatic Incubation Period

Mice inoculated intracerebrally with scrapie may have the scrapie agent isolated from their peripheral leukocytes 1 to 3 months after inoculation before clinical disease appears. Thus, Kuroda et al. (2) have demonstrated maximal scrapie infectivity in low-density mononuclear leukocytes in the blood of such infected animals 1 to 3 months after inoculation. Whether human CJD or BSE in cattle may have a transient or long-standing viremic phase has not been determined. Certainly, many CJD patients have been blood donors during their incubation periods.

Route of Infection

Furthermore, route of infection may play a critical role in pathogenesis. A long course of sporadic CJD (over 1 year), with only slowly developing dementia and early severe cerebellar signs resembling kuru, has occurred in almost all iatrogenic cases from peripheral infection of CJD-contaminated cadaveric pituitary-derived hormones. Such CJD cases are very rare among the noniatrogenic cases of CJD. In the i.c.-inoculated dura mater and con-taminated stereotactic-electrode cases, and in the corneal transplant cases, all patients have developed the classic, nonataxic, rapidly progressive CJD. This suggests that in true sporadic cases of CJD, the de novo spontaneous generation of infectious nucleating protein has occurred in the brain, since such cases are overwhelmingly of the i.c.-inoculation type of rapidly fatal classic CJD.

Conclusion

What can we conclude from all of this? That high-titer contamination of human foods should be avoided, and that no tissues from BSE-affected animals should be used for human food or animal consumption, or for pharmaceutical or cosmetic products. In the United Kingdom, incubating animals with lower titers cannot be fully avoided, but their milk, blood, and meat are most probably not infected. The ban on offal from all United Kingdom cattle for consumption by humans and animals should avoid any potential danger. If a cheap, easy, and rapid test for infectious protein (PrP^{BSE}) can be developed to screen late-stage incubating animals in the slaughterhouse, this should be done.

In the meantime, extremely thorough surveillance of all CJD cases in every country is essential. The monitoring should include attention to all features of the clinical course, pathology, and age at onset. Such data have recently led to discoveries of CJD contamination and a chain of infection, rather than the more usual spontaneous generation of the infectious agent in each individual as in all familial and in most sporadic cases of CJD. Patients under 30 years of age with a slowly dementing, prolonged, and ataxic clinical picture should suggest a peripheral route of infection.

An urgent program of developing bovine/murine transgenic and chimeric mice carrying many copies of BSE amyloid precursor protein gene should be instituted for quick isolation of BSE and probably more sensitive infectivity titration of bovine tissues. The use of human/mice transgenic and chimeric mice now available for i.c. and oral inoculation may give information suggesting susceptibility or resistance of man to BSE. We urgently need data on the relative sensitivity of these systems compared with the inoculation of susceptible primates.

References

1. Gibbs CJ Jr, Amyx HL, Bacote A, Masters CL, Gajdusek DC. Oral transmission of kuru, Creutzfeldt–Jakob disease, and scrapie to nonhuman primates. J Infect Dis 1980;142:205–8.
2. Kuroda Y, Gibbs CJ Jr, Amyx HL, Gajdusek DC. Creutzfeldt-Jakob disease in mice: persistent viremia and preferential replication of virus in low-density lymphocytes. Infect Immun 1983;41:154–61.

Part I

Host Range and Pathogenesis

2

Bovine Spongiform Encephalopathy Distribution and Update on Some Transmission and Decontamination Studies

RAY BRADLEY

A report (1) of a scrapie-like disease in a captive nyala (*Tragelaphus angasi*) in England heralded the start of the epidemic of bovine spongiform encephalopathy (BSE). In November 1986 the first cases of BSE were confirmed by histologic examination of the brain from affected cattle from two areas of southern England (2). Subsequently it was reported that clinical cases of BSE, unrecognized at the time, had probably occurred as early as April 1985 and that the origin of infection, in the first 190 cases or so, was feed in the form of ruminant-derived meat and bone meal (MBM) (3). Nothing has occurred since to alter this view (4).

An important contributory factor to the appearance of the disease was a change, in 1981/1982, away from the formerly common hydrocarbon solvent method of extracting tallow from rendered animal waste (5). As a result of an initial increasing exposure to incompletely inactivated scrapie agent from sheep offals, cattle developed clinical BSE from 1985 onward after completion of the modal incubation period of 4 to 5 years. Exposure of cattle increased from 1984–85, at the latest, as a result of recycling infection from infected cattle waste that after being processed, had been included in feed in the form of meat and bone meal. This escalation in the exposure of cattle became obvious 4 to 5 years later, from July 1989 onward, as a rapidly rising disease incidence. A large experimental study was set up to determine the effect of different rendering processes in use in member states of the European Union on the survival of BSE and scrapie agents

11

added, in the form of brain from affected animals, to animal waste before processing. The presence and titer of agents was determined by bioassay in mice (6). The interim results from this study are reported below.

Embryo transfer, conducted in accordance with the protocols of the International Embryo Transfer Society (IETS), is a safer way to move genetic material from country to country than moving live animals (7). Countries with BSE have genetically desirable breeding stock of high commercial value, much desired by other countries, including those without BSE (or scrapie), so there is a strong desire to import such animals but with minimal risk. There is no evidence that bovine embryos can transmit BSE under any circumstances, but to verify this a large experiment has been set up in Great Britain, the interim results of which are also reported.

In addition, this chapter describes the present incidence and geographic distribution of BSE, the natural and experimental host range of the BSE agent, studies on some other aspects of transmission, BSE fibrils, and research relevant to the practical investigation of this new disease.

The Present Position with BSE in Great Britain, the Rest of Europe, and Elsewhere[1]

On February 3, 1995, 143,109 cases of BSE had been confirmed in Great Britain on 32,006 farms. The comparable figures for Northern Ireland were 1,504 cases on 1,039 farms. As of January 1, 1995, there were 99 confirmed cases in the Republic of Ireland, 9 in France, 118 in Switzerland, and 17 in Portugal. The French and Swiss cases, the majority of the Irish cases, and some of the Portuguese cases were in native-born cattle. The feeding of MBM could not be excluded in most cases and is generally believed to be the source of infection. The Sultanate of Oman and Italy have had two confirmed cases each, the Falkland Islands, Denmark, and Canada one each, and Germany four. In each of these instances the animals had been imported from the United Kingdom and it is presumed they were exposed there and not in the country where the disease was identified. Most of the European countries affected by BSE had a ban in place that prevented the feeding of ruminant protein to ruminant animals (Great Britain since 1988, Ireland, France, and Switzerland since 1990). Since July 1994 there has been a ban on feeding mammalian protein to ruminant animals throughout the European Union (8). However, if a member state enforces a system that makes it possible to distinguish between protein from ruminant and nonruminant species, the commission may authorize that member state to

---

[1] The author believes the data presented for countries outside the United Kingdom and derived from official sources, to be accurate. However, he advises that the most up-to-date figures are obtained from the State Veterinary Services of the countries concerned to a particular date.

permit the feeding of nonruminant protein to ruminant animals. The United Kingdom and Switzerland also have a specified bovine offals (SBO) ban that prevents potentially infected offals from clinically unaffected cattle from entering the human food chain. (SBO comprises brain, spinal cord, thymus, spleen, and intestine, from duodenum to rectum inclusive, from cattle >6 months old.) Since September 1990 the ban in the United Kingdom prohibits the use of SBO, or protein derived from it, for feeding to humans and any species of animal or bird. In addition in Great Britain, since November 1994, the intestine and thymus from calves under 6 months old have been included in the ban, as a result of finding infectivity in the distal ileum of calves following high experimental oral exposure (see Chapter 3).

New Natural Hosts of Spongiform Encephalopathy Agents

Prior to 1986 the natural host range of spongiform encephalopathy (SE) agents included humans (for kuru, Creutzfeldt-Jakob disease, and Gerstmann-Sträussler-Scheinker syndrome), sheep and goats for scrapie, certain species of deer and elk for chronic wasting disease (CWD), and mink for transmissible mink encephalopathy (TME). TME has never been reported in the British Isles and CWD has never been reported in Europe. To this list can be added scrapie in moufflon (*Ovis musimon*) first reported in 1992 (9), although the disease, at least in the two flocks where it occurred, appeared to be endemic, indistinguishable from natural sheep scrapie, and to have no connection with the BSE epidemic.

In 1986, before BSE had been recognized, SE was reported in a captive nyala (*Tragelaphus angasi*) (1). Subsequently SE was also reported in a gemsbok (*Oryx gazella*), an Arabian oryx (*Oryx leucoryx*), six greater kudu (*Tragelaphus strepsiceros*), five eland (*Taurotragus oryx*), and a single scimitar-horned oryx (*Oryx dammah*) (10). The SE in nyala and kudu subsequently was shown to be transmissible by inoculation into susceptible strains of mice. The first cases in these species had all been exposed to the same type of contaminated feed as cattle and this is the presumed source of infection. All of the other kudu, four of the eland, and the scimitar-horned oryx were born after the introduction of the ruminant feed ban (10). The second kudu case was the offspring of the first (feed-exposed) case and theoretically may have resulted from maternal transmission (11), although the dam of four subsequent cases is still alive. Because the ruminant feed ban took some time to become fully effective after it was introduced (see Chapter 4), there is now at least some support for the view that the later-born kudu, eland, and the scimitar-horned oryx could have been exposed to infected feed despite being born after the ban. In the case of the eland only, there was a possible additional source from the enclosure contaminated by drips from a meat wagon carrying cattle heads and other cattle material unfit for human consumption and fed to lions and tigers that remain healthy. Nevertheless, there may be an increased susceptibility to disease in greater kudu and possibly a wider distribution of infectivity in the tissues than in cattle. This is being examined by additional epidemiologic and tissue infectivity studies. However, there have been no cases of SE in this small kudu herd since 1992.

In May 1990 a single case of SE in a domestic cat was reported in England (12). As of January 31, 1995, a total of 63 cases had been confirmed of which one was in Northern Ireland and one was a native-born cat in Norway. There are insufficient cases to identify the source of infection, although it is presumed to come from feed.

Cases of spongiform encephalopathy have been diagnosed in captive wild felidae, one in a puma (*Felis concolor*) in a British zoo (13), two in cheetahs (*Acinonyx jubatus*) in British zoos, one in an Australian zoo (14, 15), and one in an Irish zoo, both of which had been imported from England. It is believed that these captive wild cats were exposed to parts of raw (uncooked) cattle carcasses containing central nervous tissue that may have carried the infectious agent (10), although the precise origin cannot be established. No cases of feline SE have occurred in any wild or domestic felidae born in the United Kingdom since the SBO ban was extended to prevent these tissues being fed to any species of animal or bird.

White tigers from a zoo in England previously suggested to have been affected with SE (16) are not now regarded as having been affected with a scrapie-like disease, largely because the disease was shown not to be transmissible to a range of mammalian species, including primates, domestic cats, mice, or hamsters, although some of the studies are incomplete (CJ Gibbs, Jr, personal communication). Unfortunately, no fixed tissue or paraffin blocks remain, so further pathologic, including immunochemical, studies are not possible.

A few red necked ostriches (*Struthio camelus*) kept in German zoos showed neurologic illness and lesions of SE (17, 18), but transmissibility to rodents, the presence of specific fibrils and PrP^{Sc} have not been demonstrated (H Diringer, personal communication). Therefore, doubt still prevails in regard to the cause of this disease and whether it is a genuine member of the scrapie group.

Experimental Host Range of BSE

The following species are susceptible to BSE following experimental, parenteral exposure to brain tissue from confirmed cases of BSE: mice (19), cattle (20), pigs (21), sheep and goats (22), marmosets (23), and mink (24). Hamsters did not succumb to parenteral challenge (Table 2.1). In addition, mice fed brain and cerebrospinal fluid from confirmed cases of BSE also

TABLE 2.1. Experimental bovine spongiform encephalopathy host range (minimum incubation period in
months).

	Oral	Parenteral
Mouse	Positive (15)	Positive (9.7)
Cattle	Positive[a]	Positive (18)
Sheep	Positive (18)	Positive (14)
Goat	Positive (31)	Positive (17)
Pig	IP	Positive (16)
Marmoset	ND	Positive (46)
Mink	Positive (14)	Positive (12)
Hamster	ND	NT
Chicken	IP	IP

IP, in progress; ND, not done; NT, no transmission.
[a] Based on clinical evidence (see Chapter 3).

succumbed to the disease (25–27). Domestic fowl challenged with bovine
brain from confirmed BSE cases, either by the oral or parenteral route,
have not as yet succumbed to SE 55 months after exposure (GAH Wells,
personal communication). Likewise, pigs fed cow brain from confirmed
cases of BSE have not developed neurologic illness or SE 56 months after
challenge. Pigs similarly challenged orally with brain homogenate from
sheep with scrapie remain healthy one year after challenge (other than one
pig with intercurrent disease but without spongiform encephalopathy).

The BSE Agent Exists as a Single Strain and Maintains Its Biologic Properties Following Natural and Experimental Passage Through Various Species

The nature of the agents causing transmissible SE is unresolved. Three
main hypotheses predominate: the prion hypothesis—the agent is a modified form of a host protein (PrP); the virino hypothesis—the agent has a
genome coding only for copies of itself in close association with PrP; or it is
an unconventional virus. The prion hypothesis does not readily and satisfac-
torily explain the phenomenon of agent mutation or the occurrence of
strains with different biologic properties for which there is abundant evidence, although several ingenious, but unproven, ideas have been put forward to explain them. Recent reports by Bruce et al. (28) show that
transmissions into mice from seven unrelated cattle sources, three cats, and
a nyala, and greater kudu and from sheep goats and pigs experimentally
infected with BSE, all have similar biologic characteristics, suggesting a
single major strain of agent is involved. This argues for an agent-specific
informational molecule (possibly an undiscovered nucleic acid) and aligns

with the constancy of the clinical signs and neuropathology in cattle with BSE seen throughout the epidemic. Such a result would not have been anticipated by followers of the prion hypothesis because each of the species has a different *PrP* gene sequence, and consequently a different PrP composition and would be expected to have significantly different biologic effects in mice (incubation period length and lesion profile). Another finding from this important study was that the biologic characteristics of the BSE agent from cattle are quite different from those of contemporary or historical scrapie. However, this finding does not exclude an origin for BSE in sheep with scrapie because passage of the scrapie agent through different species can lead to a permanent change in its biologic properties (29).

Maternal and Horizontal Transmission

Horizontal transmission consequent upon scrapie infection in ewes at lambing time is important in maintaining natural scrapie as an endemic infection in a flock. Since placenta from Swaledale sheep with scrapie produced scrapie in sheep and goats after experimental oral challenge (30), it is probable that horizontal transmission to related and unrelated sheep can occur naturally. In June 1988, legislation was introduced in Great Britain to reduce any risks of this kind that there may have been for cattle from cattle placenta. Cows suspected to have BSE have to be isolated in approved premises while calving and for 72 hours after parturition, and the placenta, bedding, and discharges safely disposed of by incineration or burial, after which the premises must be cleaned and disinfected. Subsequently, it was advised that it would be a good management practice, appropriate to the control of many infections, to proceed similarly at all bovine parturitions, even though this was not a statutory requirement. These control measures were put in place in 1988 on the assumption that cattle with BSE would behave like sheep with scrapie.

However, placenta from cows affected with BSE has shown no detectable infectivity when fed to susceptible mice (27) or inoculated parenterally (31). Furthermore, no infectivity has been detected in amniotic or allantoic fluid, uterine caruncle or ovary (H. Fraser, personal communication) following bioassay in susceptible mice. Also, such placenta used to challenge cattle oronasally has not resulted in transmission of any disease over 5 years after dosing. Cattle killed at 2 years after dosing did not show histopathologic evidence of SE and no BSE fibrils were seen by electron microscopy of detergent extracts of brain. Finally, mouse bioassay of tissues in mice from these healthy 2-year-old challenged cattle show no evidence of infectivity (GAH Wells, personal communication).

Comparative titrations (using brain) and assays (spleen and lymph nodes) are being conducted in parallel in cattle and mice to evaluate any underestimate of titer by using the latter species. Results are not yet available.

The possible role of maternal transmission in BSE was recognized to be of the utmost importance (32). It is now recognized however that this form of transmission alone, even if it occurred in every single cow affected with the disease, could not sustain the epidemic under British conditions because the necessary >1:1 contact rate could not be achieved, i.e.. insufficient offspring from confirmed cases would reach adulthood and themselves produce offspring (33). A report on the epidemiologic aspects of maternal transmission is provided in Chapter 4.

The conclusions to date from all the epidemiologic and transmission studies (including those reported below on embryo transfer) are that if maternal transmission occurs at all it is at a very low and so far undetectable rate. Consequently the risk of horizontal transmission, which is unlikely to occur except in the immediate postparturient period, is also low. This is supported by the fact that the within-herd incidence has remained low and below 3% throughout the epidemic (see Chapter 4) and is now declining.

As of January 27, 1995, 424 cases of BSE had been confirmed in bulls. Only a small number of bulls in artificial insemination (AI) center studs have had BSE confirmed. Most of these have occurred during the period of layoff, which lasts until the bulls are about 6 years of age, while awaiting the results of progeny tests. Relatively few offspring from such bulls exist because only a limited number of matings take place until the bull has been proven. However, comparison has been made of the incidence of BSE in the female offspring of the first two AI bulls that succumbed to BSE and two that have not developed BSE (matched by approximate age and region of use) (34) (Table 2.2). From this it can be seen that the incidence is similar for each group. Furthermore the incidence in the progeny is not at variance with that in the general epidemic for this age group. No dams of the offspring recorded in Table 2.2 have contracted BSE. The conclusion is that bulls with BSE do not transmit infection by semen used for artificial insemination. This is supported by the finding that there is no detectable infectiv-

TABLE 2.2. The incidence of BSE in female offspring, born in 1986, of two bulls[a] with confirmed BSE and of two bulls[a] not affected by BSE.

BSE status of bulls and bull identification		Unaffected			Confirmed		
		BT	HM	Total	YC	PN	Total
BSE status of female offspring	Confirmed	2	2	4	2	1	3
	Unaffected	49	51	100	48	30	78
	Total	51	53	104	50	31	81

[a] MMB Milk Recording Scheme, December 1991.

ity in semen, testis, epididymis, seminal vesicle, or prostate from BSE-affected bulls following parenteral bioassay in susceptible mice (31; H Fraser, personal communication). A small case-control study of progeny of affected sires and dams provided no evidence of different levels of incidence in the progeny of affected as compared with unaffected sires or dams (RN Curnow, personal communication).

Embryo Transfer

Whereas in recent studies in the United Kingdom the transfer of unwashed embryos from experimentally scrapie-infected sheep to healthy sheep has resulted in the appearance of scrapie in offspring of *Sip* sAsA genotype (35), no scrapie has occurred in offspring derived from thrice-washed embryos in the United States or in the recipients of these embryos, despite adequate representation of susceptible genotypes (7, 36; AE Wrathall, personal communication).

There is no epidemiologic evidence of an association between BSE occurrence in bovine embryo recipients or offspring derived from transferred bovine embryos and the use of embryo transfer in cattle in Great Britain. A large experiment (Table 2.3) has been set up to confirm that embryos derived from BSE-affected cows (artificially inseminated with semen from either healthy bulls or bulls with BSE) and washed ten times in accordance with the protocols of the IETS do not transmit disease to the resulting

TABLE 2.3. Embryo transfer: Fate of embryos and uterine flushings.

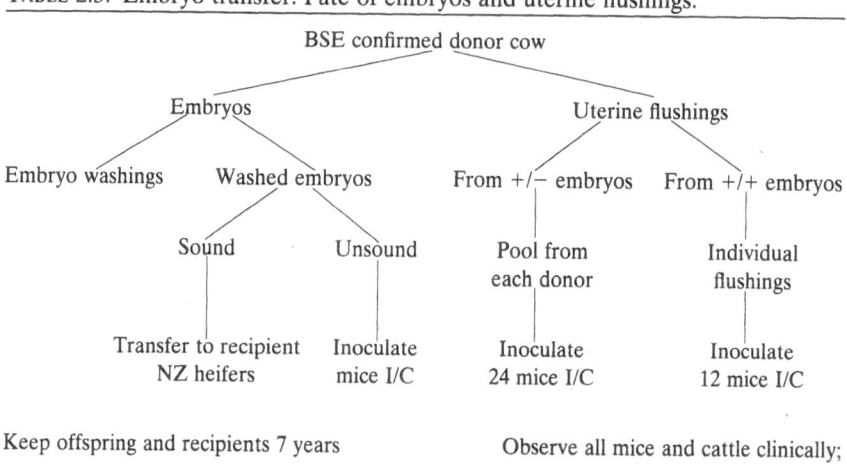

+/+, Embryos are derived from BSE affected cows inseminated with semen from BSE affected bulls.
+/−, Embryos are derived from BSE affected cows inseminated with semen collected from healthy U.K. bulls before 1980.

offspring or recipient dams (36). Matings, embryo collection, washing, and freezing were done at the Central Veterinary Laboratory, Weybridge, entirely in accordance with IETS protocols. Transfers into healthy cattle imported from New Zealand to strict criteria (and therefore devoid of exposure to, or infection with, scrapie or BSE agents) were done on a single, separate ministry farm that is quarantined. So far 256 calves have been born and reared. All these offspring and their surrogate dams will be constantly observed for clinical evidence of BSE and their brains will be examined pathologically 7 years after embryo transfer or sooner if the animal dies at a younger age. No BSE has yet been suspected or confirmed in any animal on this farm. Additionally, in separate experiments, uterine flushings and nonviable embryos are being bioassayed in susceptible mice (Table 2.3). Some experiments are still in progress but no positive results have resulted to date; 1,020 embryos or ova and 40 flushings have been inoculated.

Tissue Infectivity Studies

Bioassay in mice of tissues from clinically affected confirmed cases of BSE has demonstrated BSE infectivity only in the brain and cervical spinal cord. No detectable infectivity was found in 50 other tissues following intracerebral and intraperitoneal inoculation (Table 2.4) (31; H Fraser, personal communication). These tissues include milk, udder, skeletal muscle (red and white skeletal muscles and muscles with a cranial or spinal nerve supply), heart, liver, and kidney. Unlike in Suffolk sheep with natural scrapie (37), no infectivity was found in spleen or any lymph nodes.

Other Transmission Studies

Feeding experiments in mice have also been completed using tissues from clinically affected, confirmed cases of BSE. Transmission was achieved using a mixture of brain and cerebrospinal fluid but not with udder and milk, spleen, various lymph nodes, or placenta (27).

Two major experiments in cattle are now in progress. The first is a study to determine the attack rate and incubation period in cattle challenged orally with one of four different dose levels (1 g, 10 g, 100 g, or 3 × 100 g) of brain from BSE affected cows. Some of these cattle are showing clinical signs of disease 34 months after challenge (GAH Wells, personal communication). The second is a pathogenesis study reported separately (Chapter 3).

BSE Fibrils

Detergent extracts of brain from confirmed cases of BSE, treated with proteinase K, electron densely stained, and examined by transmission elec-

TABLE 2.4. Tissues from clinically affected cattle with no detectable infectivity by parenteral inoculation of mice grouped by anatomical system.[a]

Cerebrospinal fluid	Spleen	Esophagus
Cauda equina	Tonsil	Reticulum
Peripheral nerves	Lymph nodes	Rumen (pillar)
N. sciaticus (proximal)	Prefemoral	Rumen (esophageal groove)
N. tibialis	Mesenteric	Omasum
N. splanchnic	Retropharyngeal	Abomasum
		Proximal small intestine
		Distal small intestine
		Proximal colon
		Distal colon
		Rectum
Clotted blood	Testis	Ovary
Buffy coat	Prostate	Uterine caruncle (pregnant cow)
	Seminal vesicle	Placental cotyledon
	Epididymis	Placental fluids
Foetal calf blood	Semen	Amniotic fluid
Serum		Allantoic fluid
Midrum (mesenteric) fat	Live	Udder
Musculus (M.)	Kidney	Milk
semitendinosus	Heart	
M. longissimus	Pancreas	
M. diaphragma	Lung	
M. masseter	Trachea	
Bone marrow		
Skin		

[a] Studies on some of these tissues from different sources and other tissues not so far reported are still in progress.
Data courtesy of Dr. H. Fraser and Dr. D.M. Taylor.

tron microscopy contain BSE fibrils that are equivalent to scrapie-associated fibrils (SAF) (2, 38). There is an excellent correlation between the occurrence of BSE fibrils from specific sites in the brain stem (39) or cervical spinal cord (AC Scott, personal communication) and the confirmation of disease by histopathology.

Scott et al. (40) have recently shown that following experimental laboratory exposure to higher temperatures than are likely to be experienced under natural field conditions following death on hot summer days (e.g., 37°C for 1 week), central nervous tissue from clinically affected cattle retains detectable BSE fibrils. This offers an opportunity to confirm a positive diagnosis of BSE even in severely autolyzed brains, provided several brain/spinal cord sites are sampled.

Agent Inactivation Studies

Two major studies on inactivation of the BSE and scrapie agents are in progress. The first is a laboratory investigation of the effectiveness of porous load autoclaving, chlorine releasing agents, and 1 M and 2 M sodium hydroxide to inactivate macerates or homogenates of brain. The following brains were used: from cattle with BSE [infectivity titer $10^{3.6}$–$10^{5.2}$ mouse intracerebral (i.c.) median infective dose $(ID_{50})/g$], brain from mice terminally affected with the ME7 strain of scrapie agent (infectivity titer c. 10^8 mouse i.c. ID_{50}/g), and brain from hamsters terminally affected with the 263K strain of scrapie agent (infectivity titer $10^{9.3}$–$10^{9.5}$ hamster i.c. ID_{50}/g (41). The second inactivation study is a study to evaluate the ability of various processes used to render animal waste in member states of the European Union (EU) to inactivate the BSE and scrapie agents (6).

Laboratory Studies

Taylor et al. (41) reported that following porous load autoclaving of the macerated brains from BSE-affected cows or scrapie-affected rodents at temperatures between 134°C or 138°C for ≤1 hour, none was completely inactivated, including that from the low-titerd BSE brain, but homogenates of brain exposed to sodium hypochlorite yielding ≤16,500 ppm of chlorine for one hour showed no detectable infectivity. If dichloroisocyanurate yielding the same concentration of chlorine was used instead, and for the same time, some infectivity survived, so this chemical cannot be recommended at these concentrations and times for inactivation of transmissible SE agents. Furthermore, neither 1 M nor 2 M sodium hydroxide was completely effective either, although useful reductions of infectivity were found. Further studies are required to establish what factors impair effective inactivation by porous load autoclaving and what criteria are needed to define satisfactory cycles. This has significance for laboratories working with transmissible SE agents, for the biologic industry, which may use ovine or bovine materials during the manufacture of products, and for inactivating procedures in hospitals caring for patients affected with transmissible SEs.

One interesting and unexplained finding in the study was that attempted inactivation of some samples of BSE and scrapie agent by NaOH for 2 hours was less effective than when exposure was for 30 min or 1 hour. The result could not be explained by a variation in pH.

Rendering Studies

Pilot scale rendering equipment was used to simulate processes that are currently in use in member states of the EU or that might be of future use

TABLE 2.5. Rendering experiments (simplified).

Type	Particle diam. (mm)	Fat content	Process time (min)	Achieved end temperature (°C)	Bioassays meal (M) tallow (T)	Sample
Batch atmospheric	150	Natural (N)	150	121	M	B
Continuous atmospheric	30	N	50	*112/122*	*M*	*CE*
			125	*123/139*	*M*	*DF*
Continuous atmospheric	30	High (H) 1:1	30/120	136/137	M	GH
Continuous vacuum	*10*	*H 4:1*	*10*	*120*	*M T*	*I*
			40	*121*	*M*	*J*
Continuous wet rendering (LT)	20 1:5:4	N	120/240	101/119/72	M	KLM
Solvent extraction	20	Greaves (G)	240	53	M	N
		G + solvent	30	75	M	O
		G + steam	15	100	M	P
Batch pressure cook (raw)	30/50	N	28/30	133/136/145	M T	QRS
Batch pressure cook (processed)	2.2 Meal	<10%	20	136/145	M	TU

Raw material 861 BSE suspect cow forebrains (10%) mixed with intestines and beef bones (90%)—ratios 1:3:6 or 1:5:4.
Mouse bioassays: (1) mixed raw material, (2) meat and bone meal, (3) tallow.
Words and figures underlined indicate failures.
Modified from Woodgate (6) and Taylor (44).

(6, 42). For each of the chosen processes, the raw material (carcass bones and offal), ratio of fat to raw material, particle size, temperature/moisture profiles, pressure, and residence time in continuous systems were selected to simulate as closely as possible the conditions in the actual plants. The study was planned in three phases. In the first phase brain material from cattle clinically suspected to have BSE was titrated in mice and used to spike the raw material, 10% of brain to 90% of bones/intestines. The second phase used brains from United Kingdom sheep with scrapie, and the third was intended to study scrapie from other member states. The third phase will not proceed, at least in the form originally planned. Following cooking (rendering) to the criteria set for each process (particle size, fat content, time, temperature, and pressure), the meat and bone meal end product and some of the tallow samples were again assayed, or titrated in mice, to determine the loss in titre. In the event the initial titer of the BSE spike was low and only permitted detection of a titer loss of about 80-fold. Nevertheless, detectable BSE infectivity survived two of these processes (and in one of these, the start titer was not reduced at all in the meat and bone meal produced, although no infectivity was found in the tallow resulting from this process) (Table 2.5). These two processes were banned for processing ruminant protein in any member state of the EU (43) from January 1, 1995. The results from phase II are awaited, and phase III cannot proceed as planned as the scrapie brains required have not been collected by other member states. An alternative study is being considered.

One interesting result from the BSE study (phase I) was that one process used to prepare test material for a subsequent study (sample M, Table 2.5) and employing a temperature of 72°C for 240 min, produced greaves (precursor of meat and bone meal) in which no infectivity could be detected.

Conclusion

The measures adopted to control BSE, and to protect public (Table 2.6) and animal health (Table 2.7), were originally based partly on previous knowledge of natural scrapie, such as that in the report of Hadlow et al. (37), but

TABLE 2.6. BSE public health controls (consumer protection).

August 1988
  Carcasses of suspects destroyed
December 1988
  Milk from suspects destroyed
November 1989
  No specified bovine offal to humans

TABLE 2.7. BSE animal health controls.

June 1988
Notifiable—restrictions[a]; isolation[b]
July 1988
No ruminant protein in ruminant feed
September 1990
No specified bovine offal to any species

[a] Movement restrictions on suspect animals.
[b] Isolation of parturient cattle in approved accommodation.

increasingly on new information provided by the epidemiology and other studies on BSE. In addition, controls have been strengthened when the results of research have indicated the need to do so. For example, the prohibition on the use of specified bovine offals, or anything produced from them, in human food, was extended to cover all species of animals and birds from September 1990. This action was recommended by the independent Spongiform Encephalopathy Advisory Committee as a result of the experimental transmission of BSE to pigs by parenteral inoculation. Results of research are now increasing confidence about the effectiveness of controls. Most importantly there is definite evidence that the epidemic is in sharp decline, that the distribution of infectivity in the bovine tissues is more restricted than in scrapie in ovine tissues, and that maternal transmission either does not occur or occurs at an insignificant level. It is important that every state veterinary service carries out a risk assessment for BSE and similar conditions in other bovidae and felidae to assess whether an epidemic is likely to be initiated by exposure to infected feed, and implements the preventative measures needed to obviate any risk that is identified.

Acknowledgments. Many people have contributed information used in this chapter including Professor R.M. Barlow and Dr. D.J. Middleton of the Royal Veterinary College; Dr. D.M. Taylor, Dr. M. Bruce, Mr. J. Foster, and Dr. H. Fraser of the Institute for Animal Health, BBSRC & MRC Neuropathogenesis Unit in Edinburgh; Dr. J.K. Kirkwood and Dr. A. Cunningham of the Zoological Society of London; Mr. P. Merson of Genus and staff of the late Milk Marketing Board at Thames Ditton; Mr. J.W. Wilesmith, Mr. G.A.H. Wells, Dr. A.E. Wrathall, and Mr. M. Dawson of the Central Veterinary Laboratory; and Mr. K.C. Taylor and Dr. D. Matthews of the Notifiable Diseases Section of MAFF at Tolworth. I thank them all and also Mrs. Y. Spencer, Mrs. C. Humphries, and Mr. J. Cook for preparation of figures and tables used in the verbal presentation, Mrs. E. Davies for typing the manuscript, and Mrs. W. Bolton for secretarial assistance.

References

1. Anon. Scrapie-like disease in a captive nyala. Animal Health 1986 report of the chief veterinary officer. London: HMSO, 1987.
2. Wells GAH, Scott AC, Johnson CT, Gunning RF, Hancock RD, Jeffrey M, et al. A novel progressive spongiform encephlaopathy in cattle. Vet Rec 1987;121:419–20.
3. Wilesmith JW, Wells GAH, Cranwell MP, Ryan JBM. Bovine spongiform encephalopathy: epidemiological studies. Vet Rec 1988;123:638–44.
4. Wilesmith JW, Ryan JBM, Hueston WD. Bovine spongiform encephalopathy: case-control studies of calf feeding practices and meat and bone meal inclusion in proprietary concentrates. Res Vet Sci 1992;52:325–31.
5. Wilesmith JW, Ryan JBM, Atkinson MJ. Bovine spongiform encephalopathy: epidemiological studies on the origin. Vet Rec 1991;128:199–203.
6. Woodgate SL. Rendering systems and BSE agent deactivation. Livestock Prod Sci 1994;38:47–50.
7. Wrathall AE, Brown KFD. Embryo transfer, semen, scrapie and BSE. In: Bradley R, Savey M, Marchant B, eds. Sub-acute spongiform encephalopathies. Proceedings of a seminar held in Brussels, November 12–14, 1990. Dordrecht: Kluwer Academic, 1991:243–53.
8. Commission Decision—94/381/EC. Concerning certain protection measures with regard to bovine spongiform encephalopathy and the feeding of mammalian derived protein. Official Journal of the European Communities, No. L 172/23, July 7, 1994, Brussels.
9. Wood JLN, Lund LJ, Done SH. The natural occurrence of scrapie in moufflon. Vet Rec 1992;130:25–7.
10. Kirkwood JW, Cunningham AA. Epidemiological observations on spongiform encephalopathies in captive wild animals in the British Isles. Vet Rec 1994; 135:296–303.
11. Kirkwood JW, Wells GAH, Cunningham AA, Jackson SI, Scott AC, Dawson M, et al. Scrapie-like encephalopathy in a greater kudu (*Tragelaphus strepsiceros*) which had not been fed ruminant-derived protein. Vet Rec 1992;130:365–7.
12. Wyatt JM, Pearson GR, Smerdon TN, Gruffydd-Jones TJ, Wells GAH. Spongiform encephalopathy in a cat. Vet Rec 1990;126:513.
13. Willoughby K, Kelly DF, Lyon DG, Wells GAH. Spongiform encephalopathy in a captive puma (*Felis concolor*). Vet Rec 1992;131:431–4.
14. Office International des Epizooties. Spongiform encephalopathy in an imported cheetah in Australia. Dis Inform 1992;5:52.
15. Peet RJ, Curran JM. Spongiform encephalopathy in an imported cheetah (*Acinonyx jubatus*). Aust Vet J 1992;69:171.
16. Kelly DF, Pearson H, Wright AI, Greenham LW. Morbidity in captive white tigers. In: The proceedings of a symposium on comparative pathology of zoo animals. Washington, DC: National Zoological Park, October 1978.
17. Schoon HA, Brunckhorst D, Pohlenz J. A contribution to the neuropathology of the red-necked ostrich (*Struthio camelus*)—spongiform encephalopathy [in German]. Verh Erkrank Zootiere 1991;33:309–14.
18. Schoon HA, Brunckhorst D, Pohlenz J. Spongiform encephalopathy in an ostrich (*Struthio camelus*). A case report [in German]. Tierarztl Prax 1991; 19:263–5.

19. Fraser H, McConnell I, Wells GAH, Dawson M. Transmission of bovine spongiform encephalopathy to mice. Vet Rec 1988;123:472.
20. Dawson M, Wells GAH, Parker BNJ. Preliminary evidence of the experimental transmissibility of bovine spongiform encephalopathy to cattle. Vet Rec 1990;126:112–3.
21. Dawson M, Wells GAH, Parker BNJ, Scott AC. Primary parenteral transmission of bovine spongiform encephalopathy to the pig. Vet Rec 1990;127:338.
22. Foster JD, Hope J, Fraser H. Transmission of bovine spongiform encephalopathy to sheep and goats. Vet Rec 1993;133:339–41.
23. Baker HF, Ridley RM, Wells GAH. Experimental transmission of BSE and scrapie to the common marmoset. Vet Rec 1993;132:403–6.
24. Robinson MM, Hadlow WJ, Huff TP, Wells GAH, Dawson M, Marsh RF, et al. Experimental infection of mink with bovine spongiform encephalopathy. J Gen Virol 1994;75:2151–5.
25. Barlow RM, Middleton DJ. Dietary transmission of bovine spongiform encephalopathy to mice. Vet Rec 1990;126:111–2.
26. Barlow RM, Middleton DJ. Is BSE simply scrapie in cattle? Vet Rec 1990;126:295.
27. Middleton DJ, Barlow RM. Failure to transmit bovine spongiform encephalopathy to mice by feeding them with extraneural tissues of affected cattle. Vet Rec 1993;132:545–7.
28. Bruce M, Chree A, McConnell I, Foster J, Pearson G, Fraser H. Transmission of bovine spongiform encephalopathy and scrapie to mice: strain variation and the species barrier. Philos Trans R Soc Lond [Biol] 1994;343:405–11.
29. Kimberlin RH, Walker CA, Fraser H. The genomic identity of different strains of mouse scrapie is expressed in hamsters and preserved on reisolation in mice. J Gen Virol 1989;70:2017–25.
30. Pattison IH, Hoare MN, Jebbett JN, Watson WA. Further observations on the production of scrapie in sheep by oral dosing with foetal membranes from scrapie-affected sheep. Br Vet J 1974;130:lxv–lxvii.
31. Fraser H, Foster JD. Transmission to mice, sheep and goats and bioassay of bovine tissues. In: Bradley R, Marchant B, eds. Transmissible spongiform encephalopathies. Proceedings of a consultation on BSE with the Scientific Veterinary Committee of the Commision of the European Communities held in Brussels, September 14–15, 1993. Brussels: CEC, 1994:145–59.
32. Report of the Working Party on Bovine Spongiform Encephalopathy. Department of Health, Ministry of Agriculture, Fisheries and Food. February 1989, London (The Southwood Report).
33. Wilesmith JW. The epidemiology of bovine spongiform encephalopathy. Semin Virol 1991;2:239–45.
34. Bradley R, Wilesmith JW. Epidemiology and control of bovine spongiform encephalopathy (BSE). Br Med Bull 1993;49:932–59.
35. Foster JD, McKelvey WAC, Mylne MJA, Williams A, Hunter N, Hope J, et al. Studies on maternal transmission of scrapie in sheep by embryo transfer. Vet Rec 1992;130:341–3.
36. Bradley R. Embryo transfer and its potential role in control of scrapie and bovine spongiform encephalopathy (BSE). Livestock Prod Sci 1994;38:51–9.
37. Hadlow WJ, Kennedy RC, Race RE. Natural infection of Suffolk sheep with scrapie virus. J Infect Dis 1982;146:657–64.

38. Hope J, Reekie LJD, Hunter N, Multhaup G, Beyreuther K, White H, et al. Fibrils from brains of cows with new cattle disease contain scrapie-associated protein. Nature 1988;336:390–2.
39. Scott AC, Wells GAH, Stack MJ, White H, Dawson M. Bovine spongiform encephalopathy: detection and quantitation of fibrils, fibril protein (PrP) and vacuolation in brain. Vet Micro 1990;23:295–304.
40. Scott AC, Wells GAH, Chaplin MJ, Dawson M. Bovine spongiform encephalopathy: detection of fibrils in the central nervous system is not affected by autolysis. Res Vet Sci 1992;52:332–6.
41. Taylor DM, Fraser H, McConnell I, Brown DA, Brown KL, Lamza KA, et al. Decontamination studies with the agents of bovine spongiform encephalopathy and scrapie. Arch Virol 1994;139:313–26.
42. Taylor DM. Deactivation of BSE and scrapie agents: rendering and other UK studies. In: Bradley R, Marchant B, eds. Transmissible spongiform encephalopathies. Proceedings of a consultation on BSE with the Scientific Veterinary Committee of the Commision of the European Communities held in Brussels, September 14–15, 1993. Brussels: CEC, 1994:205–23.
43. Commission Decision—94/382/EC. On the approval of alternative heat treatment systems for processing animal waste of ruminant origin, with a view to the inactivation of spongiform encephalopathy agents. Official Journal of the European Communities, No. L 172/25, 7 July 1994, Brussels.
44. Taylor DM, Woodgate SL, Atkinson MJ. Inactivation of the bovine spongiform encephalopathy agent by rendering procedures. Vet Rec 1995;137:605–10.

3

Preliminary Observations on the Pathogenesis of Experimental Bovine Spongiform Encephalopathy

GERALD A.H. WELLS, MICHAEL DAWSON,
STEPHEN A.C. HAWKINS, ANTHONY R. AUSTIN,
ROBERT B. GREEN, IAN DEXTER, MARK W. HORIGAN,
AND MARION M. SIMMONS

Bovine spongiform encephalopathy (BSE), a scrapie-like or "prion" disease of domestic cattle first recognized in Great Britain in 1986 (1), has provided a precedent among such diseases in its occurrence as a nationwide food-borne epidemic originating from contamination of commercially processed feed with a scrapie-like agent (2–4). Although the precise routes of infection in previously described naturally occurring scrapie-like diseases of animals remain less clear, there is substantial epidemiologic evidence that transmissible mink encephalopathy (TME), a rare disease of ranch-reared mink, is also food-borne (5), and oral, or alimentary exposure, is generally considered a likely means of transmission of natural scrapie under field conditions (6–8). It seems probable also that lateral transmission, within captive populations, of chronic wasting disease of mule deer and Rocky Mountain elk (9) results from oral contamination.

Contemporaneously associated with the BSE epidemic in Great Britain has been the novel occurrence also of scrapie-like diseases in several other mammalian species. These have included the domestic cat (10–12) and nine species of captive wild animals in zoological collections, of which six were members of the family Bovidae and three were members of the family Felidae (13; I. McCandlish, personal communication). The food-borne source of infection, from common exposure to ruminant-derived protein, is considered the likely origin also of the disease in the exotic bovids and in domestic cats (3, 14). Infection of the zoo carnivores was probably also by dietary means, most likely from exposure to parts of raw, BSE-infected, bovine carcasses containing spinal cord material (13, 15).

Further evidence of a common source of infection for BSE and the attendant spongiform encephalopathies of several other species comes from the remarkable similarity of the pattern of incubation periods and pathologic profiles in a standard panel of mouse strains on transmissions of BSE, feline spongiform encephalopathy (FSE), and the scrapie-like disease of two of the exotic species of ungulates (16).

These data therefore provide considerable evidence that the oral route of infection is of major importance in BSE and in the recently described scrapie-like diseases of other species if not in all of the scrapie group of diseases in animal species. It is only very recently that experimental proof has been obtained of the infection of cattle with BSE by the oral route (17).

Evidence from studies of BSE supports the hypothesis that the major biologic variables of host and agent, which in mouse models of scrapie determine the pathologic phenotype (18–20), are, in BSE, constant. This suggested that BSE would provide the degree of predictability required to study the pathogenesis of the disease in the natural host species. Two such pieces of evidence were available prior to the initiation of the study described in this chapter. First, uniformity of the pathology defined by the pattern of distribution and severity of the vacuolar changes in the brains of affected cattle, in both natural disease and in cases experimentally induced by parenteral inoculation, provided an indication of the phenotypic stability of the infection and gave rise to the view that the epidemic in Britain involved a single strain of scrapie-like pathogen (16, 21, 22). A later finding showed further that on primary transmission, to a standard panel of isogenic mice, seven different cases of BSE presented uniformity of the pathologic phenotype in each mouse strain (16). It can be concluded that this bovine-adapted agent was established in cattle prior to clinical recognition of the disease, and since four of the seven cases strain typed were present in the national herd in 1982, this was perhaps as early as the time of putative initial exposure of cattle to a scrapie-like agent (2, 23). Second, after injection of Friesian/Holstein and Jersey calves with BSE by simultaneous intracerebral (i.c.) and intravenous (i.v.) routes, all challenged animals succumbed to the disease with a relatively small range of incubation period (24, 25), suggesting a striking lack of variation between breeds in the response to the agent. Thus, experimental exposure of cattle by the oral route might also be expected to elicit a uniform response enabling a study of the pathogenesis of BSE in the natural host species.

This chapter describes the preliminary findings in a necessarily protracted, and as yet incomplete, experiment to determine the spatial and temporal development of infectivity and pathologic changes in cattle following oral exposure to a single dose of BSE-affected brain homogenate. We have previously reported briefly the first major finding of the study, the presence of infectivity in the ileum of challenged cattle 6 and 10 months postinoculation (p.i.) (17).

Materials and Methods

Inoculum

Material for the preparation of inoculum was obtained postmortem from a total of 75 cases of BSE, from various locations in England and Wales, slaughtered in 1991 and confirmed by histopathologic examination. From each animal, killed for statutory diagnosis, the brain stem was removed from the cranial cavity through the foramen magnum by a method previously described (26). A segment including the obex was removed from the medulla oblongata and fixed in 10% formal saline for histopathologic examination to confirm the diagnosis by originally validated criteria (27). The remaining fresh tissue was frozen temporarily at $-20°C$ prior to transportation, on ice, to the Central Veterinary Laboratory and then placed at $-70°C$ for storage. Individual samples were later thawed and homogenized in batches of four to six in a Silversen blender. The homogenized batches were pooled in a large plastic container and further blended to form a single pool of inoculum.

Animals

Forty Friesian/Holstein male calves, born in 1991, were assembled when 1 to 5 weeks old, from a total of 12 farms in England. The principal criterion for the selection of calves was that at the time of their purchase they originated in herds in which there was either no history of BSE or clinically suspect cases had recovered or had not been confirmed by histopathologic examination for the statutory diagnosis of BSE.

Rigorous maintenance of the identity of the calves throughout the study was ensured in a number of ways. On arrival at the laboratory, the ear tag number of each calf was recorded. A further laboratory reference was assigned by the use of an additional ear tag, readable at a short distance from the animal. At 8 months old, a photographic record was made of each animal, to provide identification from lateral profile and facial markings should ear tags become removed.

Blood samples were collected from all calves for analysis for the PrP gene octarepeat polymorphism (28), which is the only known significant polymorphism of the bovine PrP gene. DNA was extracted directly from ethylenediaminetetraacetic acid (EDTA) blood samples, and a 400-bp fragment of the PrP gene, containing the octarepeat region, was amplified by polymerase chain reaction (PCR). The octarepeat allelism was then determined by gel electrophoresis. All of the calves were found to be homozygous for genes with six copies of the octapeptide (6:6), the allele that has been found predominant in a sample of the cattle population in Britain, but which is not a requirement for the development of BSE (29).

Calves were housed initially in individual pens and later transferred to open-ridge ventilated yards prior to dosing. Concentrate rations were formulated free of ruminant-derived protein.

Experimental Protocol

At approximately 4 months old, 10 of the calves were randomly allocated to a control group that, for 7 weeks, was housed in loose box accommodation separate from the challenged animals, but the husbandry regimen was otherwise identical in both groups. Dosing of the remaining 30 calves was carried out at 4 months old. This age was selected for dosing as it is within the period of calfhood exposure to meat and bone meal under field conditions (30) and because of the need to minimize the impact of potential intercurrent disease losses in the study, especially from enteric and respiratory diseases prevalent in conventionally reared animals up to this age.

A necessary prerequisite for the success of any temporal study of pathogenesis is that the experimental design ensures a dosage of agent that will induce disease in all challenged animals. However, there are no data available as yet on the dose-response relationships of infection of cattle with BSE. The dose used in this study was therefore one considered great enough to be certain of producing the disease in all test animals. Calves were each dosed orally with 100 g of the pool of BSE-affected brain stem homogenate. This was administered in two successive 50-mL doses using new sterile syringes and ensuring that the homogenate was deposited on the base of the tongue at the entrance to the pharynx. The 10 control calves received no treatment.

It was assumed to be inevitable that in the immediate p.i. period fecal excretion of inoculum infectivity would occur with the risk of environmental contamination and the possible consequent variable supplementation of the effective individual challenge dose. To prevent this, as far as was possible, a strict p.i. husbandry was implemented. This was based on data reviewed on the passage of digesta through the alimentary tract of cattle. In the dairy cow, particulate matter is excreted over the period from 12 hours to 10 days after ingestion; in smaller ruminants, the maximum excretion time is shorter (31). Challenged calves were kept on wood shavings, and feces were removed from the yards several times a day for 4 days, and thereafter once daily for another 10 days p.i. Thereafter, straw bedding was introduced and replenished daily to give a deep litter system. The challenged and control groups remained in separate housing until 7 weeks p.i., then the controls were returned to the yards with the challenged calves. At 14 weeks p.i. all of the calves were turned out to double-fenced pasture. They were held at pasture for a further 32 weeks and then rehoused in covered yards for the duration of the study.

Clinical observations were maintained throughout the study by a number of methods. Passive observations, during daily husbandry routines and weekly visits by veterinary staff, were made to detect behavioral changes and signs associated with BSE (1, 2, 32–34). Responses to handling and restraint were assessed during routine weighing at approximately 2-month intervals from 9 months p.i. Open-field study testing was conducted at approximately 2-month intervals, beginning 18 months p.i., in an area that permitted free expression of behavior of individual animals in isolation from their companions. This particularly enabled assessment and detection of abnormalities of kinesis and gait. Behavior studies were made over 24-hour periods by passive observation at approximately 3 month intervals from 12 months p.i. Gross behavior in terms of percentage time spent eating, ruminating, and "idling" were determined (35). Other behavioral events were recorded if deemed unusual or significant by observers. Clinical neurologic examinations were performed within 7 days of a planned necropsy.

Challenged and control animals were randomly allocated to a sequential kill protocol at the start of the experiment. Starting at 6 months old (2 months p.i.) and thereafter at intervals of 4 months, until 22 months p.i., three challenged calves and one control calf were killed at each time point. The sequential kill protocol was based on a median incubation period of 4 to 6 years in the epidemic of BSE (36), and on the assumption that the experimental dose probably represented a significantly greater exposure than that which results in disease in the field (37, 38) and, therefore, would produce disease within a shorter incubation period. Subsequently, with no evidence of the onset of clinical signs or histopathologic changes in the central nervous system (CNS) by 2 years p.i., the sequential kill protocol was extended to 6-month intervals to allow for the possibility that the incubation period might more closely reflect the median incubation period in the epidemic estimated from the epidemiologic data on age-specific incidence (36). A revised protocol was temporarily adopted to accommodate a maximum incubation period of 56 months (instead of the original 38 months) but, on the basis of subsequent experimental observations, it was modified again to give the final schedule shown in Figure 3.1.

At each of the sequential kills, a large range of tissues was sampled aseptically and stored at −70°C for mouse bioassay of infectivity (Table 3.1). With the exception of cerebrospinal fluid, feces, urine, blood, bone marrow, bone, pericardium, mitral valve, and dura mater, a similar range of tissues was sampled for histopathologic examination. The eyes, thyroid, cardiac thymus, left cranial lobe of the lung, and myocardium (right ventricle and interventricular septum) were sampled only for histo-

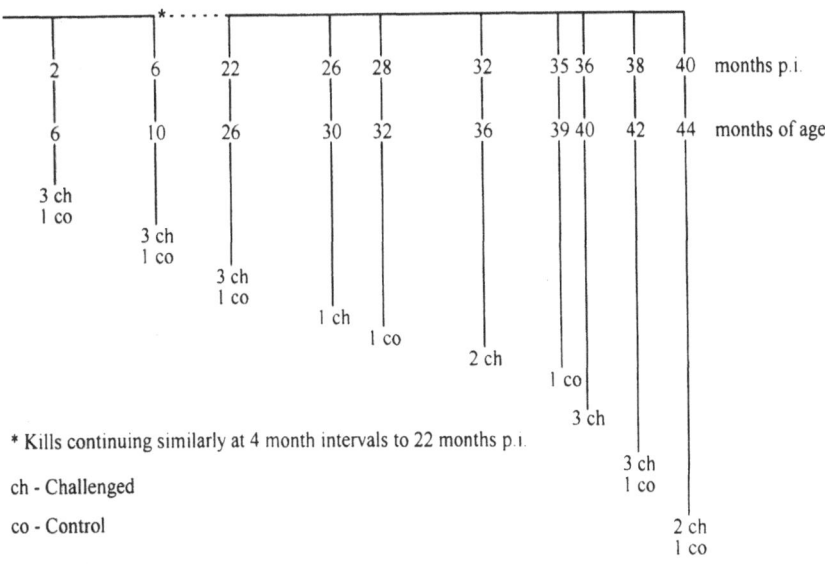

FIGURE 3.1. Schedule of sequential kill time points of challenged and control cattle.

pathologic examination. Tissues for histologic processing were collected into either phosphate-buffered, neutral, 10% formalin (BF) (viscera and skeletal muscles), or 10% formal saline (CNS).

Mouse Bioassay of Cattle Tissue Infectivity

From 2 g of the remaining original brain stem pool used to inoculate calves, a 10% homogenate in saline was prepared for assay of infectivity in RIII (Sinc[s7] genotype) isogenic mice. All inocula from organs of the experimental cattle were prepared for assay either from single tissue pools from each group of challenged calves or from single tissues from the control calf, killed at each time point in the study. To prepare inoculum, 2 g of tissue (comprising equal amounts of the identical tissue in the case of the challenged animals of each time point) were macerated and ground in a minimum volume of sterile saline, diluted to give a 10% suspension in saline, and finally passed through a gauze filter. The inoculum was stored as 4-mL aliquots at −70°C. Ampicillin (5 mg per aliquot) was added to suspensions prepared from tongue, rumen, abomasum, duodenum, distal ileum, spiral colon, tonsil, retropharyngeal lymph node, and turbinate epithelium.

Test and control tissue inocula were each assayed in 20 RIII inbred mice according to standard procedures (39). Each mouse was injected with 0.02 mL i.c. and 0.1 mL by the intraperitoneal (i.p.) route. Detailed clinical monitoring of the mice was carried out from 250 days p.i. and the clinical end point when mice show clear signs of neurologic disease was determined

TABLE 3.1. Tissues sampled fresh for mouse bioassay of infectivity.

Neural
  Brain: frontal cortex, caudal medulla
  Pituitary
  CSF
  Dura*
  Spinal cord: C2–C3, T10–T11, L3–L4
  Nodose ganglia
  Dorsal root ganglia: C3–C6, T5–T8
  Trigeminal ganglia
  Stellate ganglia
  Sciatic nerve
  Facial nerve
  Phrenic nerve
  Radial nerve*
  Semitendinosus muscle*
  Diaphragmatic muscle*
  Triceps muscle
  Masseter muscle
  Sternocephalicus muscle
  Longissimus dorsi muscle

Alimentary
  Tongue (dorsum, to include mucosa)
  Submandibular salivary gland
  Parotid salivary gland
  Cranial esophagus*
  Rumen
  Omasum*
  Abomasum (pyloric)
  Duodenum
  Distal ileum (including Peyer's patches)
  Spiral colon
  Feces*
  Pancreas
  Liver

Lymphoreticular
  Spleen
  Thymus (cervical)
  Tonsil
  Submandibular lymph node
  Retropharyngeal lymph node
  Bronchial-mediastinal lymph node
  Hepatic lymph node
  Mesenteric lymph node
  Prescapular lymph node
  Popliteal lymph node

Other
  Kidney
  Urine*
  Adrenal*
  Lung (left caudal lobe)
  Nasal mucosa (midturbinate)
  Pericardium
  Heart (left ventricle)
  Mitral valve*
  Aorta*
  Blood (buffy coat)
  Blood (serum)*
  Blood (clot)*
  Bone marrow (sternum)
  Collagen (Achilles tendon)*
  Skin*
  Bone (femoral diaphysis)*

*Tissues sampled and stored but not included in current bioassays.

according to established criteria for calculating incubation period (40). The brain of each mouse was removed, fixed in BF, and processed routinely for histopathologic examination for evidence of a scrapie-like encephalopathy. Any mice surviving to the end point for incubation and expression of the disease were killed and examined similarly, as were any mice that died or were killed due to intercurrent disease. The end point was 650 days, except in groups of mice inoculated with distal ileum and mesenteric lymph node from challenged cattle killed 2 months p.i. Here the end point was extended to 700 days in view of results obtained from tissues from challenged cattle killed 6 months p.i.

Results

While the study is as yet incomplete, a number of important observations have been made and here we report the progress to date.

Of the 12 source herds for recipient calves, 8 have subsequently (January 1995) experienced cases of BSE in homebred animals. These herds have contributed 24 of the cattle used in the study—5 control and 19 challenged animals. None of these animals belongs to the 1991-born cohorts that were used in this study. To date in the experiment, only challenged cattle have developed disease; all controls have remained healthy.

The clinical signs observed in these cattle were the same as those described previously in naturally occurring cases of BSE (1, 2, 32–35) and can be summarized under five headings: emotionality changes, expressed in increased apprehension, fear, and startle or in decreased responsiveness; sensory changes, seen as hyperesthesia or hyperreflexia; adventitial movements, comprising muscle fasciculation, tremor, or myoclonus; ataxia of gait, including hypermetria; and evidence of autonomic dysfunction in the form of decreased rumination.

Until 24 months p.i., no abnormalities were identified. The resistance shown to handling and restraint varied between animals and was considered to be within the normal range, but over the next 6 months most animals became notably more placid regarding interactions with handlers and were particularly bold in approaching observers. However, the intensity and the combination of clinical signs that indicated definite onset of clinical disease were observed at 35 months p.i. in one animal (number 277) and 37 months p.i. in another (261). The signs included changed emotionality with increased startle; sensory changes, evident as excessive nose licking, odontoprisis, and an increased sensitivity to touching the head; and an ataxia of gait. Changes in emotionality were the first indications of abnormality in these two animals but the specific features differed between them. One (277) was observed to be unusually nervous in open field testing at 33 months p.i., and this intensified to fear of handling and being moved by 35 months p.i. The other animal (261), by contrast, did not show evidence of increased fear, and although it could be startled by unexpected and unfamiliar stimuli, was generally hyporesponsive to handling and restraint. Besides these two animals with definite clinical disease, five of the other six inoculated animals remaining at 36 months p.i. exhibited occasional nose licking, head rubbing, sneezing, and yawning, indicative of sensory changes and probably an early clinical manifestation of disease (34). One animal (299) was mildly ataxic when hustled.

Hyporesponsiveness was evident to some degree in inoculated animals other than 277 and 261 surviving at 37 months p.i. (numbers 192, 296, 300) sufficient to enable them to be handled in a small pen for physiologic recordings without restraint. Kinesis in the open-field test was reduced in the two animals (261, 277) with the onset of clinical disease and there was some decline in the three other animals (192, 296, 300). There was also evidence of autonomic dysfunction from reduced time spent ruminating in the five challenged animals (192, 261, 277, 296, 300) surviving at 37 months p.i.

In summary, unequivocal clinical signs of BSE were first noted in one of the challenged cattle at 35 months p.i. Two of the three challenged cattle killed at 36 months p.i. were suspected to have early clinical features. By 37 months p.i. (status as of January 1995), clear evidence of clinical signs were present in two (261, 277) of the five remaining challenged cattle with early clinical manifestations suspected in the other three animals (192, 296, 300). At the time of writing two groups of cattle remain to be killed at 38 and 40 months p.i.

Histopathological examinations of the brain and spinal cord of cattle killed from 2 months p.i. to 26 months p.i. did not reveal any significant vacuolar changes. At 32 months p.i., when two challenged cattle were killed, vacuolar changes were detected in the CNS, but only in the vestibular nuclear complex of the pons of one animal (232). At 36 months p.i., in one of the challenged animals (299) in which early clinical signs were suspected, there was a mild spongiform encephalopathy, in which the lesion distribution pattern was typical of natural cases of BSE. No diagnostically significant vacuolar changes could be demonstrated in the other two challenged cattle. Histopathologic examinations of all other tissues sampled are as yet incomplete.

Mouse bioassay of the donor cattle brain pool confirmed infectivity with an incubation period (mean ± SEM) of 373 ± 7.6 days in RIII mice and no mice surviving to the assay end point. Mouse bioassay for infectivity of recipient tissues has provided evidence of infectivity in the distal ileum of challenged cattle killed 6, 10, 14, and 18 months p.i., but not 2 months p.i. or in any of the control animals through a similar range of time points. The current status of mice inoculated with this tissue is given in Figure 3.2. Only the assays of tissues from cattle killed 2 and 6 months p.i. are completed. No evidence of infectivity was detected at 2 months p.i. in any of the tissues, or at 6 months p.i. in any of the tissues other than the distal ileum. All other bioassays are incomplete.

Discussion

The onset of clinical disease in all challenged animals remaining in the study during the period 35 to 37 months is, given the single exposure dose, consistent with the premise and supportive data cited earlier that the major biologic variables of host and agent, which determine the pathologic phenotype in the scrapie-like diseases (41, 42), are constant in BSE of cattle. The random occurrence and low average incidence of BSE cases per herd, and indeed, several other features of the epidemic in the United Kingdom, are

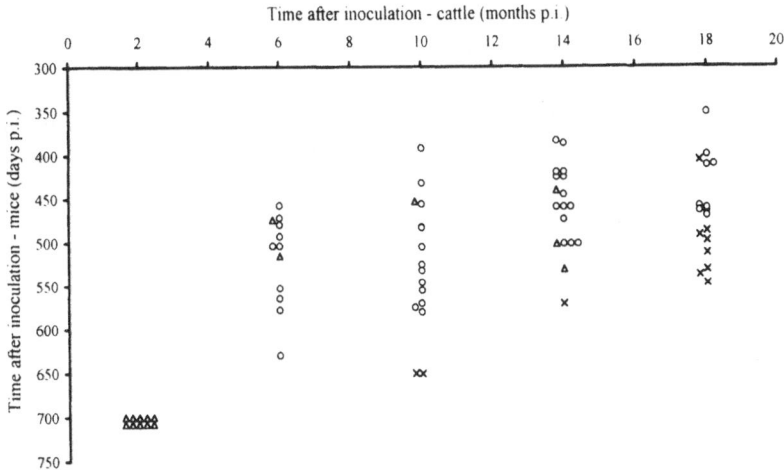

FIGURE 3.2. Mouse bioassay results (January 1995) of distal ileum inocula from challenged cattle killed 2 to 18 months p.i. Δ, SE negative; O, SE positive; ×, pending.

consistent with very low average exposure via feed and an absence of genetic variation in the cattle population (38). Given the large experimental exposure in the present experiment, it is interesting that the minimum incubation period observed in the cattle is in the lower part of the incubation period distribution, estimated from age-specific incidence in the epidemic of BSE (36). Information on the magnitude of a dose-related effect on incubation period after oral challenge with BSE is anticipated from a further experimental study that is in progress. The study examines the attack rate and incubation period of BSE in groups of cattle exposed orally to 3 × 100 g (successive daily doses), 1 × 100 g, 1 × 10 g, or 1 × 1 g of an inoculum of pooled BSE-affected brain stems.

The potential exposure of some of the recipient calves to natural BSE on their farms of origin remains a possible confounding factor in the study, although there is no evidence, which would be apparent from the development of disease in control animals, of this having occurred to date.

The observations presented here on the temporal development of clinical signs and neuropathologic changes in the CNS are incomplete. The determination of incubation period from the insidious onset of clinical signs in BSE, and other scrapie-like diseases, is problematic and particularly influenced by the criteria used and the rigor of their application. In this study clinical monitoring involved behavioral observations that may, on further analysis, indicate earlier abnormalities than those reported here. Nevertheless, the development of overt clinical signs and diagnostically significant vacuolar changes in the brain were in close temporal association. This is in contrast to observations in those rodent models of scrapie where the development of vacuolar changes relative to incubation period has been studied after infection with rodent adapted agents. Vacuolar changes in the brain have been first observed between 53% and 77% of the incubation period after i.c. inoculation (18, 43, 44) and similarly about halfway through the incubation period after intraocular inoculation (45). In Compton White (CW) mice infected i.p. with the 139A scrapie agent, vacuolar changes were detected first in the thoracic spinal cord at approximately 38% of the incubation period, and first in the brain at 65% of the incubation period (46). Comparable data for natural host species, wild type agents and natural routes of infection are not available, but, in a study of the pathogenesis of TME, mink inoculated subcutaneously with mink passaged TME agent first developed spongiform changes in the brain at about 88% of the incubation period (47).

Most of the knowledge we have on the pathogenesis of scrapie-like diseases has been obtained from studies of experimental scrapie in laboratory rodents. In experimental mouse scrapie, when infection is introduced by nonneural routes, replication of the infectious agent necessarily involves tissues of the lymphoreticular system (LRS) before it is demonstrable in the CNS (for reviews see 45, 48). The same holds true for exposure via the intragastric route when replication is initiated in Peyer's patches almost immediately (49). A very similar pattern of LRS pathogenesis before infection of the CNS has been shown in natural scrapie of sheep (50). Although in rodent models it is likely that there is initial lymphatic and, or hematogenous spread of infection to the LRS from primary sites of infection after experimental parenteral nonneural exposure, it is not the means by which infection reaches the CNS. On the contrary there is much more evidence that in such models spread to the CNS is by neural pathways, involving the peripheral nerve innervation of the LRS (48, 51).

Based on this background and the putative sheep scrapie origins of BSE, assumptions have been made in constructing a control strategy for BSE. Measures have included the removal of specified offals from all slaughtered cattle more than 6 months old and, as an extreme precautionary measure, most recently, the removal of small intestine and thymus from calves more than 2 months old to obviate potentially infected LRS tissues entering human and animal food chains (52, 53).

As briefly reported previously (17), this study has confirmed experimentally, for the first time, that cattle can be infected with BSE by oral exposure. After the single exposure of the cattle, infectivity was detectable in the distal ileum at 6 months p.i. and at all time points through to at least 18 months p.i. Replication of the agent in the tissue over this time is indicated by a progressive reduction in incubation period (Fig. 3.2). Because mouse bioassays of tissues taken from 10 months p.i. and later are incomplete, or not yet started, it is clearly premature to conclude that infectivity will not be detectable subsequently in any other tissue. In addition to the considerable amount of further information to come from the bioassay of tissues in this study there is also the potential for assay of additional tissues that have been collected and stored, should it be indicated (Table 3.1).

Further results are required to determine whether subsequent replication, spread, and persistence of agent in spleen and regional lymph nodes, as observed in the rodent models of scrapie and in natural scrapie before neuroinvasion, can be detected in these cattle. Nevertheless, the findings are consistent with previous observations on the pathogenesis of natural scrapie in sheep and of experimental scrapie in rodents when infection is by nonparenteral routes because there is early replication and persistence of infection in distal ileum in which Peyer's patches are concentrated.

Interestingly, mouse bioassay of a range of tissues, including the distal ileum, from terminally affected BSE cattle has not detected infectivity other than in the CNS (54).

This raises the possibilities of differences between the pathogenesis of BSE and natural scrapie and differences between this experimental exposure and the natural exposure of cattle to BSE. Two possible phenomena may be operating. The first relates to the potential for differences in pathogenesis of scrapie-like diseases between species and suggests that cattle with BSE may be more similar in this respect to mink with TME than to sheep with natural scrapie. In mink with TME, only very low titers of infectivity can be demonstrated in extraneural tissues (55). Study of the temporal distribution of TME agent in mink, after subcutaneous inoculation, has shown extremely limited replication of agent in lymphoid tissues prior to detection of infectivity in the CNS (47), indicating that high titers in the LRS are not a prerequisite of neuroinvasion. Furthermore, only after replication occurred in the CNS was there dissemination of agent to other extraneural tissues. Despite this latter feature, it is generally accepted that the low extraneural infectivity in mink, and therefore the low potential for shedding of agent, is consistent with the epidemiologic evidence that TME is not spread naturally and suggests that mink are not natural hosts of the TME agent (47). Clear comparisons can be drawn here with BSE, but it is probable that the relative inefficiency of titration of cattle tissue infectivity in mice, across a species barrier, as used in the present study and in assays of tissues from terminally affected cattle, could obscure the synthesis by failing to detect low concentrations of agent, whereas in the study of TME by Hadlow and colleagues (47), bioassays were optimized by titration in mink, thus avoiding species barrier effects.

The second phenomenon relates more to a possible difference in exposure between the experimental oral transmission of BSE to cattle and the natural disease and is therefore independent of species barrier effects. One illustration of this is seen when different doses of the 87V mouse scrapie strain are inoculated by the intraperitoneal route in Sinc^{p7} mice (for review see 6). While high doses result in disease, although not invariably and with a wide range of incubation periods, lower doses fail to produce scrapie in the mice, yet there is rapid replication and persistence of infectivity in spleen in both cases. In this example, the higher dose of inoculum apparently allows the agent to access cells that are necessary for neuroinvasion to occur, whereas lower doses fail to produce infection in these cells. It is clear, from the occurrence of disease, that neuroinvasion is a feature in both natural BSE and in the present study so facilitation of neuroinvasion is not the issue here; it is simply that in the experimental model of BSE, exposure to high doses of agent may enable it to gain access to a greater number of different cells in Peyer's patches than occurs with natural BSE infection, with the result that agent replicates to detectable titers.

The mean incubation period (days $\pm$ SEM) in RIII (Sinc^{s7} genotype) mice inoculated i.c. and i.p. with a 10^{-1} dilution of the donor brain stem tissue pool was greater than the mean incubation periods in previous comparable transmissions from seven cases of BSE (314 ± 3 to 335 ± 7 days) (16), suggesting that the oral inoculum pool may have contained less infectivity than in transmissions from single cases of the natural disease. Nevertheless, the dose used in this study is probably several orders of magnitude greater than that estimated to cause natural cases of BSE (38). Natural effective exposures of cattle to BSE may have an identical pathogenesis to that shown in this study but infectivity may be below the limits of detection by bioassay in mice, especially across the species barrier. Indeed, this is the most likely explanation of the failure to detect infectivity in the distal ileum of calves at 2 months p.i. in this study. The limit of detectability of infectivity in mice is estimated at approximately $10^{2.0}$ i.c. LD_{50}/g of BSE affected brain (56) but could be greater depending on the extent of the species barrier. Further experimental studies are in progress to address this in relation to BSE by a comparative titration of a BSE inoculum in mice and cattle.

Acknowledgments. The authors wish to thank Dr. M.E. Bruce, Mr. J.W. Wilesmith, and Dr. R.H. Kimberlin for helpful discussions, and Dr. T.C. Martin, Mr. S.L. Hughes, and Mr. K.J. Hughes for cattle PrP genotyping. They also acknowledge colleagues in the Veterinary Investigation Service for collection of the donor brain material and the staff of Pathology, Virology and Animal Services at the Central Veterinary Laboratory for skilled technical support.

References

1. Wells GAH, Scott AC, Johnson CT, Gunning RF, Hancock RD, Jeffrey M, Dawson M, Bradley R. A novel progressive spongiform encephalopathy in cattle. Vet Rec 1987;121:419–20.
2. Wilesmith JW, Wells GAH, Cranwell MP, Ryan JBM. Bovine spongiform encephalopathy: epidemiological studies. Vet Rec 1988;123:638–44.
3. Wilesmith JW, Wells GAH. Bovine spongiform encephalopathy. Curr Top Microbiol Immunol 1991;172:21–38.

4. Wilesmith JW. An epidemiologist's view of bovine spongiform encephalopathy. Philos Trans R Soc Lond [Biol] 1994;343:357–61.

5. Marsh RF, Hadlow WJ. Transmissible mink encephalopathy. Rev Sci Tech Off Int Epiz 1992;11:539–50.

6. Kimberlin RH. Unconventional slow viruses. In: Collier LH, Timbury MC, eds. Topley and Wilson's principles of bacteriology, virology and immunity. 8th ed. London: Edward Arnold, 1990:671–93.

7. Hadlow WJ. To a better understanding of natural scrapie. In: Bradley R, Savey M, Marchant B, eds. Proceedings of a seminar in the CEC Agricultural Research Programme held in Brussels, 12–14 November 1990, sponsored by the Commission of the European Communities, Directorate-General for Agriculture, Division for the Co-ordination of Agricultural Research. Dordrecht: Kluwer Academic, 1991:117–30.

8. Carp RI, Kascak RJ, Rubenstein R. Pathogenesis of unconventional slow virus infections. In: Liberski PP, ed. Light and electron microscopic neuropathology of slow virus infections. Boca Raton: CRC Press, 1993:33–61.

9. Williams ES, Young S. Spongiform encephalopathies in Cervidae. Rev Sci Tech Off Int Epiz 1992;11:551–67.

10. Wyatt JM, Pearson GR, Smerdon TN, Gruffydd-Jones TJ, Wells GAH, Wilesmith JW. Naturally occurring scrapie-like spongiform encephalopathy in five domestic cats. Vet Rec 1991;129:233–6.

11. Pearson GR, Wyatt JM, Henderson JP, Gruffydd-Jones TJ. Feline spongiform encephalopathy: a review. In: Raw M-E, Parkinson TJ, eds. The veterinary annual, vol 33. London: Blackwell Scientific Publications, 1993:1–10.

12. Fraser H, Pearson GR, McConnell I, Bruce ME, Wyatt JM, Gruffydd-Jones TJ. Transmission of feline spongiform encephalopathy to mice. Vet Rec 1994; 134:449.

13. Kirkwood JK, Cunningham AA. Spongiform encephalopathy in captive wild animals in Britain: epidemiological observations. In: Bradley R, Marchant B, eds. Transmissible spongiform encephalopathies. Proceedings of a consultation on BSE with the Scientific Veterinary Committee of the Commission of the European Communities held in Brussels, 14–15 September 1993. Dordrecht: Kluwer Academics, 1994:29–47.

14. Kimberlin RH. Bovine spongiform encephalopathy. Rev Sci Tech Off Int Epiz 1992;11:347–90.

15. Willoughby K, Kelly DF, Lyon DG, Wells GAH. Spongiform encephalopathy in a captive puma (*Felis concolor*). Vet Rec 1992;131:431–4.

16. Bruce M, Chree A, McConnell I, Foster J, Pearson G, Fraser H. Transmission of bovine spongiform encephalopathy and scrapie to mice: strain variation and the species barrier. Philos Trans R Soc Lond [Biol] 1994;343:405–11.

17. Wells GAH, Dawson M, Hawkins SAC, et al. Infectivity in the ileum of cattle challenged orally with bovine spongiform encephalopathy. Vet Rec 1994;135: 40–1.

18. Fraser H, Dickinson AG. The sequential development of the brain lesions of scrapie in three strains of mice. J Comp Pathol 1968;78:301–11.

19. Fraser H. The pathology of natural and experimental scrapie. In: Kimberlin RH, ed. Slow virus diseases of animals and man. Amsterdam: North-Holland, 1976:265–305.

20. Bruce ME, Fraser H. Scrapie strain variation and its implications. In: Chesebro BW, ed. Transmissible spongiform encephalopathies. Scrapie, BSE and related human disorders. Curr Topics Microbiol Immunol 1991;172:125–38.

21. Wells GAH, Hawkins SAC, Hadlow WJ, Spencer YI. The discovery of bovine spongiform encephalopathy and observations on the vacuolar changes. In: Prusiner SB, Collinge J, Powell J, Anderton B, eds. Prion diseases of humans and animals. Chichester: Ellis-Horwood, 1992:256–74.

22. Wells GAH, Hawkins SAC, Cunningham AA, Blamire IWH, Sayers AR, Harris P. Comparative pathology of the new spongiform encephalopathies. In: Bradley R, Marchant B, eds. Transmissible spongiform encephalopathies. Proceedings of a consultation on BSE with the Scientific Veterinary Committee of the Commission of the European Communities held in Brussels, 14–15 September 1993. Dordrecht: Kluwer Academics, 1994:327–45.

23. Wilesmith JW, Ryan JBM, Atkinson, MJ. Bovine spongiform encephalopathy: epidemiological studies on the origin. Vet Rec 1991;128:199–203.

24. Dawson M, Wells GAH, Parker BNJ. Preliminary evidence of the experimental transmissibility of bovine spongiform encephalopathy to cattle. Vet Rec 1990;126:112–3.

25. Dawson M, Wells GAH, Parker BNJ, Scott AC. Transmission studies of BSE in cattle, hamster, pigs and domestic fowl. In: Bradley R, Savey M, Marchant B, eds. Proceedings of a seminar in the CEC Agricultural Research Programme held in Brussels, 12–14 November 1990, sponsored by the Commission of the European Communities, Directorate-General for Agriculture, Division for the Co-ordination of Agricultural Research. Dordrecht: Kluwer Academic, 1991: 25–32.

26. Bradley R, Matthews D. Sub-acute, transmissible spongiform encephalopathies: current concepts and future needs. Rev Sci Tech Off Int Epiz 1992;11:605–34.

27. Wells GAH, Hancock RD, Cooley WA, Richards MS, Higgins RJ, David GP. Bovine spongiform encephalopathy: diagnostic significance of vacuolar changes in selected nuclei of the medulla oblongata. Vet Rec 1989;125:521–4.

28. Goldmann W, Hunter N, Martin T, Dawson M, Hope J. Different forms of the bovine PrP gene have five or six copies of a short, G-C-rich element within the protein-coding exon. J Gen Virol 1991;72:201–4.

29. Hunter N, Goldmann W, Smith G, Hope J. Frequencies of PrP gene variants in healthy cattle and cattle with BSE in Scotland. Vet Rec 1994;135:400–3.

30. Wilesmith JW, Ryan JBM, Hueston WD. Bovine spongiform encephalopathy: case-control studies of calf feeding practices and meat and bone meal inclusion in proprietary concentrates. Res Vet Sci 1992;52:325–31.

31. Church DC. Passage of digesta through the gastro-intestinal tract. In: Church DC, ed. Digestive physiology and nutrition of ruminants, vol 1. Corvallis, OR: O.S.U. Book Stores, 1969:85–99.

32. Wells GAH. Bovine spongiform encephalopathy. In: Grunsell CSG, Raw M-E, Hill FWG, eds. Veterinary annual, 29th issue. London: Wright, 1989:59–63.

33. Wilesmith JW, Hoinville LJ, Ryan JBM, Sayers AR. Bovine spongiform encephalopathy: aspects of the clinical picture and analyses of possible changes 1986–1990. Vet Rec 1992;130:197–201.

34. Austin AR, Hawkins SAC, Kelay NS, Simmons MM. New observations on the clinical signs of BSE and scrapie. In: Bradley R, Marchant B, eds. Transmissible spongiform encephalopathies. Proceedings of a consultation on BSE with the

Scientific Veterinary Committee of the Commission of the European Communities held in Brussels, 14–15 September 1993. Dordrecht: Kluwer Academics, 1994:277–87.

35. Austin AR, Simmons MM. Reduced rumination in bovine spongiform encephalopathy and scrapie. Vet Rec 1993;132:324–5.

36. Wilesmith JW, Ryan JBM. Bovine spongiform encephalopathy: recent observations on age-specific incidences. Vet Rec 1992;130:491–2.

37. Kimberlin RH. Bovine spongiform encephalopathy: an appraisal of the current epidemic in the United Kingdom. Intervirol 1993;35:208–18.

38. Kimberlin RH, Wilesmith JW. Bovine spongiform encephalopathy: epidemiology, low dose exposure and risks. Ann NY Acad Sci 1994;724:210–20.

39. Bruce ME, Chree A, McConnell I, Foster J, Pearson G, Fraser H. Agent strain variation in BSE and scrapie. In: Bradley R, Marchant B, eds. Transmissible spongiform encephalopathies. Proceedings of a consultation on BSE with the Scientific Veterinary Committee of the Commission of the European Communities held in Brussels, 14–15 September 1993. Dordrecht: Kluwer Academics, 1994:189–204.

40. Dickinson AG, Meikle MH, Fraser H. Identification of a gene which controls the incubation period of some strains of scrapie agent in mice. J Comp Pathol 1968;78:293–9.

41. Dickinson AG, Fraser H. Scrapie: pathogenesis in inbred mice: an assessment of host control and responses involving many strains of agent. In: ter Meulin V, Katz M, eds. Slow virus infections of the central nervous system. New York: Springer-Verlag, 1977:3–14.

42. Kimberlin RH, Cole S, Walker CA. Transmissible mink encephalopathy (TME) in Chinese hamsters: identification of two strains of TME and comparisons with scrapie. Neuropathol Appl Neurobiol 1986;12:197–206.

43. Masters CL, Rohwer RG, Franko MC, Brown P, Gajdusek DC. The sequential development of spongiform change and gliosis of scrapie in the Golden Syrian hamster. J Neuropathol Exp Neurol 1984;43:242–52.

44. Baringer JR, Bowman KA, Prusiner SB. Replication of the scrapie agent in hamster brain precedes neuronal vacuolation. J Neuropathol Exp Neurol 1983;42:539–47.

45. Scott JR. Scrapie pathogenesis. Br Med Bull 1993;49:778–91.

46. Cole S, Kimberlin RH. Pathogenesis of mouse scrapie: dynamics of vacuolation in brain and spinal cord after intraperitoneal infection. Neuropathol Appl Neurobiol 1985;11:213–27.

47. Hadlow WJ, Race RE, Kennedy RC. Temporal distribution of transmissible mink encephalopathy virus in mink inoculated subcutaneously. J Virol 1987; 61:3235–40.

48. Kimberlin RH, Walker CA. Pathogenesis of experimental scrapie. In: Bock G, Marsh J, eds. Novel infectious agents and the central nervous system. Ciba Foundation Symposium 135. Chichester: John Wiley, 1988:37–62.

49. Kimberlin RH, Walker CA. Pathogenesis of scrapie in mice after intragastric infection. Virus Res 1989;12:213–220.

50. Hadlow WJ, Kennedy RC, Race RE. Natural infection of Suffolk sheep with scrapie virus. J Infect Dis 1982;146:657–64.

51. Kimberlin RH. Scrapie: How much do we really understand? Neuropathol Appl Neurobiol 1986;12:131–47.

52. Wells GAH, Wilesmith JW, McGill IS. Bovine spongiform encephalopathy: a neuropathological perspective. Brain Pathol 1991;1:69–78.
53. The Spongiform Encephalopathy (Miscellaneous Amendments) Order 1994. Statutory Instrument No. 2627. London: HMSO, 1994.
54. Fraser H, Foster J. Transmission to mice, sheep and goats and bioassay of bovine tissues. In: Bradley R, Marchant B, eds. Transmissible spongiform encephalopathies. Proceedings of a consultation on BSE with the Scientific Veterinary Committee of the Commission of the European Communities held in Brussels, 14–15 September 1993. Dordrecht: Kluwer Academics, 1994: 145–59.
55. Marsh RF, Hadlow WJ. Transmissible mink encephalopathy. Rev Sci Tech Off Int Epiz 1992;11:539–50.
56. Kimberlin RH. A scientific evaluation of research into bovine spongiform encephalopathy (BSE). In: Bradley R, Marchant B, eds. Transmissible spongiform encephalopathies. Proceedings of a consultation on BSE with the Scientific Veterinary Committee of the Commission of the European Communities held in Brussels, 14–15 September 1993. Dordrecht: Kluwer Academics, 1994:455–77.

4

Recent Observations on the Epidemiology of Bovine Spongiform Encephalopathy

John W. Wilesmith

Bovine spongiform encephalopathy (BSE) became a statutorily notifiable disease in Great Britain in June 1988 following its recognition in 1986 (1). Initial epidemiologic studies provided evidence that cattle had become infected by a scrapie-like agent via infected meat and bone meal used as a protein supplement (2). A subsequent case-control study of calf-feeding practices substantiated this hypothesis (3), and action to prevent further exposure from the food-borne source was taken in July 1988 when the feeding of ruminant derived protein to ruminants was banned.

Despite this timely intervention, the epidemic of BSE in Great Britain has been the largest example of a feed-borne transmissible spongiform encephalopathy. This is because of the protracted incubation period, as a result of which the mean age at onset of clinical signs at the peak of the epidemic was 60 months. The epidemic has resulted in a considerable economic cost to the nation, major factors being the financial compensation paid to owners of suspect cases that are compulsorily slaughtered and the cost of incinerating the carcasses of such animals as well as the detriment to the export of animals and their products.

In addition, there has been the cost of staff of the State Veterinary Service, who visit reported suspect cases and conduct the clinical examinations, and for the histologic examination of brain tissue for all animals slaughtered. However, this has facilitated an unprecedented epidemiologic monitoring of the epidemic with a standard questionnaire being completed for all suspect cases slaughtered. The accumulated data have allowed analyses and enabled specific analytical studies to gain as full an understanding as possible of the epidemiology of BSE. This chapter presents a summary of the results (as of March 1995) of these analyses and specific analytical studies.

Observations on the Incidence of BSE

Mean Age at Onset of Clinical Signs and Annual Age-Specific Incidences

Table 4.1 provides the mean ages at onset of clinical signs for each 12-month (July to June) birth cohort, from 1973–74 to 1991–92, for all confirmed cases for which a complete date of birth was available, i.e., the month, day, and year.

The annual age-specific incidences of BSE in herds with homebred cases in each year 1989 to 1994 for data accumulated by December 31, 1994 are shown in Table 4.2. This indicates the first possible discernible effect of the ban on the feeding of ruminant derived protein to ruminants, which was a reduction in the incidence in 2-year-old animals in 1991. This was followed by a reduction in incidence in 3-year-old animals commencing in 1992, 4-year-old animals in 1993, and 5-year-old animals in 1994.

TABLE 4.1. Mean ages at onset (months) of clinical signs of BSE by 12-month birth cohorts.

Birth cohort[a]	No. of cases[b]	Mean age	SD	Range	Mode	Median
1973/74	1	187.0	—	—	—	—
1974/75	0	—	—	—	—	—
1975/76	0	—	—	—	—	—
1976/77	4	154.3	9.5	140–160	140	159
1977/78	10	148.7	28.0	117–203	133	138
1978/79	14	132.0	15.3	110–155	110	134
1979/80	61	116.8	18.2	91–167	95	115
1980/81	127	106.2	21.5	63–165	91	101
1981/82	384	93.0	19.5	62–158	89	89
1982/83	1,968	80.4	18.1	43–145	69	76
1983/84	5,218	72.3	16.5	30–137	64	69
1984/85	8,199	67.8	15.5	30–126	60	65
1985/86	12,280	66.0	14.2	30–114	63	64
1986/87	22,391	64.2	12.6	24–103	64	63
1987/88	36,471	60.8	10.9	21–91	60	60
1988/89	12,712	59.0	8.6	24–87	60	59
1989/90	5,593	53.8	6.4	26–85	53	54
1990/91	695	46.7	5.1	20–75	50	47
1991/92	21	35.4	2.9	29–39	36	36

[a] July to June.
[b] Data accumulated by March 31, 1995.

TABLE 4.2. Age-specific incidences of BSE in herds with homebred cases in each year 1989 to 1994.

Age (years)	1989	1990	1991	1992	1993	1994[a]
2	0.07 (24,068)[b]	0.06 (34,565)	0.02 (47,822)	0.02 (55,657)	0.01 (52,290)	0.01 (40,164)
3	0.079 (44,375)	1.32 (64,220)	1.87 (88,898)	0.66 (102,999)	0.45 (95,921)	0.20 (74,374)
4	3.42 (39,163)	3.95 (56,435)	5.77 (78,517)	6.69 (90,663)	3.29 (83,276)	2.31 (63,914)
5	3.50 (33,786)	3.47 (48,013)	3.91 (66,018)	6.56 (76,239)	8.13 (69,715)	4.66 (52,767)
6	1.57 (26,959)	1.99 (38,773)	1.98 (52,701)	2.80 (60,332)	5.03 (55,019)	6.17 (41,585)
7	0.44 (20,126)	0.70 (29,135)	0.91 (39,756)	1.04 (45,131)	1.68 (40,877)	3.20 (30,316)
8	0.18 (14,173)	0.19 (20,618)	0.35 (28,093)	0.59 (31,780)	0.73 (28,671)	1.17 (21,485)
9	0.06 (9,091)	0.08 (13,275)	0.14 (18,592)	0.28 (20,785)	0.42 (18,695)	0.52 (14,357)
10	0.02 (5,399)	0.05 (7,703)	0.07 (11,092)	0.08 (12,285)	0.35 (11,052)	0.57 (8,474)
11	0.10 (2,990)	0.05 (4,261)	0.03 (6,064)	0.06 (6,617)	0.16 (6,198)	0.34 (4,654)
12	0.00 (2,284)	0.06 (3,203)	0.04 (4,548)	0.09 (5,487)	0.06 (4,946)	0.11 (3,784)

[a] Data accumulated by March 31, 1995.
[b] Number of animals at risk.

The Epidemic Curve

The number of cases of BSE by month and year of onset of clinical signs up to August 1994 are shown in Figure 4.1; figures for the subsequent 6 months cannot yet be presented because it is possible that a proportion of the cases now being reported will have a date of clinical onset during this period. The epidemic curve indicates that the peak of the epidemic occurred at the end of 1992/beginning of 1993. This peak incidence of cases represented a maximum mean annual incidence of BSE in adult animals of less than 1%. Since 1993 the incidence has declined.

Incidence by Herd Type and Annual Within-Herd Incidences

Cumulatively, 54.0% of dairy herds have experienced at least one confirmed case of BSE compared with 14.5% of beef suckler herds. If the latter

NO. OF CASES.

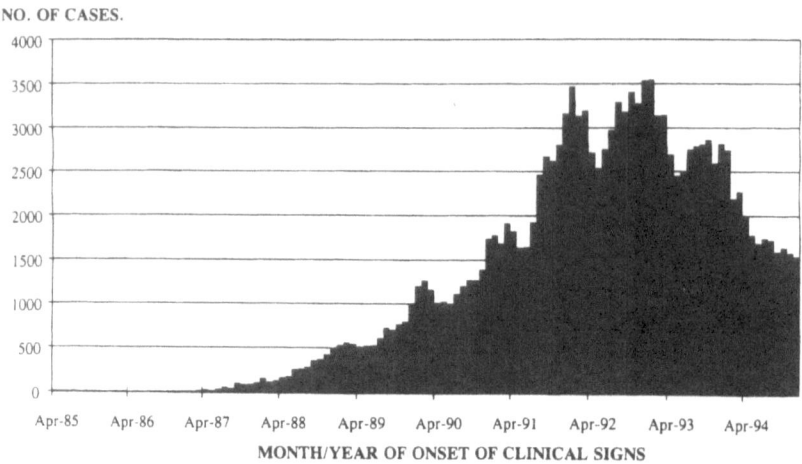

MONTH/YEAR OF ONSET OF CLINICAL SIGNS

FIGURE 4.1. Number of cases of BSE by month and year of onset of clinical signs April 1985–December 1994.

incidence is restricted to beef suckler herds with homebred cases, it becomes 10.8%.

The 6-month within-herd incidences in the two affected herd types from 1988 to 1994 are shown in Table 4.3. This indicates a generally low mean within-herd incidence that has not exceeded 3% and has shown a reduction since 1992.

Geographical Incidence of Affected Dairy Herds

Figure 4.2 depicts the cumulative incidence of BSE affected dairy herds by county in Great Britain. This indicates a marked difference in the risk of infection for herds in the southern counties of England compared with those in Scotland, with the latter having experienced a notably lower risk.

TABLE 4.3. Within-herd incidences (%) in BSE affected herds in 6-month periods 1988 to 1994.

Period	1988	1989	1990	1991	1992	1993	1994[a]
Jan–June	1.8	2.0	2.1	2.3	2.7	2.5	2.3
July–Dec	1.8	1.9	2.2	2.5	2.7	2.4	2.2

[a] Data accumulated by March 31, 1995.

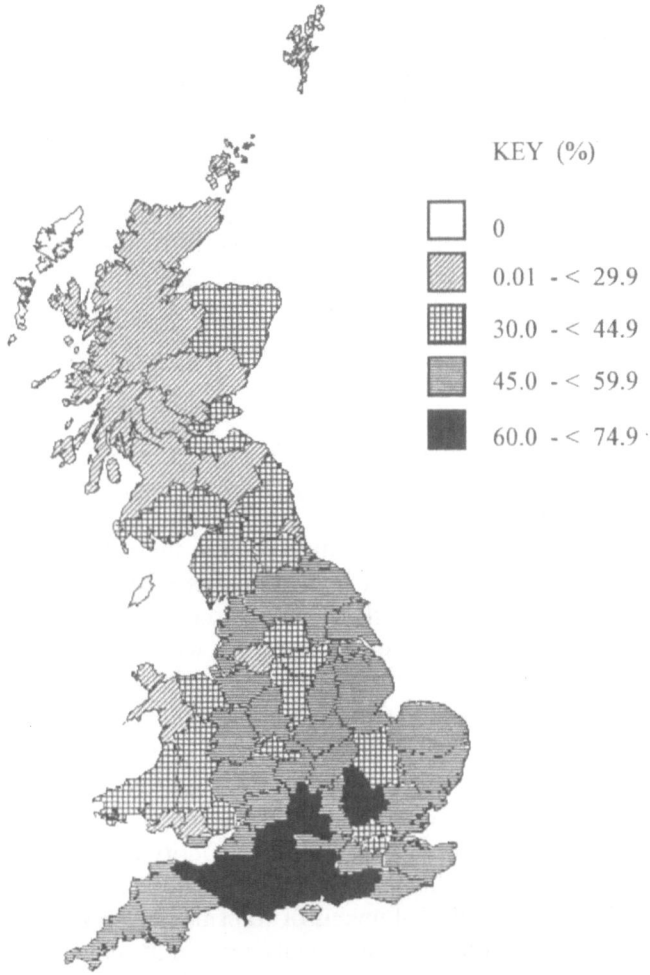

FIGURE 4.2. Cumulative incidence (%) of BSE affected dairy herds by county in Great Britain, from November 1986 to March 31, 1995.

Analytic Epidemiologic Studies

Cohort Study of Maternal Transmission

This study was initiated in July 1989 by the purchase of the animals needed for the study from naturally affected herds. The animals purchased from each affected herd comprised one or more pairs of animals, of which one member of the pair was the offspring of a case of BSE and the other was an animal born in the same calving season whose dam had reached 6 years of age and had not, at the time of purchase, succumbed to BSE. In the case of the latter member of the pair, the main epidemiologic database is used to check whether the dams of such offspring have subsequently developed BSE. If they have, irrespective of the dam's age, the pair is removed from the study. The animals in this study will be allowed to reach 7 years of age unless death or the need to slaughter an animal supervenes. The youngest animal in the study will reach 7 years of age in November 1996. As expected, because of the potential exposure of animals in the study to infected feed before purchase, cases of BSE have occurred, but it will not be possible to interpret the results until the study is completed at the beginning of 1997.

Within-Herd Case-Control Study of Cases of BSE in Animals Born on or After October 30, 1988

Detailed case studies of BSE in animals born in the months after July 18, 1988 indicated that there was a high probability of exposure for these animals to commercial feedstuffs manufactured before that date, which were already in the food chain or on farms, but which were not identifiable as such.

These case studies provided sufficient information to enable an effective case-control study of cases born some months after the statutory ban on the feeding of ruminant-derived protein to ruminants to be designed. One objective was to seek evidence of maternal transmission of BSE in advance of the availability of results of the definitive cohort study outlined above, and another was to determine the probability of lateral transmission of BSE occurring. The latter potential means of transmission was investigated by comparing the risk of infection for animals born on the day or within 3 days of the calving of a cow that subsequently developed BSE with that for animals born in the same herds at other times.

The results of this study (4) provided no evidence that the offspring of animals that are subsequently affected with BSE are at increased risk of developing BSE. A marginally statistical significant risk for animals born on the day, or between 1 and 3 days after a subsequently affected animal calved, was found, but this was unlikely to be indicative of a causal association. Even adjusting for the exposure to infected animals that calved, but had been culled from the herd before developing clinical signs of BSE, these two means of transmission, maternal and horizontal, could not account for the majority of cases in animals born on or after November 1, 1988.

In view of these findings, more in-depth analyses of the descriptive epidemiologic data were conducted. The first of these was a reexamination of the proportional distribution of home-bred cases in each region of Great Britain (i.e., cases in animals occurring in their natal herds). The results are

TABLE 4.4. Distribution of confirmed, homebred cases of BSE in Great Britain by region.

Region	1985/86 (%)	1986/87 (%)	1987/88 (%)	1988/89 (%)	1989/90 (%)	1990/91 (%)
	Birth cohort					
North	8.3	9.1	9.8	11.3	12.4	19.7
Middle and west	17.3	19.3	19.1	21.4	21.2	17.9
East	6.7	7.2	7.2	11.4	15.5	14.6
Southeast	14.1	12.0	11.9	14.7	15.2	13.3
Southwest	43.4	41.4	40.3	31.3	26.3	25.0
Wales	6.9	8.1	8.0	6.9	6.8	7.2
Scotland	3.3	2.9	3.7	3.0	2.8	2.4
Total number in cohort	10,140	18,309	29,090	10,098	4,384	543

shown in Table 4.4, indicating a notable increase in the proportional incidence of BSE in the northern and eastern regions of England.

As these regions contain a substantial proportion of the pig and poultry populations in England the second analysis examined the correlation between the incidence of BSE in animals born after the feed ban and the ratio of these two species to adult dairy cattle in each county of England. More specifically, this involved calculating the cumulative incidence of BSE in animals born after October 30, 1988, in each county using the population of adult dairy cows recorded at the Ministry of Agriculture Fisheries and Food's Annual Agricultural Census of June 1985 as the denominator. The population of total pigs and domestic avian species (referred to below as poultry) excluding ducks and geese were also derived from this census.

The results of the analysis of the correlation between the incidence of cases of BSE in animals born after October 30, 1988 and the ratio of the total pig population to that of adult cattle in each county is shown in Figure 4.3. A significant correlation was found ($r = .55, p = .00002$). Similar results (data not shown) were obtained for correlations with both the ratios of the numbers of poultry to the combined population of pigs, and poultry to the adult population of dairy cows.

Discussion

The reduction in the mean age at onset of clinical signs of BSE in successive cohorts typifies the effect of right censoring due to the protracted incubation period distribution, for which the true mean has not changed during the course of an epidemic. The 1987 and later-born cohorts are most affected by this, as the majority of the observed cases of BSE have been infected in calfhood (3). The animals in earlier-born cohorts, which developed disease

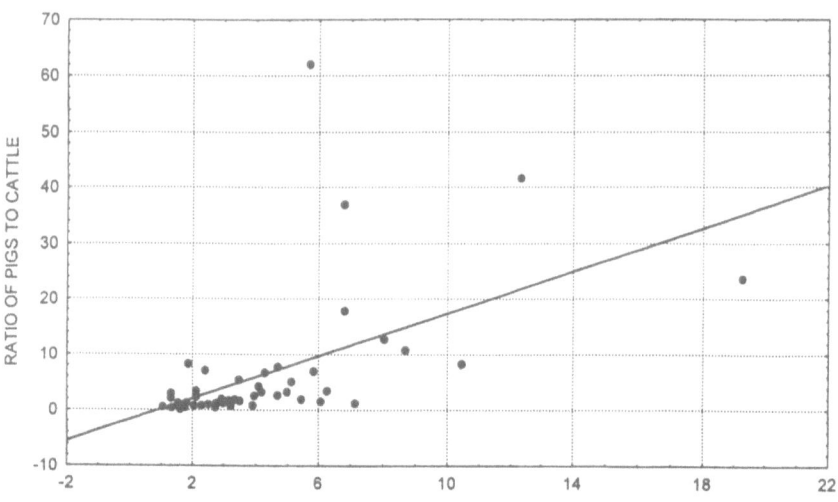

FIGURE 4.3. Correlation of the incidence of BSE in animals born after October 30, 1988 and the ratio of the total population of pigs to adult dairy cattle in each county of England.

at a much older age, are likely to have been due to infection in adulthood after 1981, when effective exposure appears to have commenced (5).

The observations on the incidence of BSE, particularly the annual age-specific incidences, during the course of the epidemic have indicated that the statutory action to prevent further food-borne exposure has had the expected effects. The reductions in incidence in 2- and 3-year-olds from 1991 to 1992 did not result in a decline in the national incidence because these cases represented a small proportion of the total. However, the subsequent reduction in the incidence in 4- and 5-year-old animals, the modal ages at onset at the peak of the epidemic, has resulted in the decline in the national incidence from 1993. This decline has been sustained and the current (March 1995) number of suspect cases of BSE being reported is approximately 50% of that for the same period in 1994.

The decline in the national animal incidence has been mirrored in the annual incidence of BSE within affected herds (Table 4.2). This has been remarkably low and is indicative of the low-dose exposure that has occurred throughout the epidemic (6). All of these expected reductions in incidence as a result of the statutorily based intervention substantiate the hypothesis that cattle became exposed via infected meat and bone meal.

The decline in the annual within-herd incidences of BSE does not provide any evidence of cattle-to-cattle transmission other than the food-borne source. However, to determine whether or not other means of transmission can occur the necessary analytical studies have been initiated or completed.

The cohort study to investigate the possibility of maternal transmission is necessarily a long-term study and it is not possible to draw any interim conclusions before its completion in 1997. In the meantime, the within-herd case-control study of cases born on or after November 1, 1988 has provided a means to examine the risk of maternal transmission (4). Although this means of transmission could not sustain the epidemic of BSE in Great Britain (7), its possible occurrence has been actively investigated, as it could be detected earlier than horizontal transmission and would indicate the possible occurrence of this means of transmission. The statistical power for the design of the case-control study in terms of the sample size was set at an 80% probability of detecting an odds ratio of 2 or greater using a 95% confidence level for an exposure that occurred at a prevalence of 5% in unaffected animals. The results of this study therefore provide convincing evidence that if maternal transmission of BSE occurs it is at a very low rate and no additional control measures on this aspect are warranted. These results are consistent with the observations on the epidemic in general, and in particular, with the results of the routinely conducted analysis to compare the expected incidence of BSE in offspring of confirmed dams with that expected from the past food-borne exposure. This analysis has failed to reveal greater incidence in the offspring of confirmed cases (8).

The within-herd case-control study provided no evidence for the horizontal transmission of BSE, leading to the general conclusion that neither maternal nor horizontal transmission could explain the very low incidence of BSE in animals born after July 18, 1988, when the legislation to prevent further exposure of the cattle population to the food-borne source was enacted. Detailed case studies of cases in animals born in 1988 after this ban and those born in 1989 have indicated that the majority could have been exposed via feed that was manufactured before July 18, 1988 and was in the supply chain or on farm for some 6 months subsequently.

The follow-up studies revealing the change in the geographic distribution of cases, and the correlation between the incidence of BSE in animals born after October 30, 1988 and the populations of pigs and poultry, provide a possible explanation for the continued exposure from feedstuffs, albeit at a very much reduced risk. This is that cross-contamination of cattle feedstuffs occurred in mills also producing pig and poultry rations, in which ruminant protein derived from specified bovine offal could have been legitimately used until September 1990. After this time it was illegal to use specified bovine offals (SBO) for the production of meat and bone for feeding to any species. These offals originally comprised the brain, spinal cord, tonsils, spleen, and intestines of animals over 6 months of age that were selected in studies on the presence of scrapie infectivity in tissues of naturally affected sheep (9). The prohibition has since been extended to include thymus and intestines from calves under 6 months old. To date infectivity has only been detected, by bioassay in mice, in brain, and spinal cord of terminal cases of natural BSE (10). The practical effect of this statutory control measure, which appears to involve a wider range of tissues than is scientifically justifiable, and so may be overcautious, should be to prevent the use of meat and bone meal produced from any potentially infected bovine tissue in any feed mill. The occurrence of BSE in cattle born since the introduction of controls in the use of SBO suggests that the measure has not been completely successful, and studies are planned to investigate further the possibility that exposure of animals born in 1989 and 1990 occurred as a result of accidental contamination of cattle feedstuffs. Potential importing countries may well need more confirmatory epidemiologic evidence to explain the occurrence of BSE in animals born in this period.

In summary, the epidemic of BSE was the result of exposure from infected meat and bone meal, with the majority of cases having resulted from the recycling of infected cattle tissues via meat and bone meal. The statutory prohibition of the feeding of ruminant derived protein resulted in an immediate marked reduction in the risk of exposure. There is no evidence of any other means of transmission capable of sustaining the epidemic, but cases of BSE in animals born after this ban are likely to continue for some years at a very low incidence as a result of accidental cross-contamination of cattle feedstuffs that has already occurred.

Acknowledgments. The studies described in this chapter were funded by the Ministry of Agriculture, Fisheries, and Food. I am indebted to Kevin Taylor and Dr. Danny Matthews for their help in facilitating the cohort and case-control studies. My thanks are also due to the staff of the epidemiology department, notably Judi Ryan for maintaining the BSE database and ensuring the validity of the results of the analyses presented. Finally, this chapter would not have appeared in a legible form without the typing skills of Lan-ea Mann.

References

1. Wells GAH, Scott AC, Johnson CT, Gunning RF, Hancock RD, Jeffrey M, et al. A novel progressive spongiform encephalopathy in cattle. Vet Rec 1987;121:419–20.
2. Wilesmith JW, Wells GAH, Cranwell MP, Ryan JBM. Bovine spongiform encephalopathy: epidemiological studies. Vet Rec 1988;123:638–44.
3. Wilesmith JW, Ryan JBM, Hueston WD. Bovine spongiform encephalopathy: case-control studies of calf feeding practices and meat and bone inclusion in proprietary concentrates. Res Vet Sci 1991;52:325–31.
4. Hoinville LJ, Wilesmith JW, Richards MS. An investigation of risk factors for cases of bovine spongiform encephalopathy born after the introduction of the "feed ban." Vet Rec 1995;136:312–8.
5. Wilesmith JW, Ryan JBM, Atkinson MJ. Bovine spongiform encephalopathy: epidemiological studies on the origin. Vet Rec 1991;128:199–203.
6. Kimberlin RH, Wilesmith JW. Bovine spongiform encephalopathy (BSE): epidemiology, low dose exposure and risks. Ann NY Acad Sci 1994;724:210–20.

7. Wilesmith JW, Wells GAH. Bovine spongiform encephalopathy. In: Chesebro BW, ed. Scrapie, Creutzfeld-Jakob disease and other spongiform encephalopathies. Current Topics in Microbiology and Immunology, vol 172. Berlin-Heidelberg: Springer-Verlag, 1991:21–38.
8. Wells GAH, Wilesmith JW. The neuropathology and epidemiology of bovine spongiform encephalopathy. Brain Pathol 1995;5:91–103.
9. Hadlow WJR, Kennedy RC, Race RE. Natural infection of Suffolk sheep with scrapie virus. J Infect Dis 1982;146:657–64.
10. Fraser H, Foster JD. Transmission to mice, sheep and goats and bioassay of bovine tissues. In: Bradley R, Marchant B, eds. Transmissible spongiform encephalopathies. Proceedings of a consultation on BSE with the Scientific Veterinary Committee of the Commission of the European Communities, Brussels, 14–15 September 1993. Brussels: European Commission, 1994:145–59.

Part II

Transmission and Pathogenesis of
Spongiform Encephalopathies

5

Scrapie: Studies on Vertical and Horizontal Transmission

JAMES L. HOURRIGAN AND ALBERT L. KLINGSPORN

Scrapie is the oldest known malady in the group of diseases known as the transmissible spongiform encephalopathies. The means by which scrapie spreads from animal-to-animal and flock-to-flock has been a subject of controversy, particularly whether transmission is vertical or horizontal, or both. The significant aspects of research efforts to define and evaluate the nature of vertical and horizontal transmission are presented. These include diagnosis and diagnostic problems, preclinical and subclinical infection, detection of scrapie agent in tissues of the reproductive tract, and the effect of different levels of exposure.

Materials and Methods

The Mission Field Trial Station is located on 450 acres of pastureland, near the city of Mission, in south Texas. It was developed to provide information for the Scrapie Eradication Program on the broad aspects of epidemiology, including diagnosis and the study of vertical and horizontal transmission. Previously exposed sheep and goats, and later, unexposed animals, were purchased and brought to the station, bred and maintained there under close observation for extended periods, usually until they died of scrapie or from natural causes. The 547 previously exposed sheep brought to the station were of the Suffolk, Cheviot, Hampshire, and Montadale breeds. The animals, purchased as field outbreaks occurred, represented 21 blood-lines in which natural scrapie was confirmed. At Mission, the animals were maintained on buffel grass pasture (and hay) supplemented with alfalfa hay pellets and mineralized salt. No meat or bonemeal was fed. Lambing and kidding occurred where the females were maintained. At weaning, young females were temporarily separated from their dams, while young males were moved to the location where only males were kept. Males had no further contact with females except for those males selected for breeding and taken temporarily to a breeding pen. Numbered duplicate ear tags were placed in both ears of all animals. Sheep were also wool-branded. Animals were observed daily for clinical signs of scrapie. Brain and other tissues from sheep and goats were submitted to the U.S. Department of Agriculture, National Veterinary Services Laboratories (NVSL), Ames, Iowa, for microscopic examination (1).

Mice were inoculated at the Mission Station Laboratory, with some inoculated at NVSL. Random-bred Swiss mice, purchased from a commercial laboratory, were inoculated intracerebrally (i.c.) when 3 weeks old. Each mouse showing signs suggestive of scrapie was examined microscopically. Central nervous system (CNS) tissues were prepared by Cajal's gold sublimate technique, and a positive finding was based on characteristic astrocytosis as described by Hadlow (personal communication). Mice found dead were not examined; however, when an inoculation experiment was terminated at the end of 2 years, at least two of any remaining mice were examined microscopically. The animals' age (months) when they died or were killed was used as the age affected, and they were not considered affected unless confirmed by microscopic examination or mouse inoculation. Sheep or goats of doubtful category, suspicious or inconclusive, were not considered "at risk" or used to calculate scrapie incidence except as otherwise noted. Sheep or goats considered to have been at risk had to have reached at least 24 months of age and be negative histologically; those brought to the station and still alive had to have reached 100 months of age, and those born at Mission, at least 60 months of age.

Diagnosis of Scrapie

Efforts to eliminate scrapie in sheep begin with an accurate diagnosis of the disease. For two centuries, observation of clinical signs was the only means to identify affected animals. Nearly a century ago vacuolated neurons were reported in the medullas of affected sheep (2). This finding was confirmed (1, 3–7) and expanded into a practical postmortem diagnostic procedure. The ability to identify infected animals was advanced in 1961 when it was reported that scrapie could be transmitted to mice by inoculation (8). Although affected sheep may exhibit a variety of syndromes, the typical clinical signs have changed little over the past 250 years (9). Early signs of scrapie are usually subtle changes in behavior or temperament. These are followed by more obvious clinical signs that are progressive and include pruritus, debilitation, rubbing against fixed objects (often accompanied by rapid movements of tongue and lips), grinding of teeth, biting of feet and limbs, locomotor incoordination, and loss of wool, condition, and weight.

Palsson (10) described clinical signs of *Rida* (scrapie) in Iceland that were similar to those generally observed, except that the Icelandic sheep usually did not exhibit pruritus; however, in recent years, pruritus has often been marked in early stages of the disease. In Japan, Sato et al. (11) reported a confirmed case in a 17-month-old Corriedale ewe with no signs of pruritus. Zlotnik and Katiyar (12) reported that sheep in India manifested a severe pruritus with absence of protracted incoordination. Dickinson et al. (13) observed that in Britain there was most often pruritus and incoordination but that either could be virtually absent in some Suffolk sheep and that extensive microscopic lesions could be present in the absence of both. Clark (14, 15) observed in a monitoring and control program that 15% of sheep exhibiting combinations of the classical signs of pruritus, emaciation, ataxia, and hyperesthesia were negative microscopically, and 26% of sheep found dead for no obvious reason were found to be positive microscopically. In contrast, clinically normal sheep found dead in the Mission flock were rarely microscopically positive. Clinical signs observed in 203 confirmed cases (200 ewes and 3 rams) in the Shetland Islands were that 39% exhibited pruritus and emaciation; 18% showed pruritus, emaciation, and hyperesthesia; and 9% showed these three signs together with ataxia. Most affected sheep were Shetland or Shetland cross with a few being Cheviot, Cheviot cross, and one was Suffolk. It was concluded that scrapie cannot be consistently diagnosed only on clinical signs.

Clark (16) described clinical signs observed in 301 natural cases of scrapie in sheep and goats at the Mission Station. Typical signs in sheep included early changes in temperament with apprehension and excitement aggravated by noise or motion stimuli, and bilateral pruritus, often beginning in the tail head or rump area. Some animals scratched or rubbed the head or sides with the rear feet and bit at the rump, sides, abdomen, and legs. Intermittent muscle tremors of the head and neck, incoordination, and a high-stepping stilted gait were also seen. Massage of the back or loins could elicit the characteristic scratch reflex action. The course varied from weeks to several months and occasionally more than one year. Appetite was usually maintained, but loss of weight and condition occurred. Signs in Targhee sheep were generally similar to those seen in Suffolks, except that only about 70% of the Targhees exhibited pruritus. Only some 31% of the Rambouillet sheep showed pruritus and a few displayed the "trotting" syndrome and would collapse to the ground and briefly remain motionless.

Comparison of Clinical Signs, Microscopic Examination, and Bioassay

The diagnosis of scrapie is usually less difficult than the discussion of diagnostic problems suggests. The signs seen in a suspect sheep are normally such that an experienced clinician can feel confident, particularly if the suspect animal is observed in familiar surroundings over a period of time to permit the disease to run its course. If this is accomplished, and diagnostic tissues are properly collected, prepared, and dispatched to the laboratory, an experienced pathologist can usually confirm the clinical diagnosis. Difficulties may arise for a variety of reasons. The suspect animal may die of other causes before clinical signs or microscopic changes are well developed. In the hot climate at Mission, autolysis occurs quickly. Conversely, in cold climates, brain tissue can be frozen in transit or because it was placed in a freezer, resulting in the tissue being too damaged for satisfactory histologic examination.

With few exceptions, animals that died or were sacrificed at Mission were found to be conclusively either positive or negative both clinically or histologically; however, some were neither clearly positive nor negative. Tissues from 158 such animals, 24 to 159 months old, were used to compare the various combinations of clinical and histologic diagnosis with the results of the mouse bioassay. These results are shown on Table 5.1, which also indicates whether the dam of a particular donor was positive, negative, or unclassified. The latter were not known to be clearly positive or negative for various reasons. The closest correlations were in the positive x positive (PxP) sheep where 43/46 (93%) were also positive in mice. In the case of clinically positive and histologically inconclusive or negative sheep 16/21 (76%) were positive in mice; and for NxN sheep 8/39 (21%) were positive in mice. The rate was 10/12 (83%) in PxN sheep and 3/5 (60%) in NxP sheep; 6/14 (43%) in clinically negative and histologically inconclusive or positive sheep. Overall, 22/29 (76%) of those that had positive dams were also positive in mice, compared with 39/69 (57%) of those with negative dams and 27/60 (45%) of those that had unclassified dams.

TABLE 5.1. Natural scrapie: inoculation of mice with tissues from adult sheep that were positive (P), suspicious (S), or negative (N) on clinical diagnosis, and histologically were positive (P), inconclusive (I), or negative (N).

Age of donors (24–159 months)			Status of dam of donor					
Status			Dam positive		Dam negative		Dam not classified	
Clinical	Histologic	Number of donors	Number of donors	Positive in mice	Number of donors	Positive in mice	Number of donors	Positive in mice
P	P	46	9	9	19	17	18	17
P	I	9	2	2	6	3	1	1
P	N	12	1	1	8	7	3	2
S	P	6	—	—	5	2	1	1
S	I	9	1	1	4	2	4	2
S	N	23	1	1	12	5	10	1
N	P	5	1	1	2	2	2	
N	I	9	4	2	3	1	2	
N	N	39	10	5	10		19	3
Total		158	29	22 76%	69	39 57%	60	27 45%

Detwiler (17) reported the mouse inoculation procedure had been discontinued by the U.S. Department of Agriculture because of the cost and excessive time required; however, the mouse inoculation procedure was very useful in helping resolve the diagnostic problems presented in Table 5.1. Only 57 of 158 (36%) of the suspects were considered affected on the basis of being positive clinically and histologically, or positive histologically only. When mouse inoculations results were included in the diagnostic procedure, the total number of confirmed cases increased to 96, a rise of 69%. This increase was particularly evident in sheep up to 35 months of age.

Detection of PrP Protein as a Diagnostic Test

More recently, additional diagnostic methods have been considered. Scrapie-associated fibrils (SAFs) reported in extracts of infected rodent brains have since been found in other animal species (18). In addition to their presence in laboratory animals and sheep, SAF and/or the major protein component (PrP or prion protein) has been detected in naturally affected goats (19), wild ruminants in zoos (20), deer and elk with chronic wasting disease (CWD) (21), domestic cats (22), puma (23), cheetah (24), and cattle with bovine spongiform encephalopathy (BSE) (25). The close correlation of SAF, or PrP, being demonstrated in sheep clinically and histologically positive for scrapie and their absence in sheep believed not to be affected encouraged using these procedures as a diagnostic test for sheep scrapie (25–30). The tissues most often evaluated were brain, spleen, and selected lymph nodes, particularly the mesenteric and prescapular nodes.

Observations of Race et al. (31) further supported a diagnostic role for PrP, which was detected in 17 of 18 brains from sheep that had clinical scrapie, and in 87% of spleens and lymph nodes from sheep that had PrP-positive brains. PrP was detected in 9 of 10 brains of the 18 sheep that were clinically positive but microscopically only suggestive or inconclusive, thus detecting cases that otherwise would have been missed.

A capillary microelectrolysis technique applied to urine in Alzheimer's disease as applicable to the diagnosis of scrapie was suggested, but no supporting data were available (32).

Preclinical/Subclinical Infection

Subclinical infection in scrapie is an extremely important issue due to the long incubation period, with lambs often being exposed and becoming infected at an early age and not developing clinical disease until reaching age 2 to 4 years or more. To determine the rate of preclinical/subclinical infection in apparently normal sheep at the Mission Station, tissues from 73 normal lambs up to 23 months of age (Table 5.2) and from 86 adult sheep 2 to 13 years old (Table 5.3) were inoculated into mice. Tissues used

TABLE 5.2. Natural scrapie: inoculation of mice with tissues from lambs considered negative clinically and histologically.

| Age of donors | Status of dam of donor | | | | | |
| | Dam positive | | Dam negative | | Dam not classified | |
	Number of donors	Positive in mice	Number of donors	Positive in mice	Number of donors	Positive in mice
At birth	1	—	5	—	—	—
<1 month	1	—	2	—	2	—
1 month	1	—	—	—	—	—
2 months	1	—	2	—	—	—
3 months	3	—	3	—	—	—
4 months	4	1	2	—	1	—
5–11 months	11	5	11	1	3	—
12–17 months	11	6	—	—	2	—
18–23 months	2	2	3	—	2	—
Total	35	14	28	1	10	
		40%		4%		

included spleen, lymph node, and brain, and in some cases additional organs. Sheep breeds represented in the study were Suffolk (70%), Hampshire (7%), Cheviot (10%), Montadale (6%), Rambouillet (4%), and Targhee (2%). All animals were considered clinically and histologically negative. As shown in Table 5.2, tissues from 14 of 35 (40%) of lambs up to 23 months old, and progeny of affected dams, were positive when inoculated i.c. into mice compared with 1 of 28 (4%) of those born to negative dams, and 0 of 10 born to unclassified dams.

Hadlow et al. (33) studied the temporal distribution of virus in certain of these animals. Virus was first detected in lymphoid tissues and intestine of normal lambs 10 to 14 months old. Detection of small amounts of virus, first

TABLE 5.3. Natural scrapie: inoculation of mice with tissues from adult sheep considered negative clinically and histologically.

| Age of donors | | Status of dam of donor | | | | | |
| | | Dam positive | | Dam negative | | Dam not classified | |
Years	Months	Number of donors	Positive in mice	Number of donors	Positive in mice	Number of donors	Positive in mice
2	24–35	4	3	5	1	—	—
3 & 4	36–59	4	—	9	—	7	—
5 & 6	60–83	7	—	10	—	12	—
7 & 8	84–107	2	—	5	—	1	—
9 & 10	108–131	2	—	2	—	9	—
11–13	132–167	—	—	—	—	7	—
Total	86	19	3	31	1	36	—
2–13 years old			16%		3%		

in the tonsil, retropharyngeal lymph node, and intestine suggested infection via the alimentary tract. Virus was also detected in the mesenteric-portal, prescapular, and prefemoral lymph nodes, spleen, ileum, and proximal colon in these sheep but not in blood clot, serum, feces, distal colon, or medulla. No virus was found in retropharyngeal, prescapular, prefemoral, or mesenteric lymph nodes or in the thymus, spleen, or ileum of six Suffolk lambs sacrificed at birth, or in the colostrum of their dams. Only one of the dams developed clinical scrapie later in life. Likewise, virus was not detected in retropharyngeal, prescapular, or mesenteric lymph nodes or in the tonsil (palatine), spleen, and ileum of 23-month-old lambs of negative dams.

The earliest isolation by the Mission Laboratory was from spleen and lymph node of a 4-month-old Suffolk ewe lamb, the progeny of a naturally affected ewe that died at 41 months of age and $4\frac{1}{2}$ months after lambing (34). This isolate was determined to be ME7 by the Moredun Laboratory in Edinburgh, Scotland. A similarly early isolation (but from brain tissue) of a $4\frac{1}{2}$-month-old Border-Leicester wether lamb born to an affected dam had previously been reported (35). The dam showed signs of scrapie at lambing, and this lamb had developed a nibbling reflex but its brain was histologically negative. The youngest sheep from which the agent was detected in brain at the Mission Laboratory was a 16-month-old Suffolk ram whose dam was affected.

There are precedents for the occurrence of natural scrapie in unusually young sheep. Foster and Dickinson (36) reported clinical scrapie in Suffolk lambs as young as 13 months. Joubert et al. (37) reported natural scrapie in sheep in France as young as 10 to 12 months. Zlotnik and Katiyar (12) reported clinical cases in India as young as 10 months. Dickinson et al. (13) observed that when all sheep more than one year old in an experimental flock of natural scrapie were examined histologically, an additional 10% of cases were found.

The 86 adult sheep 2 to 13 years old (Table 5.3) were compared with the younger lambs (Table 5.2) indicating that 3 of 19 (16%) of the adults born to affected dams were positive in mice compared with only 1 of 31 (3%) of those born to negative dams. All subclinical infections detected in adults were in the 24- to 35-month age group in which 4/9 (44%) were positive. All older adults were negative in mice, although 15 of them were progeny of affected dams, and some had other scrapie relationships: 8 of the adult rams were either twins or siblings of affected animals; 24 mouse-negative Suffolk ewes had affected progeny, and 15 were either twins or siblings of affected animals.

Detection of Scrapie Agent in Reproductive Tissues

Although the scrapie agent had been detected in many tissues of naturally affected sheep and goats, little had been done to isolate the agent from reproductive tissues. Pattison et al. (38, 39) reported that fetal membranes from each of six naturally affected Swaledale ewes produced scrapie in 9 of 11 Herdwick sheep infected orally, after incubation periods of 21 to 57 months. The incidence in orally infected sheep was essentially the same as when sheep were inoculated i.c. Palmer (40) obtained semen from a 3-year-old affected Suffolk ram showing clinical signs of scrapie over a period of 6 months and inoculated 20-day-old Dorset Horn and Suffolk lambs subcutaneously (s.c.). They were killed and examined over a period of $2\frac{1}{2}$ years with negative results. Ikegami et al. (30) detected abnormal PrP in the placenta of a naturally affected sheep.

Results from efforts to isolate the scrapie agent from reproductive tissues from Mission Station sheep by inoculating mice are shown in Table 5.4. The agent was detected in the following tissues from sheep and goats affected with natural scrapie: amniotic fluid—1/1 positive (sheep); uterus—4/10 positive (2 sheep and 2 goats); uterine caruncle—1/7 positive (sheep); fetal cotyledon—2/5 positive (1 sheep and 1 goat); ovary—3/11 positive (3 goats); fetal tissues—0/7 from 5 sheep and 2 goats, all affected, and 0/1 from a normal goat. All 10 fetuses from affected inoculated animals (4 sheep and 6 goats) were negative. The only positive result (3/5 mice inoculated with spleen were positive at 469 to 606 days postinoculation) was from twin fetuses of a Targhee ewe, which had been inoculated s.c. when 24 months old and died 21 months later, with inconclusive histology and unsuccessful transmission attempt in mice.

Scrapie Incidence in Suffolk Sheep

The scrapie incidence among 826 adult (2 to 11 years old) Suffolk sheep born in the infected Mission flock is shown in Figure 5.1. Twenty-four

TABLE 5.4. Demonstration of scrapie agent in reproductive tissues from natural cases and experimentally infected sheep and goats.

Reproductive tissue	Natural scrapie		Inoculated animals	
	Affected donor females	Negative donor females	Affected donor females	Negative donor females
Amniotic fluid	1/1	—	—	—
Uterus	4/10	—	0/2	—
Uterine caruncle	1/7	—	0/3	—
Fetal cotyledon	2/5	—	0/2	—
Ovary	3/11	—	1/2	—
Chorionic sac	0/1	—	—	—
Chorionic fluid	0/1	—	—	—
Semen	0/5	—	—	—
Testes	0/6	—	—	—
Seminal vesicle	0/1	—	—	—
Fetus	0/7	0/1	0/10	1/3

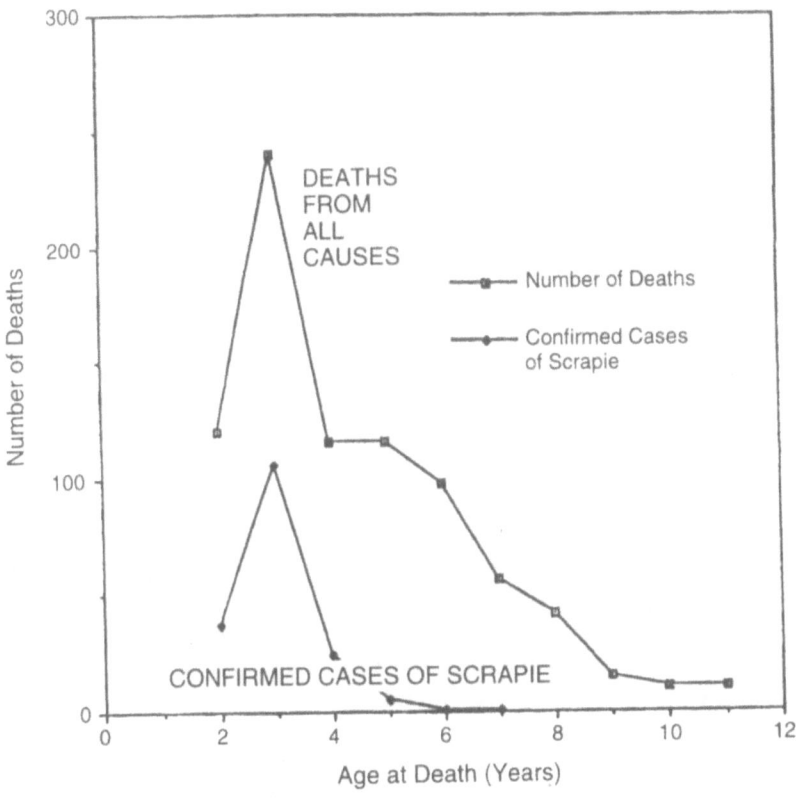

FIGURE 5.1. Natural scrapie in 826 Suffolk sheep born at Mission Station.

			AGE AT DEATH (YEARS)							
	2	3	4	5	6	7	8	9	10	11
				TOTAL DEATHS						
	121	240	116	116	98	56	42	15	11	11
				SCRAPIE CASES						
NUMBER	37	106	24	5	1	1	0	0	0	0
PERCENTAGE	31	44	21	4	1	2	0	0	0	0

percent of the ewes died with microscopically confirmed scrapie compared with 18% of rams. For both ewes and rams, the average age at death was 41 months. In 2-year-old sheep, 31% of deaths from all causes were confirmed cases of scrapie. The rate increased to 44% in 3-year-old sheep and decreased to 21%, 4%, and 2% as sheep became progressively older. Total deaths each year were about equal for both rams and ewes, but the maximum age at death was about 2 years older for ewes. Both sexes were believed to have been subjected to similar exposure until weaning, when the male lambs were moved to facilities occupied by rams only; thus, the males probably received less cumulative horizontal exposure than did the females.

Foster and Dickinson (36) reported that in a selectively bred flock of 120 Suffolk sheep, the average age at death from scrapie during 11 years' surveillance decreased from $3\frac{1}{2}$ to $1\frac{1}{2}$ years, with earliest cases as young as 13 months. Selective breeding involved 62 sires that were clinical cases or had affected parents. It was suggested that the flock died out from a progressively increasing level of infection from exposure to affected flockmates with an associated shortening of the incubation period.

The familial patterns of scrapie in Suffolk sheep at Mission are shown in Table 5.5. The scrapie status of the parent had a marked effect on the rate of scrapie in the progeny, as 78% of progeny were affected if both sire and dam were affected, and 42% if the dam only was affected. If the scrapie status of the sire was unknown or ignored, 34% of the progeny of scrapie-free dams were affected, compared with 62% of the progeny of affected dams. If the scrapie status of the dam was unknown or ignored, 26% of the progeny of scrapie-free sires were affected, compared with 45% of the progeny of affected sires. Essentially the incidence of scrapie in progeny doubled if either dam or sire was affected.

Concentrating on the progeny of affected females might appear to offer an effective scrapie-control measure; however, these 27 affected progeny of affected dams represented only 26% of the 103 total affected progeny. The remaining 74% had dams considered not to have been affected. Although it was possible that some unaffected dams were misclassified, or were "nonclinical or carrier" dams, in this Suffolk group 100 scrapie-free dams of affected progeny ranged from 25 to 160 (average 97) months of age. The unaffected dams of 50% of the affected progeny were more than 100 months old.

TABLE 5.5. Natural scrapie in Suffolk sheep (familial patterns).

		Fate of progeny		
Parents	Number at risk	% affected	Number affected	Average age at death
Dam free × sire free	105	25%	26	44
Dam free × sire positive	129	39%	50	43
Dam positive × sire free	31	42%	13	37
Dam positive × sire positive	18	78%	14	35
Total	283	36%	103	42
If sire status unknown				
Dam free	342	34%	—	—
Dam positive	82	62%	—	—
If dam status unknown				
Sire free	172	26%	—	—
Sire positive	170	45%	—	—

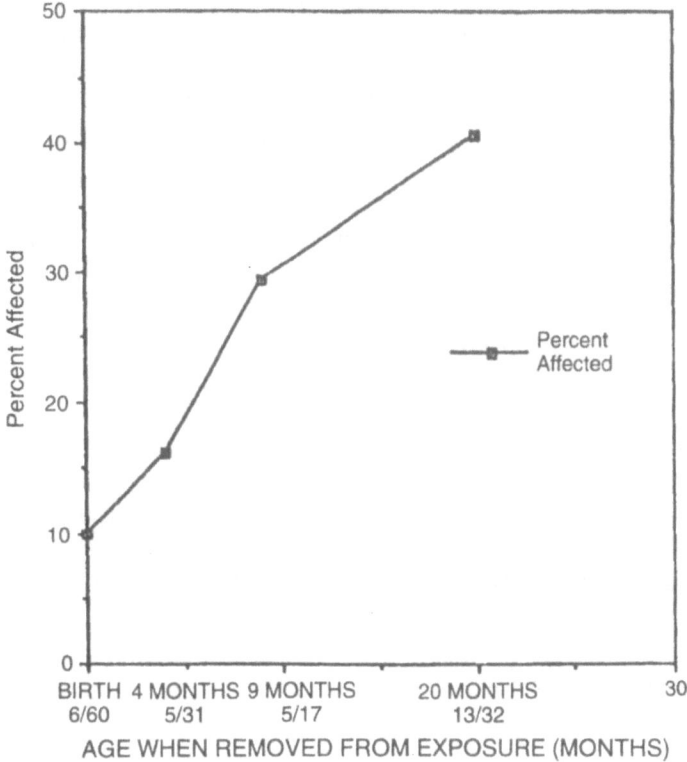

FIGURE 5.2. Natural scrapie in sheep removed from exposure to affected animals at various ages.

Hadlow et al. (33) found the virus widespread in natural cases 34 to 57 months of age, and in highest concentration in the brain. It was also detected in the spinal cord and sciatic nerve. Infection was not detected in blood clot, saliva, mandibular and parotid salivary glands, thyroid, heart, lung, kidney, or skeletal muscle. Nor was it detected in the placenta (cotyledons) from two ewes, ovary, and uterus of three ewes, or the testes and seminal vesicles of one ram.

Natural Scrapie in Sheep Subjected to Different Levels of Exposure

A total of 244 sheep (127 Suffolk, 59 Rambouillet, and 58 Targhee) were removed from scrapie exposure at birth (within a few hours of birth) or at 4, 9, or 20 months of age and placed in isolation pens (Fig. 5.2). These ages were selected as being representative of flock operations when sheep might be sold at weaning or the first or second autumn following birth. Ten percent (6/60) of sheep removed at birth, and deemed at risk, developed the disease. Four of the six were Suffolks and two were Targhees. Although the isolation of these young lambs was substantial and in accordance with general concepts of scrapie contagion at that time, it may not have been absolute.

Tissues from 44 lambs removed at birth and dead from various causes were inoculated into mice, with negative results. Inoculated tissues from lambs up to 1 year of age were spleen and lymph node, and for lambs more than 1 year old, also included brain. Only 10 of these lambs lived more than 4 months; of these 10, only three were progeny of affected dams. These negative findings were thus of limited scope. Sheep that remained in the infected flock until they were older, and thus subject to a more cumulative exposure, developed scrapie at higher rates. Sixteen percent of these removed at weaning (4 months), 29% of those removed at 9 months, and 41% of these removed at 20 months were affected. The average age at death decreased from 46 months for those removed at birth to 39 months for those removed at 20 months. Scrapie incidence in 36 at-risk Suffolk sheep isolated from birth was 11%, considerably less than would have been expected had they remained in the infected flock.

Horizontal Transmission of Natural Scrapie

Transmission to Animals Exposed from Early Age

Many studies have shown that scrapie can spread horizontally from naturally affected sheep to previously unexposed sheep and goats (Table 5.6). Greig and Russell (41) observed indirect spread at pasture. Brotherston et al. (42) observed spread from a succession of naturally affected sheep to goats housed with them from one week or less of age, and to progeny of pregnant goats and Blackface sheep exposed similarly when 1 to 3 days old. Scrapie incidence in the exposed animals was 66%, 55%, and 43%. Dickinson et al. (43–45) reported 28% of Scottish Blackface sheep reared in lifetime pasture contact with natural cases, developed scrapie as did 57% of Suffolks reared in maximum contact, and 19% if reared in partial isolation. They also found that 75% of the progeny of naturally affected dams developed scrapie if reared on the dam, compared with 29% if the progeny were bottle-fed. There were additional observations of scrapie in goats reared in infected flocks of sheep (46–51).

Transmission to Animals Not Exposed Until After Weaning

Emphasis has been on the importance of the exposure of lamb to dam (in utero or postnatally) and/or exposure of the lamb at a very early age as

TABLE 5.6. Scrapie: evidence of horizontal transmission to previously unexposed animals from naturally affected sheep.

	Exposure	Results	Age affected
Greig and Russell (41) Scotland	Biweekly pasture rotation with affected sheep for 3 years	Cheviot: 1/4 (25%) Half-bred: 7/12 (58%) Progeny: 2/9 (22%)	6–8 years 6–8 years $1\frac{1}{2}$–$3\frac{1}{2}$ years
Brotherston et al. (42) England	6 goats exposed from 1 week of age in a covered pen to a succession of affected sheep	Goats: 4/6 (67%) (observed for 4 years)	39–47 months
Brotherston et al. (42) England	6 pregnant goats, and their progeny, exposed in a covered pen to a succession of affected sheep	Goat progeny: 6/11 (55%) (observed for 4 years)	29–37 months
Brotherston et al. (42) England	11 Blackface lambs 1–3 days old exposed in a covered pen, to a succession of affected sheep	Blackface lambs: 3/7 (43%) (observed for >5 years)	45–51 months
Dickinson et al. (43, 44) Scotland	Scottish Blackface sheep reared in lifetime contact with affected sheep at pasture	Blackface sheep: 28% developed scrapie	Observed 5–6 years
Dickinson et al. (43, 44) Scotland	Progeny of "off free" sires and dams (1 dam affected later)	Maximum contact: 8/14 (57%) Partial isolation: 5/27 (19%)	Observed 4 years
Dickinson et al. (45) Scotland	Progeny of affected dams	Progeny reared on dam: 6/8 (75%) Bottle fed: 2/7 (29%)	Observed $3\frac{1}{2}$ years

being an essential part of scrapie epidemiology. In an experiment to test this premise, 140 previously unexposed sheep and goats were selected from scrapie-free flocks in the United States and New Zealand. These animals were 3 to 9 months old when they arrived at the Mission Station and were placed in infected pastures and corrals. The male or female arrivals mixed freely with animals of their respective sex in the infected flock. The previously unexposed sheep were bred only within their own groups. Animals closely related to the purchased animals remained in the flocks of origin as controls. The results of this experiment are shown in Table 5.7. Natural

TABLE 5.7. Natural scrapie in previously unexposed sheep and goats brought to Mission, Texas, at 3 to 9 months of age and in their first generation progeny born at Mission.

Unexposed dams brought to Mission

Breed	Total	Age first exposed (months)	Scrapie		Progeny born at Mission				
			Number	Age (months)	At risk	Affected		Age at death (months)	
						Number	%	Average	Range
Angora goat	11	3	—	—	27	5	19	43	40–46
Dairy goat	6	5	—	—	24	15	63	51	39–113
Rambouillet	31	3	1	88	94	10	11	40	31–50
Targhee	31	6	1	89	95	18	19	44	35–52
Hampshire	33	8	1	89	24	3	13	40	38–41
Suffolk (US)	8	7	—	—	8	1	13	32	32
Suffolk (NZ)	20	9	3	73, 102, 107	28	10	36	48	35–67
Total	140		6		300	62	21	43	31–113

scrapie occurred in 6 of the 140 previously unexposed animals. One case each occurred in Rambouillet, Targhee, and Hampshire ewes at 88, 89, and 89 months of age, respectively. Three cases were confirmed at 73, 102, and 107 months of age in the Suffolks brought from New Zealand when 9 months old. It was difficult to calculate scrapie incidence in the previously unexposed animals, as many had died of various causes prior to the late-appearing cases. If only the sheep living longer than the earliest case in each breed were considered as being at risk, the incidence was 25% in the New Zealand Suffolks, 10% in the Hampshires, 7% in the Targhees, and 5% in the Rambouillets. This work was planned before the scrapie incubation-period (SIP) gene concept was developed. However, it was suggested that the ewes developed the disease at an advanced age because they were not exposed from birth and therefore received a later and a lesser cumulative dose of the infectious agent than they normally would have. Although the dams were all affected at advanced age, their progeny born in the infected-flock environment at Mission developed scrapie at the usual average age of 43 months.

Scrapie was confirmed in 62 of 300 (21%) of the first generation progeny born at Mission and included cases in progeny of all breeds of previously unexposed sheep and goats taken there. Incidence in the progeny ranged from 11% in Rambouillets to 63% in dairy goats, and 36% in progeny of the Suffolks from New Zealand. The familial distribution of cases usually observed in infected flocks was less apparent in these sheep. The rate in first generation progeny of the three affected New Zealand ewes was 2/8 (25%) compared with 8/20 (40%) in the progeny of 11 apparently normal New Zealand ewes. The rate in progeny of the one affected Targhee ewe was 3/7 (43%) and the Targhee ewes that remained normal had 15 affected progeny. The one affected Hampshire ewe had no affected progeny and there were 3/23 (13%) affected progeny of 16 normal Hampshire ewes. The one affected Rambouillet ewe had 1/6 (17%) progeny affected; the 22 normal dams had 9/88 (10%) progeny affected. Ewes as young as $1\frac{1}{2}$ years gave birth to lambs that developed the disease and normal ewes that lived as long as $14\frac{1}{2}$ years had scrapie progeny.

Experimental Scrapie

Negative Evidence of Horizontal Transmission

Gordon (52) (Table 5.8) reported no scrapie transmission to 37 goats and 18 sheep in contact in indoor pens with a succession of inoculated affected sheep and goats over a period of more than 55 months. Brotherston et al. (42) found no evidence of scrapie transmission to 7 goats, kept in a yard with a succession of 978 inoculated Herdwick sheep, observed for 7 years. Pattison (53) observed no scrapie in 192 goats housed in indoor pens for 5 years with a succession of inoculated affected goats, and concluded that the

TABLE 5.8. Experimental scrapie: negative evidence of horizontal transmission.

	Exposure	Results
Gordon (52) Compton	20 Cheviot lambs 6 months of age and 37 goats mingled with a succession of inoculated affected sheep and goats in indoor pens for 55 months	0/37 goats 0/18 sheep (observed for 55 months)
Brotherston et al. (42) Moredun	Dams before kidding and kids from birth yarded with 978 inoculated cases in Herdwick sheep	0/7 goats (observed for 7 years)
Pattison (53) Compton	192 goats were housed in indoor pens for 5 years with a succession of inoculated affected goats	0/192 goats (observed for 7 years)

spread of scrapie between and among sheep and goats was a rare occurrence. Although sheep and goats developed scrapie when housed or pastured with sheep affected with the natural disease (Table 5.6), this did not occur when sheep and goats were similarly exposed to clinical cases that had been inoculated with the agent (Table 5.8).

Inoculated Dams and Their Progeny

Gordon (52) (Table 5.9) inoculated 140 Cheviot ewes s.c. using SSBP (a relatively high-passage inoculum largely of Cheviot origin) 20 days after

TABLE 5.9. Experimental scrapie: evidence of vertical transmission in progeny of inoculated dams.

	Inoculated dams	Progeny left on dam
Gordon (52) Compton	140 Cheviot ewes and 4 rams inoculated s.c. (SSBP) 20 days after breeding began; rams, 2/4 (50%); ewes, 45/140 (32%)	45 positive dams 8/40 (20%) 95 negative dams 2/72 (3%) (affected at 7 months >)
Dickinson et al. (45) Moredun	5 Border Leicester ewes s.c. (SSBP) just before or after breeding	2/4 (50%) (affected at 8–13 months)
Pattison (53) Compton	30 goats were inoculated i.c. with goat brain at mating; 30/30 (100%) $7\frac{1}{2}$–17 months p.i.	0/33 goats (observed for 44 months)
Dickinson et al. (45) Moredun	Embryo transfer to Border Leicester dam inoculated s.c. SSBP/21 17 days posttransfer; scrapie signs at weaning	1/1 (at 4 years)

breeding began. He observed that 8 of 40 (20%) of the progeny of 45 inoculated ewes that developed scrapie were affected compared with 2 of 72 (3%) of the progeny of 95 inoculated ewes that remained normal. All progeny were left on dams. Most affected progeny developed the disease at 7 to 18 months of age. Dickinson et al. (45) inoculated five Border Leicester ewes s.c. (SSBP) just before or after breeding and observed that two of four (50%) of progeny left on dam were affected at 8 to 13 months of age. A Border Leicester embryo transfer recipient was inoculated s.c. (SSBP/21) 17 days posttransfer and showed clinical signs at weaning; the lamb was affected at 4 years of age. Pattison (53) did not observe scrapie in 33 progeny of affected goats inoculated s.c.

Inoculated Animals at Mission Station

W.W. Clark inoculated brain, spleen, and lymph node of Suffolk origin into 121 female sheep and goats at the Mission Station. Results are shown in Table 5.10. Scrapie was confirmed in the inoculated females as follows: Suffolk 15/28 (54%); Rambouillet 14/29 (48%); Targhee 5/24 (21%); dairy goats 23/28 (82%); and Angora goats 12/12 (100%). There were 185 progeny born to the inoculated dams. The incidence was 7/61 (11%) in the progeny of 69 inoculated affected dams and 6/124 (5%) in the progeny of 52 inoculated unaffected dams. A similar comparison was made with animals left with dams when 9/76 (12%) developed scrapie compared with 4/109 (4%) of those removed at birth.

Progeny left with the dams were affected at an average age of 33 months (range 21–47) and those removed averaged 38 (range 32–44) months of age.

TABLE 5.10. Experimental scrapie in inoculated dams and their progeny at Mission, Texas.

Inoculated dam			Fate of progeny				
Breed	Positive	Negative	Left on dam		Removed at birth		Total
Suffolk	15		2/7		—	—	2/7 —
		13		2/5		0/13	2/18
Rambouillet	14		1/9		1/14		2/23
		15		2/23		1/25	3/48
Targhee	5		0/1		0/7		0/8
		19		0/9		0/20	0/29
Dairy goats	23		1/7		1/15		2/22
		5		0/14		1/15	1/29
Angora goats	12		1/1		—		1/1
		0	—		—		—
	69		5/25 (20%)		2/36 (6%)		7/61 (11%)
	—	52		4/51 (8%)		2/73 (3%)	6/124 (5%)
Total			9/76 (12%)		4/109 (4%)		13/185 (7%)
69/121 (56%)							

The progeny of inoculated sheep (Table 5.9) were left with the dams and developed scrapie at a much younger age (7 to 18 months) and SSBP inoculum and Cheviot sheep were used in the experiment. At Mission, the inoculum was from Suffolk sheep and none of the inoculated sheep were Cheviot. The majority of the Suffolk, Rambouillet, and Targhee sheep shown in Table 5.10 had been subjected to natural exposure for years, without showing clinical signs and averaged 40 to 67 months of age when inoculated. When Rambouillet and Targhee ewes were inoculated once s.c., 10/37 (27%) were affected, compared with 9/16 (56%) when similar Rambouillet and Targhee ewes were inoculated s.c. on a monthly basis.

Inoculation of Sheep and Goats Before and at Birth

Pattison (53) inoculated three goat fetuses i.c. (goat-passaged SSBP/1) through the uterine wall 57 days before birth. All developed the disease at $5\frac{1}{2}$ to 7 months of age. Hadlow et al. (54) inoculated 21 fetuses intramuscularly (i.m.), in the neck, through the uterine wall with brain of naturally affected Suffolks. The fetuses were 76 to 109 days gestation age. The lambs were sampled at 47 to 322 days postinoculation. Trace amount of virus was detected at 6 to 9 months of age in mesenteric, suprapharyngeal, or prescapular lymph node from three lambs. Also, seven lambs were inoculated into the palatine tonsils when 3 to 12 days old. Virus was not detected in the lambs killed when 5 to 7 months old. Stamp (55) reported that 11 of 14 Cheviot lambs inoculated when less than 1 day old developed typical signs when $3\frac{1}{2}$ to 5 months of age.

Attempts to Isolate Scrapie Agent from Feed
or Environmental Contaminants

Hadlow et al. (33) hypothesized, on the basis of virus distribution studies, that scrapie spreads from sheep-to-sheep via the alimentary tract. Attempts were made to detect the scrapie agent in materials the lamb might normally ingest. The fetus could receive early exposure from amniotic fluid or perhaps other reproductive tissues from which the scrapie agent has been detected (reported elsewhere in this chapter). Pattison (56) (citing Fitzsimmons and Pattison) failed to detect scrapie agent in internal parasites: *Haemonchus contortus* in sheep and goats, *Nippostrongylus brasiliensis* in rats, and *Sypacia obvelata* in mice. Pattison (53) using material from experimental scrapie cases in goats, dosed six goats orally with 10 mL saliva in 50 mL goat's milk, and 2.5 g feces in 110 mL goat's milk, three times/week (142 doses) over a period of 1 year. Goats were observed for 40 months and remained normal. Gordon (52) inoculated goats i.c., with brain, saliva, milk, urine, and feces from inoculated affected goats. Results were negative except for the brain tissue. Hadlow et al. (33) did not detect virus in the colostrum or mammary glands of one naturally affected and five normal Suffolk ewes, even though virus was present in the supramammary lymph nodes of two naturally affected Suffolk ewes (one lactating and one nonlactating). The virus was not found in the feces of Suffolks 10 to 14 months old from which the virus was detected in other tissues. The agent was detected in 0/6 feces; 0/7 urine; 0/2 milk; 0/1 colostrum from sheep or goats; or in pasture grass, trough water, or manure from hospital pens (De Camp, personal communication). There was low level (1/10 mice inoculated i.c.) evidence in 1/17 attempts to detect the agent in *Haemonchus contortus*. Palsson (10) reported no conclusive results from microscopic studies and transmission experiments using field mice (*Apodemus sylvaticus*) from *Rida*-infected farms in Iceland. Taylor (57) noted that almost all of an oral dose of scrapie given to mice was excreted in the feces.

Embryo Transfer and Scrapie Transmission

Foote et al. (58) reported no evidence of scrapie transmission when embryos were transferred from inoculated donor ewes to scrapie-free recipient ewes. None of the 60 resultant lamb progeny developed scrapie. Foster et al. (59) reported that when embryos were transferred from inoculated donor ewes to scrapie-free recipient ewes 6/20 (30%) of the resultant progeny were affected with scrapie at about age 2 years. The research differed in several respects. Foote used both Cheviots and Suffolks, SSBP/1 or Suffolk inoculum, oral and s.c. routes, collected embryos 6 to 36 months postinoculation, and washed all embryos three times. Foster used only SSBP/1, s.c. inoculation, collected embryos $1\frac{1}{2}$ to 6 months postinoculation, and did not wash the embryos. It was not possible to independently evaluate the effect of washing embryos. All were washed in one study and none in the other.

Discussion

Stockman (60) concluded in 1913 that scrapie spread by contagion although many farmers believed the disease was hereditary, or that it was both hereditary and contagious. He thought some farmers might not distinguish between hereditary transmission and congenital infection. Kimberlin (61) concluded in 1990 that the infection was most commonly transmitted from ewe to lamb, both up to parturition and afterward when ewe and lamb run together, with maternal transmission reflected in cases peaking in the fourth year, and horizontal spread accounting for some cases in older sheep.

Scrapie spreads from flock-to-flock, state-to-state, and country-to-country by the movement of infected but clinically normal sheep incubating the disease. Often these preclinical cases had passed veterinary inspection and were certified by regulatory officials as being free of scrapie and exposure thereto. These animals carried the infection into receiving flocks and to additional breeds of sheep establishing new centers of infection. The mechanisms of natural spread from animal-to-animal are less clear.

Although there is much circumstantial evidence, but no direct proof, that lambs are naturally infected in utero, historical perception and practical experience stress the role of the dam in scrapie epidemiology. Fetal lambs and kids are susceptible to experimental inoculation, and lambs and adult sheep can be infected at essentially any age.

The scrapie agent has been detected by mouse bioassay in reproductive tissues from natural cases in sheep or goats including amniotic fluid, uterus, uterine caruncle, fetal cotyledon, ovary, and placenta. Although these findings suggest the fetus may be at risk, there was only one example of the agent being detected in fetal tissue, from an inoculated ewe (M. DeCamp, personal communication). Lambs and kids first exposed when a week or less of age to naturally affected sheep developed scrapie at relatively normal ages and rates. When progeny were reared with their dams that had been inoculated s.c., 20% of progeny of dams developing scrapie also developed the disease, compared with 6% of the progeny of inoculated dams remaining normal. When the scrapie status of the inoculated dams was ignored, 12% of all progeny left with the dams developed scrapie, compared with 4% of progeny removed from the dams at birth. It would have been advantageous but was not feasible to maintain each of the inoculated dams in separate isolation pens, isolate the lambs at the moment of birth, and hold each of these in separate isolation pens.

Horizontal transmission of scrapie also occurred at advanced ages of 73 to 107 months in six ewes from scrapie-free flocks, first exposed after weaning and at 3 to 9 months of age. Their first generation progeny, born in the infected flock at Mission, were affected at an average age of 43 months. Ten percent of sheep, removed from exposure shortly after birth, developed scrapie. The rate increased progressively in sheep removed from exposure at 4, 9, or 20 months of age, and they were affected at a younger age. It was suggested that the differences in rate and age affected were largely dose-related, reflecting cumulative exposure to the scrapie agent.

Clinical scrapie may follow a single exposure, the usual when animals are inoculated. In the natural infection, the exposure may be random and the animals may be subject to multiple exposures of variable dosages. Other factors involved include origin of agent, its passage history, route of exposure, and species and genotype of the recipient. It was suggested that horizontal transmission occurred via the alimentary tract. The agent was detected in amniotic fluid and placenta that could contaminate lambing areas. The agent would likely persist in the environment because it resists chemical and physical abuse, and because it survives in soil (62). Other possible sources of exposure include ingestion of colostrum and milk or urine, feces, saliva, or internal parasites. Except for one detection in internal parasites, all attempts to demonstrate the agent in these possible sources were negative. The agent could have been present at levels too low to detect but sufficient to be a factor due to extensive dissemination and multiple exposures.

Scrapie epidemiology is a complex phenomenon encompassing flocks as well as individual sheep, and a flock may have been infected months or years before the disease is confirmed. During this period, infected but apparently normal sheep, are sold to slaughter or for breeding purposes or die from various causes before clinical signs appear, or the signs are not recognized or not reported. Under such circumstances, the first case of scrapie confirmed in a flock is often not the first case or the only infected animal in the flock. Eliminating only the sheep related to this "index case" leaves other family lines in which the disease has occurred in the flock or in other flocks into which they have moved. Preclinical infection was detected in apparently normal lambs as young as 4 months and at progressively higher rates in sheep up to 25 months old. These animals would not be expected to show clinical signs of natural scrapie until reaching some 2 to 4 years of age. Close observation of sheep allowed to live out their full life span revealed that most of the confirmed cases were progeny of ewes that lived up to 13 years of age with no evidence of scrapie. Although the scrapie rate was much higher in the progeny of affected ewes, the apparently normal ewes lived longer, had more progeny, and a greater total number of affected progeny.

Clearly scrapie is a difficult, but not impossible, disease to control. For many years it was hoped that technology would be developed to eradicate scrapie by slaughtering only selected animals in the infected flock. Presently no practical method is known that will eliminate the scrapie agent from infected flocks except slaughter of all exposed animals in the flock and those moved into other flocks. Palsson (10) described an eradication program in Iceland in which sheep in a quarantined area were destroyed during 1945 to 1957 and restocked with 4- to 5-month-old lambs brought from an area where *Rida* (scrapie) had not occurred. Two to 4 years after the area was restocked, *Rida* was again observed on a limited number of farms, some of which had been left free of sheep for 3 years. *Rida* recurred only on farms previously infected, but not on all of such farms. The author concluded that the infection persisted for 3 or more years.

Sigurdarson (63) updated the Icelandic control program, reporting in 1978 that another attempt at eradication was made. Repopulation was not permitted for at least 2 years, and prior to restocking, heroic efforts were made to clean and disinfect premises and facilities, destroying certain wood structures and skimming away surface soil in heavily exposed areas. Results have been encouraging. Only two of the 397 farms restocked have been reinfected, and these were believed due to carelessness. Some farms had remained free for up to 11 years since restocking.

Acknowledgments. Dr. W.W. Clark, director of the Mission Field Trial, Mission, Texas (1965–1984), conducted the sheep, goat, and cattle experiments. Dr. M. deCamp, director from 1984 to her retirement, also conducted the mouse inoculation and laboratory studies. Dr. W.J. Hadlow (retired), Department of Health and Human Services, PHS, NIH, Rocky Mountain Laboratories, Hamilton, Montana, provided valuable counsel and conducted laboratory studies on samples from Mission animals.

References

1. McDaniel HA, Morehouse LG. The diagnosis of scrapie. Report of Scrapie Seminar, Washington, DC, January 27–30, 1964. ARS91-53, ARS USDA: 41–52.
2. Besnoit C, Morel C. Note sur les lesions nerveuses de la tremblante du mouton. Rev Vet 1898;23:397–400.
3. Cassirer R. Sur la maladie du trot du mouton. Virchow Arch 1898 (July 15); 153:89–110.
4. Bertrand I, Carre H, Lucam F. La "tremblante" du mouton. Ann Anat Pathol Med Chir (Paris) 1937;14(7):565–86.
5. Brownlee A. Histopathological studies on scrapie: an obscure disease of sheep. Vet J 1940;96:254–64.
6. Holman HH, Pattison IH. Further evidence of the significance of vacuolated nerve cells in the medulla oblongata of sheep affected with scrapie. J Comp Pathol Ther 1943;53:231–6.
7. Zlotnik I. The pathology of scrapie. Report of scrapie seminar, Washington, DC, January 27–30, 1964. ARS91-53, ARS USDA:213–24.
8. Chandler RL. Encephalopathy in mice produced by inoculation with scrapie brain material. Lancet 1961;1:1378–9.
9. Parry HB. In: Oppenheimer DR, ed. Scrapie disease in sheep. London: Academic Press, 1983.
10. Palsson PA. Rida (scrapie) in Iceland and its epidemiology. In: Prusiner SB, Hadlow WJ. Slow transmissible diseases of the nervous system, vol 1. New York: Academic Press, 1979;357–66.
11. Sato N, Matsui T, Nakagawa M, Muramatsu Y, Shinagawa M, Kubota M. [A case of scrapie without pruritus and skin lesions.] J Jpn Vet Med Assoc 1991; 44:901–4.
12. Zlotnik I, Katiyar RD. The occurrence of scrapie disease in sheep of the remote Himalayan foothills. Vet Rec 1961;73:543–4.
13. Dickinson AG, Young GB, Stamp JT, Renwick CC. An analysis of a natural scrapie in Suffolk sheep. Heredity 1965;20:485–503.
14. Clark AM. Diagnosis of scrapie. Vet Rec 1991;128:214.
15. Clark AM, Moar JAE. Scrapie: a clinical assessment. Vet Rec 1992;130:337–8.
16. Clark WW. Clinical signs of scrapie in sheep and goats. Proceedings of the 84th Annual Meeting, U.S. Animal Health Association, Nov 2–7, 1980;479–82.
17. Detwiler LA. Scrapie: an overview. American Association of Small Ruminant Practice, Symposium on Diseases of Small Ruminants, Corvallis, OR, June 7–9, 1990;32–7.
18. Merz PA, Somerville RA, Wisniewski HM, Isobal K. Abnormal fibrils from scrapie-infected brain. Acta Neuropathol 1981;54:63–74.

19. Perrin GG, Perrin GL, Benoit C. Detection of scrapie-associated fibrils in goats. Vet Rec 1991;129:432.
20. Kirkwood JK, Wells GAH, Cunningham AA, et al. Scrapie-like encephalopathy in a greater kudu (*Tragelaphus strepsiceros*) which had not been fed ruminant-derived protein. Vet Rec 1992;130:365–7.
21. Williams ES. Spongiform encephalopathy (chronic wasting disease in cervids). 94th Annual Meeting of the U.S. Animal Health Association 1990;541–2.
22. Pearson GR, Wyatt JM, Gruffydd-Jones TJ, et al. Feline spongiform encephalopathy: fibril and PrP studies. Vet Rec 1992;131:307–10.
23. Willoughby K, Kelly DF, Lyon DE , Wells GAH. Spongiform encephalopathy in a captive puma (Felis concolor). Vet Rec 1992;131:431–4.
24. Peet RL, Curran JM. Spongiform encephalopathy in a cheetah. Aust Vet J 1992;99:171.
25. Scott AC, Wells GAH, Stack MJ, White H, Dawson M. Bovine spongiform encephalopathy: detection and quantitation of fibrils, fibril protein (PrP) and vacuolation in brain. Vet Microbiol 1990;23:295–304.
26. Gibson PH, Somerville RA, Fraser H, Foster JD, Kimberlin RH. Scrapie-associated fibrils in the diagnosis of scrapie in sheep. Vet Rec 1987;20:125–7.
27. Scott AC, Done SH, Venables C, Dawson M. Detection of scrapie-associated fibrils as an aid to the diagnosis of natural sheep scrapie. Vet Rec 1987;120:280–1.
28. Dawson M, Mansley LM, Hunter AR, Stack MJ, Scott AC. Comparison of scrapie-associated fibril detection and histology in the diagnosis of natural sheep scrapie. Vet Rec 1987;121:591.
29. Rubinstein R, Merz PA, Kascsak RJ, et al. Detection of scrapie-associated fibrils (SAF) and SAF proteins from scrapie-affected sheep. J Infect Dis 1987;156:36–42.
30. Ikegami Y, Ito M, Isomura H, et al. Preclinical and clinical diagnosis of scrapie by detection of PrP protein in tissues of sheep. Vet Rec 1991;128:271–5.
31. Race R, Darwin E, Jenny A, Taylor W, Sutton D, Caughey B. Diagnostic implications of detection of proteinase K-resistant protein in spleen, lymph nodes, and brain of sheep. Am J Vet Res 1992;53:883–9.
32. Brugère H, Banissi C, Brugère-Picoux J, Chatelain J, Buvet R. [Contribution of urinary examination to the diagnosis of spongiform encephalopathy of sheep.] Bull Mensuel Soc Vet Pratigue France 1991;75:277–9, 282–8.
33. Hadlow WJ, Kennedy RC, Race RE. Natural infection of Suffolk sheep with scrapie virus. J Infect Dis 1982;146:657–64.
34. Hourrigan JL, Klingsporn AL, Clark WW, DeCamp M. Epidemiology of scrapie in the United States. In: Prusiner SB, Hadlow WJ. Slow transmissible diseases of the nervous system, vol 1. New York: Academic Press, 1979: 331–56.
35. Renwick CC, Zlotnik I. The transmission of scrapie to mice by intracerebral inoculation of brain from an apparently normal lamb. Vet Rec 1965;77: 984–5.
36. Foster JD, Dickinson AG. Age at death from natural scrapie in a flock of Suffolk sheep. Vet Rec 1989;125:415–7.
37. Joubert L, Lapras M, Gastella J, Prave M, Laurent D. Un foyer de tremblante du mouton en Provence. [An epidemic of sheep scrapie in Provence.] Bull Soc Sci Vet Med Compare 1972;74:165–84.

38. Pattison IH, Hoare MN, Jebbett JN, Watson WA. Spread of scrapie to sheep and goats by oral dosing with foetal membranes from scrapie-affected sheep. Vet Rec 1972;90:465–8.

39. Pattison IH, Hoare MN, Jebbett JN, Watson WA. Further observations on the production of scrapie by oral dosing with foetal membranes from scrapie-affected sheep. Br Vet J 1974;130:65–7.

40. Palmer AC. Attempt to transmit scrapie by injection of semen from an affected ram. Vet Rec 1959;71:664.

41. Greig J, Russell T. Scrapie: observations on the transmission of the disease by mediate contact. Vet J 1940;96:203–6.

42. Brotherston JG, Renwick CC, Stamp JT, Zlotnik I. Spread of scrapie by contact to goats and sheep. J Comp Pathol 1968;78:9–17.

43. Dickinson AG, Young GB, Stamp JT, Renwick CC. Scrapie in Suffolk sheep: a genetic investigation using experimental matings. Report of Scrapie Seminar, held at Washington, DC, January 27–30, 1964; ARS91-53, USDA:228–44.

44. Dickinson AG, Stamp JT, Renwick CC. Maternal and lateral transmission of scrapie in sheep. J Comp Pathol 1974;84:19–25.

45. Dickinson AG, Young GB, Renwick CC. Scrapie: experiments involving maternal transmission in sheep. Report of Scrapie Seminar held in Washington, DC, on January 27–30, 1974;ARS91-53, USDA:244–8.

46. Chelle PL. Un cas tremblante chez la chevre. Bull Acad Vet France 1942; 15:294–5.

47. Hourrigan JL, Klingsporn AL, McDaniel HA, Riemenschneider MN. Natural scrapie in a goat. J Am Vet Med Assoc 1969;154:538–9.

48. Hourrigan JL. Epidemiology of scrapie in the United States. Proceedings of the 92nd Annual Meeting of the U.S. American Health Association 1988;386–401.

49. Stemshorn BW. Un cas de tremblante naturelle chez une chevre. Can Vet J 1975;16:84–6.

50. Harcourt RA, Anderson MA. Naturally occurring scrapie in goats. Vet Rec 1974;94:504.

51. Wood JNL, Done SH, Pritchard GC, Wooldridge MJA. Natural scrapie in goats: case histories and clinical signs. Vet Rec 1992;131:66–8.

52. Gordon WS. Review of work on scrapie at Compton, England (1952–1954). Report of Scrapie Seminar held at Washington, DC, January 27–30, 1964; ARS91-53, USDA:19–40.

53. Pattison IH. The spread of scrapie by contact between affected and healthy sheep, goats, or mice. Vet Rec 1964;76:333–6.

54. Hadlow WJ, Jackson TA, Race RE. Experimental infection of fetal and newborn Suffolk sheep with scrapie virus. Am J Vet Res 1984;45:2637–9.

55. Stamp JT. Scrapie disease of sheep. A review of the contradictory evidence as to the nature of the disease. Vet Rec 1958;70:50–5.

56. Pattison IH. Fifty years with scrapie: a personal reminiscence. Vet Rec 1988;23:661–6.

57. Taylor DM. Bovine spongiform encephalopathy and human health. Vet Rec 1989;125:413–5.

58. Foote WC, Clark WW, Maciulis A, et al. Prevention of scrapie transmission in sheep using embryo transfer. Am J Vet Res 1993;54:1863–8.

59. Foster JD, McKelvey WAC, Mylne MJA, et al. Studies on maternal transmission of scrapie in sheep by embryo transfer. Vet Rec 1992;130:341–3.

60. Stockman S. Scrapie: an obscure disease of sheep. J Comp Pathol Ther 1913;26:317–27.
61. Kimberlin RH. Transmissible encephalopathies in animals. Can J Vet Res 1990;54:30–7.
62. Brown P, Gajdusek DC. Survival of scrapie virus after 3 years internment. Lancet 1991;337:269.
63. Sigurdarson S. Epidemiology of scrapie in Iceland and experience with control measures. In: Bradley RM, Savey M, Marchant BA, eds. Subacute spongiform encephalopathies. Proceedings of a seminar in the CEC Agricultural Research Programme; 1990 Nov 12–14; Brussels, Belgium. Dordrecht, Boston, & London: Kluwer Academic, 1991:233–42.

6

Transmission of Sheep and Goat Strains of Scrapie from Experimentally Infected Cattle to Hamsters and Mice

CLARENCE J. GIBBS, JR., JIRI SAFAR,
MICHAEL P. SULIMA, ALFRED E. BACOTE,
AND R. ANDRES SAN MARTIN

Epidemiologic data indicate that bovine spongiform encephalopathy (BSE) probably originated in cattle by feeding rendered meat and bone meal that was accidentally contaminated by a scrapie-like agent (1). However, there remain several important questions related to the origin and biologic properties of BSE. First, what is the origin of the agent? Is it a previously unrecognized strain of BSE agent rarely inducing sporadic disease in cattle and artificially spread or is it scrapie transmitted from sheep, goat, or possibly other species and adapted to cattle? Second, how was the disease spread? Third, does the large-scale spread of a possibly new or newly adapted scrapie-like agent represent any risk to humans or nonhuman species? The biologic substrate or carrier of the agent's strain-like behavior in spongiform encephalopathies is unknown (2) and their biologic properties are defined only indirectly. The possibilities to study them are limited to the characteristics obtained by comparative pathology (3), following the anatomic distribution (4) of PrP^{sc}, comparing their species barrier behavior and by following the incubation time of each strain in genetically defined hosts (3).

Earlier, we reported the successful transmission of sheep and goat strains of scrapie to cattle (5). At that time, we stated that confirmation of the experimental transmission of scrapie to cattle must be made by serial passage of the agent in cattle, passage of scrapie from the brains of affected cattle back into Suffolk sheep and Angora goats, and passage of scrapie from affected cattle into mice and hamsters. We also reported that homogenates of brain from affected cattle had already been injected into the LVG/LAK strain of golden Syrian hamsters and two strains of mice

(NZW/LacZ and NIH Swiss) both of which have a short incubation period for sheep scrapie isolates and both of which are genetically similar to the strains used to isolate the scrapie-like agent from cases of BSE in the United Kingdom. We now report the successful primary transmission and serial passage of scrapie from the brains of two of the three affected cattle into mice and hamsters.

Materials and Methods

Bovine Specimens

Ten cattle (nine females and one steer), 8 to 11 months old, were injected with scrapie-infected brain homogenate by the intracerebral (i.c.), intramuscular (i.m.), and subcutaneous (s.c.) routes as well as being "dosed" orally (p.o.). Five of these cattle were inoculated with second-passage brain from Suffolk sheep (cows 1 through 5), and five were inoculated with the first goat-to-goat passage of scrapie from a natural case of the disease in an Angora goat (cows 6 through 10). One of the five cattle inoculated with Suffolk sheep brain homogenate (cow 5), and two of the cattle inoculated with Angora goat brain homogenate (cows 8 and 10) developed signs of neurologic disease 27 to 48 months after inoculation; we previously reported (5) the isolation and identification of scrapie amyloid protein (PrP27-30) from the brain of each of these animals. Homogenates of the brains of cows 5, 8, and 10 were used for inoculation into hamster and mice in these studies.

Bioassay in Hamsters and Mice

The LVG/LAK strain of golden Syrian hamsters and the NIH outbred strain of Swiss-Webster mice were received from the Animal Product Area, Frederick Cancer Research and Development Center, Frederick, Maryland. The NZW/LacJ strain of mice were from Jackson Laboratories, Bar Harbor, Maine. Groups of hamsters and mice were inoculated i.c. with 0.03 mL of individual 10% homogenates of brain from cows 5, 8, and 10, respectively, suspended in phosphate-buffered saline (PBS), pH 7.4. Groups of four hamsters and six mice were inoculated i.c. with 0.03 mL of serial 10-fold dilutions of homogenates of hamster brains and mouse brains, respectively, from animals that were positive on primary transmission. Titers were calculated by the method of Reed and Muench (6). The clinical diagnosis of scrapie was confirmed by histopathologic examination of the brains of randomly selected ill hamsters and mice and by sodium dodecyl sulfate–polyacrylamide gel electrophoresis (SDS-PAGE) and Western immunoblots of the abnormal isoform of scrapie amyloid protein (PrP27-30) extracted from hamster and mouse brains by the scaled-down method of Hilmert and Diringer (7). The final extract was separated in 12.5% SDS-PAGE silver-stained, and transferred to nitrocellulose. Nitrocellulose strips were immunostained with rabbit polyclonal antibodies against hamster or mouse PrP^{sc} as described (8).

Results

Transmission of Scrapie to Cattle

In 1979, ten cattle (nine females and one steer), 8 to 11 months of age, were inoculated i.c., i.m., s.c., and were "dosed" orally (p.o.) with homogenates of brain tissue infected with scrapie. The results of the experiment are summarized in Table 6.1. Five cattle (cows 1 through 5) were inoculated with second-passage brain from Suffolk sheep (2nd/Sc/Sh) and five (cows 6 through 10) were inoculated with the first goat-to-goat passage of scrapie from a natural case in an Angora goat (1st/Sc/G). One animal (cow 5) that had received scrapie-infected sheep brain and two animals (cows 8 and 10) that had received scrapie-infected goat brain presented with neurologic signs 27 to 48 months following inoculation, consisting of progressive difficulty in rising, a stiff-legged stilted gait, incoordination, abnormal tail position, disorientation, and terminal recumbency. From onset to terminal stages, the disease lasted 1 to 2.5 months. Histopathologic examination, presented in detail elsewhere (9), of the brains of affected animals revealed mild diffuse gliosis and few vacuoles, changes that were considered insufficient to confirm a clinical diagnosis of scrapie. Early attempts to transmit the disease by inoculating homogenates of brain from affected cattle to mice at the U.S. Department of Agriculture's Field Station at Mission, Texas, were unsuccessful.

TABLE 6.1. Summary of transmission of scrapie to cattle.

Cow	Breed/sex[a]	Pass/inocula[b]	Clinical	Pathology	PrP27-30
1	B/F	2nd/Sc/Sh	Traumatic death	Negative	Negative
2	H/F	2nd/Sc/Sh	Negative	Vacuoles?	Negative
3	H/F	2nd/Sc/Sh	Negative	Vacuoles	Negative
4	J/F	2nd/Sc/Sh	Negative	Negative	Negative
5	JxH/F	2nd/Sc/Sh	Neurologic signs	Vacuoles	Positive
6	B/F	1st/Sc/G	Negative	Negative	Negative
7	H/F	1st/Sc/G	Tetany?	Not Done	Negative
8	H/F	1st/Sc/G	Neurologic signs	Negative	Positive
9	J/F	1st/Sc/G	Negative	Negative	Negative
10	J/SM	1st/Sc/G	Neurologic signs	Mild gliosis	Positive

[a] B/F, Brahman/female; H/F, Hereford/female; J/F, Jersey/female; JxH/F, Jersey/Hereford/female; J/S/M, Jersey Steer/male.
[b] 2nd/Sc/Sh, second passage scrapie Suffolk sheep; 1st/Sc/G, first passage scrapie Angola goat.

Based upon the failure of early histopathologic findings to confirm a diagnosis of scrapie, frozen brain tissues from each of the ten cattle were examined for the presence of the abnormal isoform of the scrapie amyloid protein (PrP27-30). As shown in Table 6.1, PrP27-30 was detected in the brains of each of the three animals with clinical signs suggestive of scrapie (cows 5, 8, and 10); none was detected in the other seven animals. The yield from the positive brains varied greatly. The pattern of the PrP27-30 on Western immunoblots, reacted with polyclonal rabbit anti-PrP antibodies, consisted of three bands of molecular weight 27–30, 23–26, and 18–20 kd, similar to mouse PrP27-30 where the bands are thought to represent different degrees of glycosylation of the same protein. The pattern was also similar to that reported for BSE (10).

Primary and Serial Passage in Hamsters and Mice

The successful primary transmission of scrapie from cattle to laboratory rodents and the serial passage of these strains in hamsters and mice are summarized in Figure 6.1. Strain C-5 from cow 5 infected with second-passage Suffolk sheep scrapie on primary passage in hamsters had incubation periods in two attempts of 182 days and 128 days, respectively. On serial passages in hamsters, the incubation period shortened significantly to

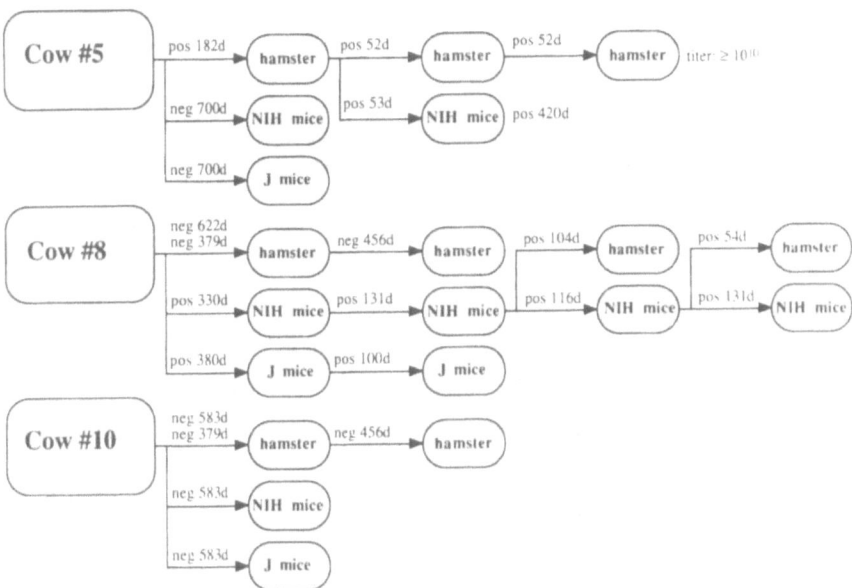

FIGURE 6.1. Illustrative data on the transmission of scrapie from PrP27-30 positive cow brains to golden Syrian hamsters, NIH Swiss-Webster mice (NIH) and NZW/LacJ mice.

52 days, and on the second hamster-to-hamster passage the infectivity titer was $\pm 10^{10} LD_{50}/mL$ as determined by the i.c. route. Strain C-5 failed to induce disease on primary passage in NIH Swiss mice and NZW/LacJ mice during the 700 days they were on test. Attempts to transmit the disease from first passage in hamsters to NIH Swiss mice were successful following an asymptomatic incubation period of 420 days. In contrast, although strain C-8 from cow 8 infected with first-passage Angora goat scrapie failed to transmit to hamsters on primary passage and serial blind passage in hamsters following observation period ranging from 379 days to 622 days, it did produce disease in both NIH Swiss mice and NZW/Lac-J mice following incubation periods of 330 days and 380 days, respectively, on primary passage. Serial passages of C-8 in both strains of mice were associated with a reduction in incubation periods to 131 days and 100 days, respectively, but

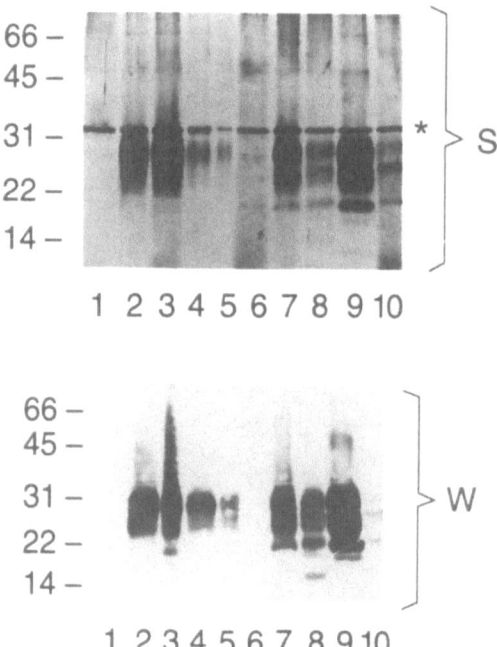

FIGURE 6.2. Detection of PrP27-30 on silver-stained SDS-PAGE gels (S) and Western immunoblots (W) of different passages of scrapie in hamsters and mice inoculated with brain homogenates of cows 5 and 8, and controls. Lane 1, normal hamster brain; lane 2, scrapie 263K-infected hamster brain; lane 3, C-5 scrapie-infected second-passage hamster brain; lane 4, C-8 scrapie-infected hamster brain following two passages in mice; lane 5, normal mouse brain; lane 6, Suffolk sheep strain C-506 scrapie-infected mouse brain; lane 7, Fujisaki strain Creutzfeld-Jakob disease (CJD)-infected mouse brain; lane 8, C-8 scrapie third-passage infected mouse brain; lane 9, human brain infected with CJD. Molecular weight of the marker is in kilodaltons. The asterisk indicates the band of proteinase K.

additional passages did not further reduce the incubation period in Swiss mice. Brain homogenates prepared from first Swiss mouse-to-mouse passage and from second mouse-to-mouse passage were used to successfully transmit the disease to hamsters following incubation periods of 104 days and 54 days, respectively. Strain C-10 (cow 10) repeatedly failed to transmit disease to hamsters and both strains of mice, although it gave the strongest reactivity of PrP27-30 on Western immunoblots (Fig. 6.2).

Discussion

In this chapter, we have extended our preliminary report (5) by presenting background date on the transmissibility, incubation time, and species barrier behavior of two distinct, wild strains of sheep and goat scrapie following passage through cattle and their transmissibility to mice and hamsters. From the initial observation it was clear that cattle are susceptible to wild strains of scrapie of sheep and goat origin and have relatively short incubation periods following combined intracerebral, peripheral, and peroral routes of inoculation. Clinical symptoms of the disease were similar to those reported for BSE in the United Kingdom (10, 11), and are presented together with neuropathology details elsewhere (9). The observed high susceptibility may be due to the 98% homology between sheep and bovine PrP gene (12). In contrast to the neuropathologic findings in British BSE, the brain pathology of the three cows that became clinically ill and showed neurologic signs initially revealed only marginal spongiform or reactive glial changes that originally precluded a definite pathologic diagnosis (9). However, on reexamination more than 15 years later, Clark et al. (9) reported that the neurohistologic changes were alike, slight and subtle, and were the same as those of scrapie, mainly astrocytosis and neuronal degeneration. Despite the limited pathologic changes, the brains of all three clinically ill animals were positive for PrP27-30, but with large variations in the PrP27-30 content (5).

During the reisolation of both scrapie strains from cows 5 and 8 into mice and hamsters, the wild scrapie strains maintained their distinct characteristics. The original sheep scrapie after one passage in a cow (C-5) was transmissible only to the hamsters; mice became susceptible only after one passage through golden Syrian hamsters and the asymptomatic incubation period was 420 days in contrast to the 53-day incubation period of the hamster material with which the mice were inoculated. The goat scrapie strain after passage in a cow (C-8) was transmissible on primary passage only to mice and transmission to hamsters occurred only after two passages in NIH mice. The second cow (C-10) inoculated with the identical goat strain of scrapie that caused disease in C-8 and which had the highest concentrations of PrP27-30 as noted on SDS-PAGE gels and Western immunoblots has failed to transmit the disease despite repeated attempts.

Both sheep (C-5) and goat (C-8) scrapie strains, when adapted to golden Syrian hamsters, displayed progressive shortening of incubation time with incubation periods on serial passage of the final 52 days (C-5) and 54 days (C-8) after i.c. inoculation, respectively, and high final infectivity titer of $\pm 10^{10} LD_{50}/g$ of brain. The incubation time of mice-adapted inocula was shorter in NZW/LacJ mice than in NIH Swiss mice, as expected (13). The comparative pathology of all transmitted cases in mice and hamsters is the subject of a separate article in preparation.

The results presented here demonstrate the susceptibility of cattle to scrapie strains of sheep and goat origin. Both strains maintained their distinct biologic characteristics after single passages in cows. However, the repeated lack of transmission to rodents inoculated with a brain homogenate of cow C-10 contrasted to the repeatedly positive transmission from C-8 inoculated with the same goat-passaged strain of scrapie suggests the possibility of a stochastic event with an abrupt change in biologic properties and resulting in a different transmission pattern to new species of animals. A similar phenomenon was previously reported and could be interpreted as a host-induced change in the ability to target and replicate in the new host species (3).

Acknowledgment. The authors thank Devera G. Schoenberg, M.S., for editing the manuscript and document production assistance.

References

1. Wilesmith JW, Ryan JBM. Bovine spongiform encephalopathy: observation on the incidence during 1992. Vet Rec 1993;132:300–1.
2. Prusiner SB. Chemistry and biology of prions. Biochemistry 1992;31:12277–88.
3. Hecker R, Taraboulos A, Scott M, et al. Replication of distinct scrapie prion isolates is region-specific in brains of transgenic mice and hamsters. Genes Dev 1992;6:1213–28.
4. Kimberlin RH, Walker CA, Fraser H. The genomic identity of different strains of mouse scrapie is expressed in hamsters and preserved on reisolation in mice. J Gen Virol 1989;70:2017–25.
5. Gibbs CJ Jr, Safar J, Ceroni M, et al. Experimental transmission of scrapie to bovine species. Lancet 1990;335:1275.
6. Reed LJ, Muench H. A simple method of estimation fifty per cent endpoints. Am J Hyg 1938;27:493–7.
7. Hilmert H, Diringer H. A rapid and efficient method to enrich SAF-protein from scrapie brains of hamsters. Biosci Rep 1984;4:165–70.
8. Safar, J. Ceroni M, Piccardo P, et al. Scrapie-associated precursor proteins: antigenic relationship between species and immunocytochemical localization in normal, scrapie, and Creutzfeldt-Jakob disease brains. Neurology 1990;40:513–7.

9. Clark WW, Hourrigan JL, Hadlow WJ. Encephalopathy in cattle experimentally infected with the scrapie agent. Am J Vet Res 1995;56:606–12.
10. Hope J, Reekie LJ, Hunter N, et al. Fibrils from brains of cows with new cattle disease contain scrapie-associated protein. Nature 1988;336:390–2.
11. Scott AC, Wells GA, Stack MJ. Bovine spongiform encephalopathy: detection and quantitation of fibrils, fibril protein (PrP) and vacuolation in brain. Vet Microbiol 1990;23:295–304.
12. Prusiner SB, Fuzi M, Scott M, et al. Immunologic and molecular biologic studies of prion proteins in bovine spongiform encephalopathy. J Infect Dis 1993;167:602–13.
13. Carlson GA, Hsiao K, Oesch B, et al. Genetics of prion infections. Trends Genet 1991;7:61–5.

7

Experimental Transmission of Scrapie to Cattle

Randall C. Cutlip, Janice M. Miller,
Richard E. Race, Allen L. Jenny,
Howard D. Lehmkuhl, and Mark M. Robinson

Scrapie is a slowly progressing neurodegenerative disease of sheep and goats that is the prototype of the spongiform encephalopathies of humans and animals (1). Following an incubation period of 2 or more years, progressive signs of neurologic disease lead to death with spongy changes, neuronal necrosis, and astrocytosis in the brain. Scrapie has been known in Europe for more than 250 years (2). The disease was first recognized in the United States in 1947 (1). From then until September 1989, scrapie was diagnosed in 505 flocks with an increased incidence of 30% in the last 10 years. This increase and epidemiologic evidence that bovine spongiform encephalopathy (BSE) in Great Britain was acquired through feeding-rendered by-products of sheep to cattle (3–5) caused great concern in the livestock and related industries in the United States. Therefore, with the assistance and encouragement of the National Renderers Association, the National Cattlemen's Association, the National Milk Producers Federation, the American Sheep Industry Association, and the Food and Drug Administration, this project was undertaken to determine if the U.S. scrapie agent(s) could be transmitted to cattle and cause a disease resembling BSE.

Methods

For this purpose, a group of 18 young Jersey and Holstein calves was inoculated intracerebrally (i.c.) (6) and a second group of 18 young calves was inoculated orally with a pooled suspension of raw brain that was prepared from 9 black face sheep with signs and lesions of scrapie. A third group of 24 calves was fed rendered meal and tallow from four flocks of black face sheep having a high incidence of scrapie. Of the 82 sheep that were rendered, one of two emaciated ewes that were examined histologically had lesions that were typical of scrapie. Carcasses of the sheep were rendered by the batch method at 129°C for 3.5 hours with expression of tallow by pressure without organic solvents. The rendered meal and tallow were fed at the rate of 6% and 3%, respectively, at the earliest time the calves would eat solid ration containing the rendered product. The fourth and fifth groups were composed of control calves that either were inoculated with a suspension of normal sheep brain or were not inoculated (Table 7.1).

To assess possible transmission of the scrapie agent, one-half of each group of calves was euthanized one year after exposure to the scrapie agent. Brain from each calf was examined for microscopic lesions using hematoxylin and eosin (H&E) and glial fibrillary associated protein (GFAP) (7) stains; for scrapie-associated prion protein (PrP-res) using immunohistochemical (IHC) (8) and immunoblotting (9) techniques; and for infectious agent by assay in C57Blk/6 mice. Calves that developed signs of neurologic disease and two control calves were similarly examined for lesions, PrP-res, and infectious agent. Calves with clinical signs were euthanized when permanent recumbency occurred except for two that died before recumbency became permanent. Those calves that have not developed disease will be observed daily for clinical signs until they become ill or for a total of 8 years.

TABLE 7.1. Preliminary results of inoculating calves with the scrapie agent.

Calf group	Type inoculum	Route inoculum	Signs (NP/NT)	Lesions (NP/NT)	IHC-PrP (NP/NT)	IMB-PrP (NP/NT)	Agent (NP/NT)
1A	RSB	IC	0/8	0/8	7/8	7/8	0/8
B	RSB	IC	9/9	0/9	9/9	9/9	1/9
2A	RSB	OR	0/8	0/8	0/8	0/8	0/8
B	RSB	OR	0/9	ND	ND	ND	ND
3A	RMT	OR	0/12	0/12	0/12	0/12	0/12
B	RMT	OR	0/12	ND	ND	ND	ND
4A	NSB	IC	0/1	0/1	0/1	ND	ND
B	NSB	IC	0/5	ND	ND	ND	ND
5A	NON	ND	0/1	0/1	0/1	ND	ND
B	NON	ND	0/1	ND	ND	ND	ND

IHC-PrP, immunohistochemistry for scrapie-associated prion protein; IMB-PrP, immunoblot for scrapie-associated prion protein; Agent, infectious agent as determined by bioassay in C57Blk/6 mice; NP/NT, number positive/number tested; A, part of group killed 1 year after inoculation; B, part of group retained for clinical observation; RSB, raw scrapie brain; RMT, rendered meal and tallow; NSB, normal sheep brain; NON, none; IC, intracerebral; OR, oral; ND, not done.

Results

Preliminary results of this study are presented in Table 7.1. Clinical signs were seen in all calves that had been inoculated i.c. with raw scrapie brain and held for more than 1 year after inoculation (6). Consistent lesions were not seen. PrP-res was seen in the brains of all calves that were inoculated i.c., including those without clinical signs that were euthanized 1 year after inoculation. Calves that were inoculated orally with raw scrapie brain or fed rendered meal and tallow had no signs of disease, no lesions, and no detectable PrP-res. Infectious agent was detected by bioassay in the brain of one i.c. inoculated, symptomatic calf but in none of the asymptomatic calves or in spleen-mesenteric lymph node pools of any calf. The mice used for bioassay were examined by IHC for PrP-res in the brain. Results of examination for microscopic lesions are pending.

Of those calves that developed clinical signs of disease after i.c. inoculation, the incubation period was 14 to 18 months and the course of the disease was 1 to 5 months. There was no correlation between length of incubation and course of the disease. Signs were insidious in onset and progressive until death occurred. In most calves, stiffness was the first sign noticed. This was followed by or accompanied by decreased appetite and lethargy. Lethargy was severe in all calves. An abnormal gait, associated with incoordination and weakness of the rear quarters, occurred in all calves and some developed transient lameness of 24 to 48 hours' duration. The posterior weakness and loss of weight progressed until recumbency became permanent or death occurred. Strabismus accompanied by exophthalmia was a common feature. Clonic spasms of muscle fascicles in the rear limbs and paddling in lateral recumbency were each seen in one calf. There was no evidence of aggressiveness, hyperexcitability, hyperesthesia (tactile or auditory), or hypermetria of limbs as has been reported for BSE (3).

Microscopic examination of H&E-stained sections of the brain from affected or unaffected calves did not reveal the primary lesions of BSE (3), i.e., vacuolation of neurons or neuropil, loss of neurons, and gliosis. Although some vacuolation of neuropil was seen in the brain stem and spinal chord of some affected calves, it did not differ greatly from that seen in two unaffected control calves. Neither were there differences in astrocytes between infected and control calves.

All calves that were inoculated i.c. with raw scrapie brain had PrP-res in the brain that was detectable either by IHC or immunoblotting. Except for two calves that were killed 1 year after inoculation, PrP-res was detected with each method. The IHC technique showed that PrP-res was widely distributed in the brain stem within the neuronal cytoplasm. In contrast to the distribution in sheep, where PrP-res was common in perivascular locations and abundant in gray matter neuropil (8), little PrP-res was located outside neurons in the calves. Subjective estimates indicated that the quantity of PrP-res was much greater in calves with clinical disease than in calves that were killed before clinical signs appeared.

Discussion

In conclusion, after 4 to 4.5 years of an 8-year study, we have demonstrated that the U.S. sheep scrapie agent will transmit to cattle by direct i.c. inoculation but not by oral exposure. Because of the limited incubation period and small number of calves exposed, the possibility of oral transmission is not precluded by these results. In cattle that were affected, neither clinical signs nor lesions were characteristic of BSE. Particularly noticeable was the decreased sensory responses, lack of aggressiveness and absence of vacuolation of neurons and neuropil. The route of inoculation or the amount of inoculum may have influenced the nature of the illness and type of lesion that was induced. The scrapie agent in other hosts can induce variable signs and lesions depending on the strain of the agent, strain and species of the host, route of inoculation, and number of passages in the host (10, 11). It is possible that additional passages through cattle will change the character of the disease resulting in changes that are more like those described for BSE.

Acknowledgment. The authors thank Dr. Richard Rubenstein, New York State Institute for Basic Research, Staten Island, NY 10314, for providing the immunoglobulin to conduct the immunohistochemical tests for PrP-res.

References

1. Detwiler L. Scrapie. Rev Sci Tech 1992;11:491–537.
2. Parry HB. Recorded occurrences of scrapie from 1750. In: Oppenheimer DR, ed. Scrapie disease in sheep. London: Academic Press, 1983:31–59.
3. Wells GAH, Scott AC, Johnson, CT, et al. A novel progressive spongiform encephalopathy in cattle. Vet Rec 1987;121:419–20.
4. Wilesmith JW, Wells GAH, Cranwell MP, Ryan JB, et al. Bovine spongiform encephalopathy: epidemiological studies. Vet Rec 1988;123:638–44.
5. Wilesmith JW, Ryan JBM, Atkinson MJ. Bovine spongiform encephalopathy: epidemiological studies on the origin. Vet Rec 1991;128:199–203.
6. Cutlip RC, Miller JM, Race RE, Jenny AL, Katz JB, Lehmkuhl HD, et al. Intracerebral transmission of scrapie to cattle. J Infect Dis 1994;169:814–20.
7. Jeffrey M, Wells GAH, Bridges AW. An immunohistochemical study of the topography and cellular localization of three neural proteins in the sheep nervous system. J Comp Pathol 1990;103:23–35.
8. Miller JM, Jenny AL, Taylor WD, Marsh RF, Rubenstein R, Race RF. Immunohistochemical detection of prion protein in sheep with scrapie. J Vet Diagn Invest 1993;5:309–16.

9. Race R, Ernst D, Jenny A, Taylor W, Sutton D, Caughey B. Diagnostic implications of detection of proteinase K–resistant protein in spleen, lymph nodes,
and brain of sheep. Am J Vet Res 1992;53:883–9.
10. Outram GW, Fraser H, Wilson DT. Scrapie in mice. Some effects on the brain
lesion profile of ME7 agent due to genotype of donor, route of injection and
genotype of recipient. J Comp Pathol 1973;83:19–28.
11. Fraser H, Bruce ME, McBride PA, Scott JR. The molecular pathology of
scrapie and the biological basis of lesion targeting. Prog Clin Biol Res
1989;317:637–44.

8

An Assessment of Transmissible Mink Encephalopathy as an Indicator of Bovine Scrapie in U.S. Cattle

MARK M. ROBINSON

A search of the literature concerning transmissible mink encephalopathy (TME) outbreaks in the United States and a postal survey of the feeding practices of Wisconsin mink producers were conducted to estimate the maximum prevalence and incidence of a hypothetical bovine transmissible encephalopathy within the context of the following hypothesis: the source of agent for each TME outbreak in Wisconsin was food-borne exposure of mink to a cow with an unrecognized transmissible encephalopathy, and every exposure to such a cow resulted in a TME outbreak. Within the context of this hypothesis, it was estimated that the maximum prevalence of the hypothetical transmissible encephalopathy in nonambulatory adult cattle could have been no more than 1:27,500. By extrapolation, the hypothetical disease occurred with an annual incidence of 1:975,000 adult cattle per year in Wisconsin, if it occurred at all.

Transmissible mink encephalopathy is a rare, but devastating neurodegenerative disease of ranch-reared mink (1–3). There have been five documented outbreaks of TME in the United States involving a total of 11 mink ranches.

Morbidity and mortality were between 60% and 100% of the adult mink population at most of the ranches (1, 3). TME is considered a "dead-end" disease because there was no vertical transmission observed during any of the outbreaks, and horizontal transmission occurred only through cannibalization of affected mink.

Epidemiologic analyses of four of the TME outbreaks indicated that the etiologic agent was food-borne (1, 3); however, identification of the source of the TME agent was problematic. Initially, the source was thought to be sheep or sheep offal that were naturally infected with the scrapie agent and incorporated into mink feed. Scrapie is a transmissible encephalopathy of sheep and goats that is similar to TME.

The association of scrapie with TME was conjectural (1, 4, 5), but persisted for more than 20 years in spite of a lack of supportive epidemiologic or experimental evidence. There was no documentation that sheep were used in mink feed prior to any of the TME outbreaks in the United States (1–3), and experiments to reproduce TME by oral exposure of mink to sheep scrapie isolates were unsuccessful (5).

The most recent outbreak of TME in the United States occurred in 1985 (3). The investigators hypothesized that the source of infection for this outbreak was one or more of the nonambulatory cattle incorporated into the mink feed produced at the ranch.

By definition, nonambulatory cattle are those that cannot rise or remain standing without support. This condition may be due to trauma, metabolic disorder, neurologic dysfunction, or a variety of other causes (6). Generally, nonambulatory cattle are slaughtered for nonhuman consumption or rendered.

Mink ranchers either make their own feed from a combination of cereal, animal offal, and whole animals, or obtain the feed as a wet or dry mix from cooperative or commercial suppliers. The fresh meat portion of the mink feed used prior to the 1985 TME outbreak cited above was derived from nonambulatory cattle obtained from local farms and incorporated into the feed mix at the ranch (3). All other feed components were obtained commercially. No sheep or sheep offal was used in the feed mix, according to the owner. The anecdotal evidence that nonambulatory cattle, but not sheep, were used in mink feed prior to this outbreak gave rise to the hypothesis that an unrecognized transmissible encephalopathy of cattle exists in the United States (3, 7).

It is important to note that this hypothesis predated the recognition of bovine spongiform encephalopathy (BSE) in the United Kingdom (8) and, further, that no instance of BSE or any other form of transmissible encephalopathy has been reported to date in either imported or native cattle in the United States. Since a bovine transmissible encephalopathy has not been reported in the United States, the possibility that nonambulatory cattle were the source of infection for any TME outbreak remains conjectural, although worrisome. To better understand how a hypothetical bovine transmissible encephalopathy could be responsible for a TME outbreak yet remain undetected itself, data were gathered through a literature search of TME in the United States and a postal survey of the feeding practices of Wisconsin mink producers, and used to derive an estimate of the maximum prevalence and incidence of a hypothetical bovine transmissible encephalopathy within the context of the following hypothesis: the source of agent for each TME outbreak in Wisconsin was food-borne exposure of mink to a cow with an unrecognized transmissible encephalopathy, and every exposure to such a cow resulted in a TME outbreak.

Materials and Methods

Literature Search

Index Medicus, Medline, CAB Abstracts, and BIOSIS Previews were searched for the years 1960 to 1994 inclusive using the keywords *epidemiology*, *mink*, *encephalopathy*, *transmissible mink encephalopathy*, and *TME*.

Postal Survey

Nine questions were developed in consultation with members of the USDA-APHIS-VS Centers for Epidemiology and Animal Health (Ft. Collins, CO) to obtain information concerning the current and past use of nonambulatory cattle in mink feed. The questions addressed (a) the longevity of the mink business, (b) the average size of the business, (c) any history of mixing mink feed on the premises, and (d) any history of using nonambulatory cattle in the mink feed.

The survey distribution was restricted to Wisconsin for three reasons: (1) historically, Wisconsin has produced at least one third of the commercial mink pelts grown in the United States (9), (2) four of the five documented TME outbreaks in the United States occurred there (1, 3), and (3) the adult milk cow inventory in Wisconsin has ranged between 1.8 and 2.4 million, and averaged approximately 2 million milk cows per year since 1945 (10). This last factor is of particular significance for two reasons. First, transmissible encephalopathies generally occur in adult animals such as milk cows. For example, the majority of cases of BSE in the United Kingdom have occurred in 3- to 5-year-old dairy cattle (11). By analogy, the adult dairy cow population in Wisconsin represented a large group of cattle in which a transmissible encephalopathy would have had an opportunity to mature to clinical expression because the majority of cows were between 3 and 7 years old. Second, the relatively large and constant number of adult dairy cows would have provided a constant source of nonambulatory adult cattle.

A single-blind distribution technique was used. The questionnaires were numerically coded, addressed, and distributed through the mail by an associate of the Great Lakes Ranch Service (Thiensville, WI). The list that correlated the code with the potential respondent was kept by the Ranch Service. Every known mink rancher in Wisconsin was included in the mailing. Distribution occurred at the end of breeding season and before whelping season in 1993, a period of relatively low management activity, to maximize the possibility of obtaining responses. Each questionnaire included a self-addressed, stamped return envelope. Questionnaires were returned by the respondents to the Great Lakes Ranch Service and forwarded unopened to the Animal Disease Research Unit (Pullman, WA) for collation and hand tabulation.

Assessment

Responses to the survey questions were analyzed for information about the recent and historical use of nonambulatory cattle in mink feed. For both analyses, it was assumed that responses to questions about averages (i.e., average number of breeder females per year, average number of nonambulatory cattle used per week) were representative of the entire period of continuous business for each respondent.

Several additional assumptions were made to facilitate an assessment of the hypothetical association between the use of nonambulatory cattle in mink feed and the occurrence of TME. It was assumed that (a) a transmissible encephalopathy occurred in adult cattle and was the immediate or underlying cause for their inclusion in the nonambulatory population; (b) nonambulatory cattle affected with a transmissible encephalopathy were randomly distributed among the nonambulatory population; (c) on average, a nonambulatory cow was used entirely in one day's ration; and (d) the use of one nonambulatory cow afflicted with a transmissible encephalopathy was sufficient to cause a TME outbreak among several thousand mink.

Results

Literature Search

More than 30 peer-reviewed publications from the years 1960 to 1994 were identified that dealt with investigations of the biochemical or molecular nature of the TME agent or experimental transmissions of the TME agent to other species, but only two were found that contained specific information about the epidemiology of TME outbreaks in Wisconsin (1, 3). The first peer-reviewed publication describing the diseased appeared in 1965 (1). Several nonrefereed publications provided ancillary information, but this was not relied on for the present analysis unless it was verified through cross-reference with peer-reviewed publications. Parenthetically, there were no publications of either type that included epidemiologic information about the only TME outbreak in the United States outside of Wisconsin, which occurred during 1963 in Idaho (2).

According to the literature, there have been four TME outbreaks in Wisconsin, occurring in 1947, 1961, 1963, and 1985 (1, 3). The 1947 outbreak involved one ranch, the 1961 outbreak involved five ranches, the 1963 outbreak involved two ranches, and the 1985 outbreak involved one ranch. The epidemiology of these outbreaks has been reported (1, 3) and only a summary will be presented here. The common features for three of the four outbreaks (1947, 1963, 1985) were (a) between 60% and 100% of the adult mink at each ranch died, (b) the probable time of infection could be identified to within several weeks, (c) dead or dying cattle were used as a feed component at each of the ranches during the probable period of infection, and (d) no other potential food source of an agent of transmissible encephalopathy was used in common at these ranches during the probable period of infection (1, 3). In contrast, only several hundred mink were affected with TME at each of five ranches during the 1961 outbreak, with a mortality level between 10% and 30% of the adult mink at each ranch. The feed components could not be identified beyond the fact that the ration was prepared and supplied from a common facility (1). The use of sheep or cattle in this ration could not be confirmed or denied.

Postal Survey

A total of 179 questionnaires were distributed to mink ranchers in Wisconsin. It was not known at the time of the mailing if all of the addresses represented current businesses since the mailing list was several years old and the mink industry had been in a state of economic depression. Seventy-one questionnaires were returned with responses, 14 were returned undelivered, and 94 were not returned. The 14 returned undelivered were assumed to represent ranches that no longer existed and were subtracted from the total of potential respondents. Therefore, the response rate was 43%.

Of the respondents, 10 (14.1%) were no longer in business and provided no responses other than that information; two (2.8%) were not in business, but had used nonambulatory cattle for a specific period of time while in business; 12 (16.9%) were currently in business, but had never mixed their own feed; seven (9.9%) were currently in business and had mixed, but were not currently mixing their own feed, and had never used nonambulatory cattle; 28 (39.4%) were currently in business and were mixing their own feed, but had never used nonambulatory cattle; four (5.6%) were currently in business and had used, but did not currently use, nonambulatory cattle; and eight (11.3%) were currently in business and currently used nonambulatory cattle. Seven of the eight ranchers in the last category used adult nonambulatory cattle (>2 years old) (Table 8.1) while the eighth had used only calves and only for a short period of time prior to the survey. Based on this information, 13 (18.3%) of the respondents had a history of using nonambulatory adult cattle in their mink feed.

The relative success of the postal survey was facilitated by the use of the Great Lakes Ranch Service as an intermediary. Previous experience had shown that mink ranchers were reluctant to divulge specific information about their operations to unknown individuals, particularly those associated with governmental agencies. Due to the death of our contact at the Ranch Service, Dr. G.R. Hartsough, it was determined that an effort to contact nonrespondents to improve the response rate would not be possible. Furthermore, it was determined that any selection bias present in the

TABLE 8.1. Current nonambulatory cattle use in mink feed.

Farm	Mink/year[a]	Cattle/year[b]
A	15,600	468
B	20,800	624
C	7,800	364
D	17,680	312
E	10,400	780
F	5,200	60
G	10,400	468
Total	87,880	3,076

[a] Average number of adult mink and kits per year.
[b] Average number of nonambulatory cattle used per year.

responses obtained from the first mailing would not significantly affect a "worst case" assessment because an increase in reported nonuse would have a neutral effect on the estimates and an increase in reported usage would actually decrease the estimates of maximum prevalence and incidence of the hypothetical disease (see below).

Assessment

Epidemiologic information indicated that infection of mink with the foodborne agent occurred during the year prior to the TME outbreak (1, 3). Therefore, data from respondents who reported use of nonambulatory cattle between the years 1946 and 1984 were used to estimate the maximum prevalence and incidence of the hypothetical bovine transmissible encephalopathy. Nine of the respondents reported the use of nonambulatory cattle in their mink feed during part or all of this period (Table 8.2). In addition, information from the literature and investigator contacts about the use of nonambulatory cattle on the premises of the 1985 TME outbreak was used in the calculations (3; R.F. Marsh, personal communication). No information about the number of nonambulatory cattle used at ranches affected with TME prior to 1985 was available.

During the period in question, there were at least ten producers who used nonambulatory cattle in their mink feed for a total of 253 ranch-years (Table 8.2). An average of 428 nonambulatory cattle were fed to mink each ranch-year, or more than one per day. It should be noted that this is a conservative estimate for the use of nonambulatory cattle in mink feed because (a) it is possible that some of the nonrespondents used nonambulatory cattle, (b) the number of cattle used at ranches that had a TME outbreak prior to 1985 could not be determined, and (c) a large number of mink ranchers went out of business between 1946 and 1984 due

TABLE 8.2. Nonambulatory cattle use in mink feed, 1946–1984.

Farm	Years[a]	Mink/year[b]	Downers/year[c]
1	34	15,600	468
2	6	15,600	104
3	4	20,800	624
4	39	7,800	364
5	7	4,160	52
6	39	17,680	312
7	33	10,400	780
8	37	5,200	60
9	24	10,400	468
10[d]	30	24,820	780
Total	253	132,460	4,012

[a] Number of years in business between 1946 and 1984, inclusive.
[b] Average number of adult mink and kits per year.
[c] Average number of nonambulatory cattle used per year.
[d] Mink production and nonambulatory cattle use levels obtained from the literature (3) and personal communication with RF Marsh, Madison, WI.

to economic and social pressures (12) and could not be included in the survey group because their present location could not be determined. It is probable that some of these producers used nonambulatory cattle, also.

It has been estimated that an average of 2% to 3% of adult cattle develop clinical problems resulting in nonambulatory conditions each year (6). If this is correct, of the approximately 2,000,000 adult cows in Wisconsin (10), mink ranchers use between 5% and 8% of the nonambulatory cattle available [(3076)/(2,000,000 × 2%–3%)] (Table 8.1). For the purposes of a "worst case" analysis, the lower level of 5% was used as the historical use level for 1946 to 1984. Therefore, it is possible that there was a total of 80 cows (1 cow per TME outbreak × 4 TME outbreaks/5% use level) in the nonambulatory cattle population between 1946 and 1984 that had transmissible encephalopathies and the potential to cause TME outbreaks. Thus, there were 80 cattle out of a cumulative population of approximately 2.2 million nonambulatory cattle (428 nonambulatory cattle used per ranch-year × 253 ranch-years/5% use level), or 1:27,500 nonambulatory adult cattle that would have had a transmissible encephalopthy during the 39-year period.

The estimate of 2.2 million cumulative nonambulatory cattle made above, which was derived from extrapolation of the postal survey data, compares favorably with that obtained by using 2% to 3% as an annual average of the incidence of nonambulatory cases among the Wisconsin adult cattle population (2,000,000 × 2%–3% × 39 years = 1.5 to 2.3 million nonambulatory cattle). If a use level higher than 5% is chosen, the estimate of the prevalence of the hypothetical bovine transmissible encephalopathy among nonambulatory cattle decreases. For example, if mink ranchers used 10% of the available nonambulatory cows, the number of clinical cases of a transmissible bovine encephalopathy would have been 40 in 39 years and the prevalence of the clinical-stage transmissible encephalopathy would have been 1:55,000 nonambulatory adult cattle.

Returning to the scenario in which 80 clinical cases of the hypothetical bovine transmissible encephalopathy occurred in 39 years, the average incidence of this disease would have been 2.05 clinical cases per year. If this average incidence is applied to the average annual inventory of 2,000,000 milk cows, the incidence of the hypothesized disease in Wisconsin dairy cows was no greater than 1:975,000 adult cows per year.

Discussion

The assumptions on which the assessment was based were examined to reduce them to lowest number of supported or unsupported variables. The first assumption, that a transmissible encephalopathy occurred in adult cattle and was the immediate or underlying cause for their inclusion in the nonambulatory population, allowed the hypothesis that a bovine transmissible encephalopathy existed.

The second assumption, that nonambulatory cattle affected with a transmissible encephalopathy were randomly distributed among the nonambulatory population, allowed for an estimate of the maximum prevalence and incidence of the hypothetical bovine disease to be derived and generalized to the cattle population of Wisconsin. If the disease exists, it is possible that its distribution is not random, but the lack of evidence for or against this possibility makes the assumption of random distribution the more logical to incorporate.

The third assumption, that on average a nonambulatory cow was used entirely in one day's ration, was based on information obtained from informal contacts with industry specialists and mink ranchers. Nonambulatory cattle are used in mink feed primarily at large mink ranches, and are collected from local sources. Following euthanasia, most or all of the bovine carcass is butchered, ground, and mixed with other feed components. Smaller mink ranchers generally rely on feed cooperatives or commercial feed suppliers instead of preparing their own feed in the described manner because the cost/benefit of collecting and processing nonambulatory animals is not favorable. This assumption was supported by data from the postal survey in that the seven ranchers who were currently in business and using nonambulatory adult cattle in their feed had large operations and used an average of 3,076 nonambulatory cattle per year, or approximately 8.5 nonambulatory cattle per ranch per week (Table 8.1). Additionally, the historical data indicated that approximately 428 nonambulatory cattle were fed to mink each ranch-year, or more than one per day (Table 8.2).

The fourth assumption, that the use of one nonambulatory cow afflicted with a transmissible encephalopathy was sufficient to cause a TME outbreak among several thousand mink, was based on a combination of experimental observations and literature findings. It has been shown that consumption of approximately 0.6 g of BSE-infected brain tissue of United Kingdom origin was sufficient to transmit a spongiform encephalopathy to a mink (13). A 500-kg dairy cow contains approximately 1 kg of central nervous system (CNS) tissue (14). Therefore, it is theoretically possible that the uniform distribution in feed of 1 kg of CNS tissue from a nonambulatory cow with a transmissible encephalopathy would provide sufficient exposure to the transmissible agent to kill at least 1,600 mink. If other tissues contained the transmissible agent, the number of affected mink could be higher. In fact, the TME outbreaks of 1947, 1963, and 1985 involved approximately 1,300 (>99%), 2,000 (>99%), and 4,400 (approximately 60%) of the adult mink present, respectively (1, 3). The study of the oral transmission of BSE to mink coupled with the number of deaths recorded for each outbreak supports the assumption that one cow was the source of the transmissible agent for each outbreak.

The focus of this report has been the use of nonambulatory cattle in mink feed and the relationship of this practice to TME outbreaks. If ranch-reared mink have served as sentinels for a transmissible encephalopathy in the Wisconsin cattle population, estimates from this analysis and the fact that there have been only four documented TME outbreaks suggest that the cattle disease has occurred infrequently. Since neither BSE nor any other transmissible encephalopathy of cattle has been reported to date in the United States, this conclusion seems intuitively obvious. However, an effort to define the potential prevalence and incidence of the disease within the context of a specific hypothesis has not been made before. This effort is relevant because it provides a quantitative basis for policy decisions at the national level concerning problems of perceived risk.

All estimates in this analysis were made in terms of the clinical expression of a transmissible bovine encephalopathy because the hypothesis that TME outbreaks were associated with the use of nonambulatory cattle in mink feed was based on the implicit assumption that there was a direct relationship between the hypothetical disease and movement of an animal from normal to nonambulatory cattle populations (3). No effort was made to include preclinical cattle in the estimates because (a) the clinical disease has not been described in the United States, (b) the relationship between the pathogenesis and potential for transmission is unknown even for BSE in the United Kingdom, and (c) the assumptions and variables associated with such an analysis were determined to be unmanageable.

The estimates in this analysis depend in part on the accuracy of the record of TME outbreaks in Wisconsin. Due to the vigilance of interested scientists, veterinarians associated with the mink industry, and producers themselves, it is unlikely that there have been undocumented TME outbreaks in Wisconsin. However, it is possible that other potential TME outbreaks have existed, but were curtailed because the time between exposure of mink and the complete depopulation of the ranch or scaling back of operations through elimination of all adult mink was insufficient to allow clinical expression and detection of the disease.

Similarly, the analysis depends on a one-to-one relationship between the use of an affected cow in mink feed and a TME outbreak, as well as the random distribution of affected cattle within the nonambulatory population so that mink ranchers did not experience a bias in obtaining these animals. An alternative scenario would include the assumption that it required concurrent exposure of mink to more than one affected cow to establish the TME infection, or that the hypothetical disease in cattle could have been masked by other diseases that resulted in a nonrandom distribution of affected cattle and preferential exclusion or inclusion in the mink food chain. The proper development of these variables would require documentation that the disease exists in this country and the ability to identify live but affected cattle.

The purpose of this assessment was to establish boundaries and define the potential magnitude of an hypothesized disease problem of major potential importance without making unjustifiable assumptions or overstepping the available data. Epidemiology and risk assessment are achieving increased attention in light of advances in information dissemination and increased public scrutiny of scientific communications. Public or media misinterpretations of research involving transmissible encephalopathies may have severe economic consequences in terms of international trade and government regulation. In some cases, misinterpretations can create unfounded health concerns and public anxiety.

A central tenet of epidemiology and risk assessment is the complete yet responsible use of all available data to try to answer specific health questions. The proponents of epidemiologic risk assessment have the added responsibility to clarify their assumptions and logic to accurately communicate their findings. In this chapter, an attempt has been made to fulfill these goals in a field where assumptions are more numerous than facts, and the economic consequences of inaccurate assumptions and unprovable anecdotes are potentially disastrous.

Finally, if a transmissible encephalopathy of adult cattle does exist in Wisconsin or the United States, but occurs with a frequency of one clinical case per 975,000 adult cattle per year, is it appropriate to consider it an infectious disease by current definitions?

Acknowledgments. The author thanks Drs. A.M. Bleem, D.B. Francy, J.R. Gorham, D.D. Hancock, G.R. Hartsough, W.D. Hueston, and R.F. Marsh for assistance in experimental design, developing the postal survey, gathering data, and for comments about the presentation of this work. This chapter is dedicated to the memory of G.R. Hartsough.

References

1. Hartsough GR, Burger D. Encephalopathy of mink: I. Epizootiologic and clinical observations. J Infect Dis 1965;115:387–92.
2. Hadlow WJ, Race RE, Kennedy RC. Experimental infection of sheep and goats with transmissible mink encephalopathy virus. Can J Vet Res 1987;51:135–44.
3. Marsh RF, Bessen RA, Lehmann S, Hartsough GR. Epidemiological and experimental studies on a new incident of transmissible mink encephalopathy. J Gen Virol 1991;72:589–94.
4. Hanson RP, Eckroade RJ, Marsh RF, ZuRhein GM, Kanitz CL, Gustafson DP. Susceptibility of mink to sheep scrapie. Science 1971;172:859–61.
5. Marsh RF, Hanson RP. On the origin of transmissible mink encephalopathy. In: Prusiner SB, Hadlow WJ, eds. Slow transmissible diseases of the nervous system, vol 1. New York: Academic Press, 1979:451–60.
6. Centers for Epidemiology and Animal Health. Nonambulatory cattle and transmissible spongiform encephalopathy in the United States. Ft. Collins, CO: USDA:APHIS:VS, 1993.
7. Marsh RF, Hartsough GR. Evidence that transmissible mink encephalopathy results from feeding of infected cattle. In: Murphy BD, Hunter DB, eds. Proceedings of the Fourth International Scientific Congress in Fur Animal Production. Toronto: Canada Mink Breeders Association, 1988:204–7.
8. Wells GAH, Scott AC, Johnson CT, et al. A novel progressive spongiform encephalopathy in cattle. Vet Rec 1987;121:419–20.
9. Marsh RF, Hadlow WJ. Transmissible mink encephalopathy. Rev Sci Tech Off Int Epiz 1992;11:539–50.
10. Wisconsin Agricultural Statistics Service. Wisconsin 1991 Dairy Facts. Madison, WI: Wisconsin Department of Agriculture, Trade, and Consumer Protection, 1992.
11. Wilesmith JW, Ryan JBM. Bovine spongiform encephalopathy: recent observations on the age-specific incidences. Vet Rec 1992;130:491–2.
12. Bridges V, Bleem A, Walker K. Risk of transmissible mink encephalopathy in the United States. In: Animal health insight, fall. Fort Collins (CO): USDA-APHIS-VS Animal Health Information, 1991.
13. Robinson MM, Hadlow WJ, Huff TP, et al. Experimental transmission of the agent of bovine spongiform encephalopathy (BSE) to mink. J Gen Virol 1994;75:2151–5.
14. Sisson S, Grossman JD. The anatomy of the domestic animals. 4th ed. Philadelphia: WB Saunders, 1953:859.

9

Experimental Infections of Cattle and Mink with the Agents of Transmissible Mink Encephalopathy, Scrapie, and Bovine Spongiform Encephalopathy

Mark M. Robinson

The agents of transmissible mink encephalopathy (TME), scrapie, and bovine spongiform encephalopathy (BSE) were used to infect cattle and mink to determine the interspecies transmissibility of each agent and the characteristics of any resulting disease. While each of the agents tested caused disease in cattle (TME, scrapie) or mink (BSE, scrapie), the time to death and the severity of clinical signs and neurohistologic changes varied between the different agents within a species.

The advent of the BSE epizootic in the United Kingdom coupled with the continued sporadic appearance of TME in the United States led to the speculation that an undetected form of bovine scrapie exists at a low level of incidence in the United States and has been the cause of the TME outbreaks (1). For this and other reasons, we decided that interspecies transmission studies of TME, sheep scrapie, and BSE were necessary to provide fundamental information about the comparability of these diseases and their respective agents. Moreover, this information could help to identify undetected transmissible encephalopathies in cattle and other farm animals should they exist. This chapter describes experimental challenges of cattle with the agents of TME and scrapie, and of mink with the agents of BSE and scrapie. Some of the results presented here are incomplete because the experiments are in progress, and are so indicated in the text or tables.

Materials and Methods

Inocula

A 10% (w/v) filtered suspension of mink brain homogenate from the Hayward, Wisconsin outbreak of TME (2) and a 10% (w/v) homogenate of mink brain from the first experimental passage of the Blackfoot, ID outbreak of TME (3) were provided by Dr. William J. Hadlow (Hamilton, MT). A 10% (w/v) homogenate of bovine brain from a steer inoculated with mink brain homogenate from the Stetsonville, Wisconsin outbreak of TME (1) was provided by Dr. Richard F. Marsh (Madison, WI). Three 10% (w/v) homogenates of sheep brain (Phelan, Kruegor, Swaim) from scrapie-affected sheep collected from three different flocks in two states were provided by Dr. Randall C. Cutlip (Ames, IA). Four 10% (w/v) homogenates of sheep brain (TN1591, W6, WA1989, 89–247) from scrapie-affected sheep collected from four different flocks in three states were prepared at this laboratory. A 5% (w/v) pool of bovine brain homogenates was made from three bovine brain homogenates prepared from cattle inoculated with scrapie-affected sheep or goat brain homogenates at Mission, Texas (4) and provided by Dr. Clarence J. Gibbs, Jr. (Bethesda, MD) with the permission of Dr. James L. Hourrigan (Vienna, VA). Two homogenates of BSE-affected bovine brain (PG30/90, PG45/90) were provided by Drs. Gerald A.H. Wells and Michael Dawson (Central Veterinary Laboratory, United Kingdom) and used undiluted or prepared as 10% (w/v) homogenates.

Inoculations

Ten-week-old castrated, weaned Holstein calves were anesthetized with xylazine. A 1.5-mm hole was made with an electric drill through the calvarium in the right parietal region of each calf, and 0.2 mL of one inoculum (Hayward, Blackfoot, or cattle-passaged Stetsonville isolate of TME; Phelan, Kruegor, Swaim, TN1591, W6, WA1989, or cattle-passaged Mission isolate of scrapie) was injected through a 22-g, $1\frac{1}{2}''$ needle. Each inoculum was used in four calves. A separate group of 12 calves was inoculated subcutaneously with 2 mL of the cattle-passaged Stetsonville TME isolate in the right prescapular region.

One-year-old male and female standard dark mink were anesthetized with a xylazine/ketamine solution. A hole was drilled through the calvarium in the left parietal region of each mink with an electric drill and #12 wire bit, and 0.1 mL of inoculum (PG30/90 or PG45/90 isolate of BSE; Phelan, Kruegor, Swaim, TN1591, W6, WA1989, or 89–247 isolate of scrapie) was injected through a 23-g, $1\frac{1}{2}''$ needle.

Similar mink were used for oral dosing with the BSE and scrapie agents. Pooled undiluted brain homogenate from BSE-affected cattle was mixed with normal mink food and fed under observation while 10% (w/v) brain homogenates from scrapie-affected sheep were placed in water containers from which the mink drank the homogenates under observation.

Results

All of the calves challenged intracerebrally with the Hayward, Blackfoot, and cattle-passaged Stetsonville isolates of TME developed clinical signs of encephalopathy and were euthanized or found dead between 16 and 28 months postinoculation (Table 9.1). The clinical signs were subtle and included apprehension, inappetence, and decreased water consumption. Excitability and auditory hyperesthesia became evident over a period of several days to weeks. Some steers exhibited bruxism, some lapped water, and some exhibited nystagmus. Hypermetria and hindlimb ataxia became evident in all of the steers and progressively worsened. None of the steers inoculated subcutaneously with the cattle-passaged Stetsonville TME isolate have developed clinical signs after 4.5 years.

The neurohistologic changes in steers from each group were similar and extensive. Spongiform degeneration of the neuropil and astrocytosis were the predominant features. Pyknotic neurons were present as a regular feature, but vacuolated neurons were sparse. The most severely affected structures included the septal area, diencephalon, and midbrain.

Studies of the experimental infection of cattle with field isolates or cattle-passaged isolates of the scrapie agent are incomplete at this writing (Table 9.2). All of the steers in three groups (TN1591, W6, Mission) have been euthanized at advanced stages of disease. In the other groups, one or more steers remain alive. Clinical signs exhibited by these animals were more subtle than the steers inoculated with TME agent (Table 9.3), and consisted primarily of hypermetria and ataxia. The time to death for scrapie and cattle-passaged scrapie inoculants appeared to be longer when compared with TME inoculants, but this impression cannot be confirmed until the end of the transmission studies. To date, the neurohistologic changes in representatives from five of the seven groups do not appear to be significant,

TABLE 9.1. Challenge of cattle with the TME agent.

Inoculum	Route	A/I[a]	TTD[b]
Hayward	i.c.	4/4	584 (days) (89)
Blackfoot	i.c.	4/4	654 (144)
Stetsonville	i.c.	4/4	557 (47)
Stetsonville	s.q.	0/12	NA

[a] Number affected/number inoculated.
[b] Time to death (s.e.m.).
i.c., intracerebral inoculation; s.q., subcutaneous inoculation; NA, not applicable.

TABLE 9.2. Challenge of cattle with the scrapie agent.

Inoculum	Route	A/I[a]	TTD[b]
TN1591	i.c.	3/4[c]	628 (days) (67)
W6	i.c.	4/4	628 (42)
WA1989	i.c.	2/4[c]	625 (1 alive)
Phelan	i.c.	2/4	NA (4 alive)
Kruegor	i.c.	4/4	601 (2 alive)
Swaim	i.c.	3/4	NA (4 alive)
Mission	i.c.	4/4	724 (56)

[a] Number affected/number inoculated.
[b] Time to death (s.e.m.).
[c] One intercurrent death.

especially when compared with those of either BSE or TME in cattle (Table 9.3). This observation is similar to previous findings (5–7). Adaptation of the scrapie agent to cattle at second cattle passage was not evident when the time to death and neurohistologic changes of the sheep-to-cattle transmissions were compared to those of the cattle-passaged (Mission) scrapie isolate.

Brain homogenates from BSE-affected cattle were used to infect mink by intracerebral and oral inoculations. Specific details of these experiments have been described (8). A summary of the results is presented in Table 9.4. Intracerebral (i.c.) inoculation produced clinical signs of encephalopathy within an average of one year. Oral inoculation with less than 1 g of pooled BSE-affected bovine brain homogenate produced clinical signs at approximately 3 months on average following the intracerebral inoculants. In all cases, these signs included reduced aggression, inappetence, lethargy, and mild to moderate hindlimb ataxia. None of the mink exhibited the classic signs of TME, i.e., tail curling, scattering of feces, hyperexcitability, and hyperaggressiveness. While the clinical signs in these mink were not comparable to classical TME, the neurohistologic changes were similar (Table 9.3).

I.c. and oral challenges of mink with field isolates of sheep scrapie are in progress. The preliminary results of these experiments are presented in

TABLE 9.3. Relative severity of clinical signs and neuropathology.

Recipient	Donor	Route	Clinical signs	Neuropathology
Cattle(BSE)	?	Oral	+ + to + + + + +	+ + + +
Cattle	Mink	i.c.	+ +	+ + + + + +
Cattle	Sheep	i.c.	+ (in progress)	+/−
Mink(TME)	?	Oral	+ + + +	+ + +
Mink	Cattle	i.c.	+ +	+ + to + + +
Mink	Cattle	Oral	+ +	+ + to + + +
Mink	Sheep	i.c.	+ (in progress)	+ + to + + +
Mink	Sheep	Oral	− (in progress)	

TABLE 9.4. Challenge of mink with the BSE agent.

Inoculum	Route	A/I[a]	TTD[b]
PG30/90	i.c.	8/8	364 (days) (22)
PG45/90	i.c.	8/8	375 (14)
Pool	oral	10/10	462 (11)

[a] Number affected/number inoculated.
[b] Time to death (s.e.m.).

Table 9.5. Several of the i.c. inoculated mink remain alive either with or without clinical signs, and none of the orally dosed mink have developed clinical signs of encephalopathy in the 20 months since their exposure. Compared with classic TME or experimental infection with BSE, the clinical signs in affected mink were much reduced (Table 9.3). Inappetence and lethargy predominated, with a relatively short clinical period. However, preliminary examination of brain sections from affected mink indicate that the neurohistologic changes are similar to those described previously (8, 9).

Discussion

A qualitative comparison of the clinical signs and neurohistologic changes observed in natural BSE and TME with those observed in these studies is presented in Table 9.3. While this is an oversimplification of the available data, three points can be illustrated by this chapter's tables. First, the clinical signs and neurohistologic changes of the steers inoculated with the scrapie agent were more subtle than those inoculated with the TME agent (Table 9.3) and not comparable to those of natural BSE. Second, the time between experimental exposure and death of cattle inoculated with the

TABLE 9.5. Challenge of mink with the scrapie agent.

Inoculum	Route	A/I[a]	TTD[b]
TN1591	i.c.	6/6	402 (days) (16)
	Oral	0/10	NA
W6	i.c.	6/6	389 (27)
	Oral	0/10	NA
WA1989	i.c.	5/6	422 (1 alive)
89–247	i.c.	2/6[c]	493 (2 alive)
	Oral	0/6	NA
Phelan	i.c.	6/6	446 (42)
Kruegor	i.c.	4/6	442 (2 alive)
Swaim	i.c.	1/6	524 (5 alive)

[a] Number affected/number inoculated.
[b] Time to death (s.e.m.).
[c] Two intercurrent deaths.

scrapie agent (Table 9.2) appears to be longer than those inoculated with the TME agent (Table 9.1). Third, to date the BSE agent is the only agent tested here or elsewhere that produced disease in mink following oral inoculation (Tables 9.4 and 9.5) (1, 9).

These observations suggest that the TME agent is somewhat more virulent in cattle than the scrapie agent following i.c. inoculation, and show that the BSE agent is capable of causing a TME-like disease in mink following oral exposure. While this information does not prove or disprove the hypothesis that a transmissible encephalopathy exists undetected in United States cattle, it serves to provide a foundation for understanding the interspecies communicability of scrapie, TME, and BSE.

Acknowledgment. The author thanks Drs. R.C. Cutlip, M. Dawson, C.J. Gibbs, Jr., J.R. Gorham, W.J. Hadlow, T.P. Huff, D.P. Knowles, and R.F. Marsh for providing materials for these experiments and advice on the experimental design and execution.

References

1. Marsh RF, Bessen RA, Lehmann S, Hartsough GR. Epidemiological and experimental studies on a new incident of transmissible mink encephalopathy. J Gen Virol 1991;72:589–94.
2. Hartsough GR, Burger D. Encephalopathy of mink. I. Epizootiologic and clinical observations. J Infect Dis 1965;115:387–92.
3. Hadlow WJ, Race RE, Kennedy RC. Experimental infection of sheep and goats with transmissible mink encephalopathy virus. Can J Vet Res 1987;51:135–44.
4. Gibbs CJ Jr, Safar J, Ceroni M, Di Martino A, Clark WW, Hourrigan JL. Experimental transmission of scrapie to cattle. Lancet 1990;335:1275.
5. Hourrigan JL. Experimentally induced bovine spongiform encephalopathy in cattle in Mission, TX, and the control of scrapie. J Am Vet Med Assoc 1990;196:1678–9.
6. Cutlip RC, Miller JM, Race RE, et al. Intracerebral transmission of scrapie to cattle. J Infect Dis 1994;169:814–20.
7. Clark WW, Hourrigan JL, Hadlow WJ. Encephalopathy in cattle experimentally infected with the scrapie agent. Am J Vet Res 1995;56:606–12.
8. Robinson MM, Hadlow WJ, Huff TP, et al. Experimental infection of mink with bovine spongiform encephalopathy. J Gen Virol 1994;75:2151–5.
9. Hanson RP, Eckroade RJ, Marsh RF, ZuRhein GM, Kanitz CL, Gustafson DP. Susceptibility of mink to sheep scrapie. Science 1971;172:859–61.

10

BSE-Free Status:
What Does It Mean?

Richard F. Marsh

It has now been 10 years since bovine spongiform encephalopathy (BSE) first appeared in Great Britain. Over this period BSE has spread to 10 other countries thereby having worldwide impact on the export and import of livestock as well as the use of bovine tissues in drug manufacture. Perhaps it is now time to examine what we have learned and consider how this knowledge can be used to prevent other epizootics of BSE.

What Have We Learned?

First, for decades we knew of the heat resistance of the sheep scrapie agent, but no one foresaw the implication of this physicochemical property for the origin and amplification of BSE. Regardless of whether BSE began by the transmission of a sheep or bovine agent, it was the long-standing practice of feeding animal protein to cattle that allowed the disease to become well established before recognized. Many countries have learned this lesson. The European Union presently has a ban on the feeding of ruminant animal protein.

Second, BSE has lasted longer and had a greater economic impact than expected. While Great Britain discontinued feeding ruminant animal protein in 1988 with the Specified Offal Ban, only now has there been a significant decrease in the incidence of disease. Prion diseases are not like conventional diseases that can be controlled or eradicated over a relatively short period of time by testing for infected individuals followed by quarantine and/or treatment. The long incubation periods with BSE allowed for years of exposure to subclinically infected animals, then additional years to subside while preventative measures took their slow course.

Third, the magnitude of the public response has been unexpected. There is great public concern over possible human health risks with BSE. Despite assurances from the British government and epidemiologic studies showing no evidence of transmission of prion diseases from animals to man,

there has been a constant public perception that no one really knows whether humans can be infected with BSE. This perception initially translated into a 30% reduction in beef consumption in Great Britain, and even a total ban on British beef from school menus in London for a few weeks immediately following the revelation that BSE could be naturally transmitted to cats. It would seem that BSE must be dealt with both emotionally as well as scientifically, and that perception is indeed reality when unequivocal facts are not available. (1)

Finally, the BSE epizootic in Great Britain has stimulated new interest in the epidemiologic interrelationships of prion diseases in animals and new research on species susceptibility and prion strain variation. This chapter discusses how these new findings may be applied to surveillance programs for bovine prion diseases in BSE-free countries.

New Epidemiologic Perspectives

Sheep to Cattle

Does BSE prove that sheep scrapie can be naturally transmitted to cattle? Perhaps, but not for certain. BSE may have been caused by the discontinuation of an organic solvent extraction step in the rendering process that allowed for greater exposure to a *cattle* pathogen. However, this question can never be answered and, furthermore, is quite meaningless.

We now know that a single brain from a prion disease–affected animal may contain several other strains (e.g., biotypes or subpopulations) of these transmissible agents that are capable of producing disease if passaged in other species (2, 3). Since scrapie-affected sheep brains are also likely to contain strains other than the one producing disease in that particular animal, it would be unwise to feed this material to other ruminants and allow for selection and amplification of a potential pathogen. Therefore, in countries that have not yet banned the practice of feeding ruminant animal protein, it remains important to reduce the prevalence of scrapie infection in sheep.

Spontaneous Disease

The epidemiology of human prion diseases is in many ways better understood than those in animals. Ten percent of patients with Creutzfeldt-Jakob disease (CJD) have a familial history of disease that can be traced to abnormalities in their prion protein gene on chromosome 20. An additional 5% to 10% of CJD cases are iatrogenic in origin due to exposure to contaminated human growth hormone or tissue grafts. But it is the remaining 80% to 85% of CJD cases that should be of interest to animal epidemiologists because they appear to be spontaneous with no evidence of natural transmission.

If spontaneous cases of prion disease can occur in humans, they likely also occur in animals. Normally not naturally transmitted, these spontaneous incidents can still pose a danger by the unnatural act of cannibalism as seen in kuru in humans, or by the intervention of man and the feeding of animal protein to ruminants. Therefore, even scrapie- and BSE-free countries such as Australia and New Zealand remain at risk if they continue to feed ruminant animal protein.

Strain Variation

Accompanying the increased research on scrapie and BSE has been a rapid expansion of knowledge on the molecular basis of prion disease. Stimulated principally by studies on prion protein gene abnormalities in familial human diseases, new recombinant DNA techniques have produced transgenic and ablated mouse models that have furthered our understanding of the role of the prion protein in producing these neurodegenerative diseases. Experiments are currently focused on three main areas of research: (1) What is the posttranslational modification responsible for producing the protease-resistant form of the protein? (2) How does the modified protein convert its normal counterpart? and (3) How do these changes account for strain variation?

Prions produced in the same species can have distinct biologic properties affecting incubation period, clinical signs, neuronal cell targeting, and host range (4). One important example of this relevant to BSE is a study by Cutlip et al. (5) testing the susceptibility of cattle to intracerebral inoculation of American sources of sheep scrapie. These animals developed progressive neurologic disease but the clinical signs and histopathologic lesions were atypical of BSE. Affected animals were not aggressive but gradually became more debilitated until unable to stand, resembling more a downer cow than a "mad" cow. More importantly, the brains of these animals showed no histologic evidence of spongiform degeneration. These findings indicate that there are strains of prions capable of producing clinicopathologic disease in cattle that is different from BSE.

Relevance to Surveillance Programs

Surveillance programs for BSE-like diseases in countries that continue to feed ruminant animal protein must take into account strain variation as well as different possible sources of exposure. In the United States, there are four possible sources for a prion disease of cattle.

Exposure to Imported BSE-Infected Cattle

From 1981 to 1989 the United States imported 499 head of cattle from Great Britain (6). At least 290 of these animals have since died and their carcasses rendered into meat and bone meal. In 1993 a case of BSE was diagnosed in a Canadian cow imported from Great Britain.

The Canadian government responded by killing all other British imports. However, 22 of these animals had already been purchased by American buyers and transferred to the United States.

If the United State's exposure is due only to feeding animal protein from BSE-infected imports, a latent period of 20 to 30 years can be expected based on three to four animal passages with 6- to 8-year incubation periods before the disease reaches a prevalence level likely to be detected by current surveillance practices. The good news is that the BSE strain that caused the epizootic in Great Britain produces spongiform degeneration of the brain that can easily be recognized by diagnostic pathologists.

Exposure to Sheep Scrapie

The Cutlip study showed that cattle are susceptible to intracerebral inoculation of American sources of sheep scrapie. Cattle exposed orally to the same inoculum remain unaffected after 4 years.

The United States has had endemic scrapie for almost 50 years, a period during which there has been continuous feeding of ruminant animal protein to cattle. Recognizing the importance of reducing or eliminating scrapie infection in American sheep, the government has attempted several eradication programs, none of which has been completely successful because of lack of producer participation and the failure of some state regulatory agencies to enforce federal rules. The low participation in the current voluntary Scrapie Control Program leaves little doubt that the prevalence of scrapie infection is increasing in American sheep.

A prion disease in American cattle exposed to a similar strain of scrapie seen in the Cutlip study would produce distinct difficulties in diagnosis. First, the clinical presentation would likely be different than BSE. Instead of most affected animals showing behavioral changes of aggression, the main clinical feature would be progressive debilitation. Therefore, the major focus for surveillance programs would need to move from rabies-negative cattle brains to animals categorized as "downer cows," a nebulous disease classification with multiple etiologies. Second, histopathologic diagnosis would not be possible since there would be no detectable spongiform degeneration. Diagnosis would need to be based on detection of the modified form of the prion protein by immunoblot analysis or immunohistochemical staining of affected brain tissue.

Exposure to a Rare Cattle Pathogen Causing Transmissible Mink Encephalopathy (TME)

Epidemiologic studies have indicated that TME may be caused by feeding ranch-raised mink tissues from down cattle (7). If mink represents a sentinel species for a rare cattle pathogen, it is likely that this agent could also be transmitted to cattle by feeding animal protein. Experimental studies testing bovine susceptibility to various sources of TME have shown that TME-affected cattle have clinical signs similar to those described in the Cutlip et al. (5) study. However, contrary to cattle experimentally infected with American sheep scrapie, the brains of TME-affected cattle have spongiform degeneration, and this microvacuolation can be distinguished from that seen in BSE-affected cattle by topographic distribution (8). Therefore, a prion disease of American cattle produced by the TME agent could be differentiated from the previous two sources of exposure by histopathology and by mink pathogenicity. While BSE is experimentally transmissible to mink by either oral or intracerebral inoculation, the clinicopathologic presentation is different from TME (9).

Exposure to a Spontaneous Incident of Prion Disease

Although the least likely source of exposure for American cattle, spontaneous cases of prion disease remain a threat as long as we continue to feed ruminant animal protein. The clinicopathologic presentation of this form of prion disease would be dependent on the conformation of the prion protein (4), thereby determining the neuronal cells affected.

"It Means What I Say It Means"

What does BSE-free status really mean? As discussed above, prion diseases in cattle may have different clinical signs and the brains of affected animals may have histopathologic features atypical of spongiform encephalopathies. Therefore, surveillance programs should carefully define the population of animals tested and their criteria for a negative finding.

Wisconsin Downer Cow Study

Every country needs to develop a bovine prion disease monitoring program designed to test the oldest animals in populations fed ruminant animal protein for the longest time. In the United States, Wisconsin has a large dairy industry with no restriction on feeding ruminant animal protein. Furthermore, Wisconsin has designated processing plants for nonambulatory (downer) cattle. The following proposed testing program for this population of animals is one example of how surveillance could be established and interpreted.

Description of Test Animals

A recent survey by the Wisconsin Department of Agriculture, Trade, and Consumer Protection, that defined a "downer cow" as any animal that was down for 24 hours or longer, found an estimated annual incidence of 4.5% (10), or 67,500 animals based on a 1.5 million population. Of these, 39% died and were rendered or sold for mink food, 23% recovered and were kept, 10% recovered and were sold, and 28% were sent to slaughter.

We have recently completed a study of the characteristics and disposition of nonambulatory (NAMB) cattle processed by one Wisconsin slaughter-house over a 1-week period (Olander et al. in review). Since this was a designated NAMB plant, 69% of the 150 animals processed by the slaughter-house that week were NAMB. Of these 104 animals, nine were con-demned on antemortem inspection and there were 33 whole carcass postmortem condemnations. Therefore, these heads (brains) were not available for prion disease testing. Aging of the remaining 62 heads by dental eruption and wear indicated that approximately half of the NAMB brains available for testing were from animals 4 years of age or older.

Testing Procedures

In the Wisconsin Downer Cow Study, it is proposed that brain tissue be collected from the foramen using a method and instrument initially devel-oped by the Central Veterinary Laboratory in Weybridge, England. Some of this tissue will be fixed in formalin for possible later histopathologic and immunohistochemical examination. The remainder of the brain tissue will be extracted for the disease-specific, modified form of the prion protein (PrP^{BSE}). These enriched PrP^{BSE} samples will first be screened by immunoblot analysis using a dot blot procedure, then positive or question-able samples further tested by Western blot analysis.

Interpretation

Since it is not possible to test every bovine showing clinical signs suggestive of prion disease, it is necessary to sample from some defined population. It is hoped that this is a high-risk population based on the feeding of ruminant animal protein, the age of the animals tested, the presence of clinical signs consistent with neurodegenerative disease, and their possible exposure to various sources of prion infectivity. In the Wisconsin Downer Cow Study,

TABLE 10.1. Estimated probabilities of finding a PrP^BSE-positive animal in the Wisconsin Downer Cow Study.

Disease frequency	Probability of finding one positive animal	Probability of finding no positive animals	Maximum number of positive animals detectable
1:13,750	0.25	0.75	4
1:10,000	0.33	0.67	5
1:8,000	0.39	0.61	5
1:6,000	0.51	0.49	6
1:4,000	0.63	0.37	7
1:2,000	0.86	0.14	9
1:1,300	0.95	0.05	12
1:1,000	0.98	0.02	14

we have made the following assessment of the meaning of negative findings based on the following parameters and estimates:

1. If a prion disease is currently present in American cattle, it likely exists at a prevalence of 1 per 27,500 animals or less (11).
2. Downer cows are a population of animals that may contain a prion disease–affected individual.
3. Testing for PrP^BSE is the best available method for detection.

Based on the above assumptions, and testing 2,000 cattle brains a year from animals 5 years of age or older from a designated NAMB plant processing 17% of down cows in the State of Wisconsin available for testing, Table 10.1 estimates that the probability of finding no positive animals would be less than 5% if the frequency of infection is greater than 1 in 1,300. If the frequency is 1 in 6,000 or 1 in 13,750, the probability of finding no positives would be 50% and 75%, respectively.

Summary

Because of the diversity in clinicopathologic features of prion diseases in cattle due to strain variation, it is important to carefully define surveillance programs regarding the population of animals examined and the method of testing. The Wisconsin Downer Cow Study describes an example of one possible program that tests older nonambulatory cattle delivered to a designated processing plant for PrP^BSE. However, whichever surveillance program is established in any given country, it must be adaptable to changes in animal population, feeding practices, and exposure to different possible sources of prion infectivity, and it must be ongoing from year to year and decade to decade. Because of long incubation periods and the need for several animal passages before the infection becomes amplified to a prevalence level sufficient for diagnosis, these diseases take decades to become established.

The alternative is to simply stop feeding ruminant animal protein.

References

1. Marsh RF. Mad cow disease. In: 1996 Yearbook of Science and Technology. New York: McGraw-Hill, 1995:173–4.
2. Fraser H, Dickinson AG. Scrapie in mice: agent strain differences in the distribution and intensity of grey matter vacuolation. J Comp Pathol 1973;83:29–40.
3. Bessen RA, Marsh RF. Identification of two biologically distinct strains of transmissible mink encephalopathy in hamsters. J Gen Virol 1992;73:329–34.
4. Bessen RA, Marsh RF. Distinct PrP properties suggest the molecular basis of strain variation in transmissible mink encephalopathy. J Virol 1994;68:7859–68.
5. Cutlip RC, Miller JM, Race RE, et al. Intracerebral transmission of scrape to cattle. J Infect Dis 1994;169:814–20.
6. Anonymous. Bovine spongiform encephalopathy update. In: Foreign Animal Disease Report. USDA/APHIS/VS 1994 (Winter);22–23:4–6.
7. Marsh RF, Bessen RA, Lehmann S, Hartsough GR. Epidemiologic and experimental studies on a new incident of transmissible mink encephalopathy. J Gen Virol 1991;72:589–94.
8. Robinson MM, Hadlow WJ, Knowles DP, et al. Experimental infection of cattle with the agents of transmissible mink encephalopathy and scrapie. J Comp Pathol 1995;113:241–51.
9. Robinson MM, Hadlow WJ, Huff TP, et al. Experimental infection of mink with bovine spongiform encephalopathy. J Gen Virol 1994;75:2151–5.
10. Anonymous. Downer cow characteristics. In: Wisconsin Farm Reporter. Wisconsin Agricultural Statistics Service (WDATCP) 1994;26(4):1.
11. Robinson MM. Transmission studies with transmissible mink encephalopathy and BSE, and a survey of mink feeding practices. In: Symposium on Risk Assessment of the Possible Occurrence of BSE in the United States. J Am Vet Med Assoc 1993;204:72.

11

Differing Neurohistologic Images of Scrapie, Transmissible Mink Encephalopathy, and Chronic Wasting Disease of Mule Deer and Elk

WILLIAM J. HADLOW

Although bearing different names, the several examples of transmissible spongiform encephalopathy in animals (1) are but variations in the expression of a common pathologic process—a process that results in degeneration of neurons (2). The variations are evident not only in the natural history of each disease but also in their neuropathologic patterns. Given the individual peculiarities of the diverse host species and the varied behavior of the closely related causative agents, these diseases are expected to differ clinically and neurohistologically. Defining them in terms of their similarities and differences would provide a clearer picture of their distinguishing features in comparative studies that include the related human diseases and all the experimental counterparts (3).

Three of the animal diseases, as they occur naturally, illustrate the differing neurohistologic images: scrapie of domestic sheep and goats (4), transmissible encephalopathy of ranch-raised mink (5), and chronic wasting disease of captive mule deer and Rocky Mountain elk (6). As in all examples of the encephalopathy in animals (1), the essential lesion is evident by light microscopy as a spongiform change in the neuropil, dark shrunken or otherwise altered neuronal cell bodies, and astrocytosis.

The spongiform change in gray matter, arising mainly from focally dilated dendrites (7), varies from scattered patches of small holes to diffuse porosity. It may be absent or barely detectable in scrapie, as it often is in Suffolk sheep (8, 9) and in some dairy goats (10), prompting thoughts about the suitability of the term *spongiform encephalopathy* for that disease (Fig. 11.1). Even so, it can be prominent in other breeds of sheep (11, 12) and in several breeds of goats (11, 13). But it is never as severe as it is in experi-

FIGURE 11.1. Scrapie. Mild spongiform change with astrocytosis and neuronal degeneration in the dorsolateral thalamus of a Saanen goat. Hematoxylin and eosin (H&E) stain. Bar = 60 μm.

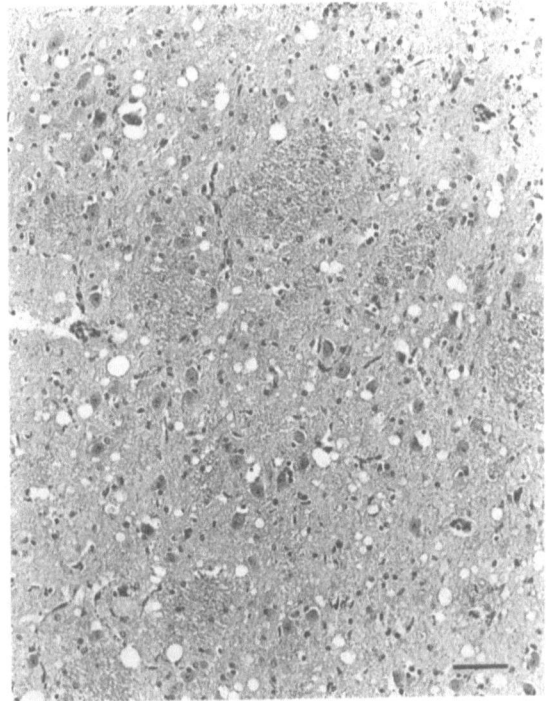

mental caprine scrapie in which the thalamus often appears honeycombed (14). In contrast, spongiform change is a constant component of the degenerative lesion in chronic wasting disease of cervids (15); it is aptly called a spongiform encephalopathy (Fig. 11.2). So, too, is transmissible mink encephalopathy, although the severity of the spongiform change varies widely among individual mink (16). Yet, it can be more diffusely severe in that disease than it ever is in either of the other two (Fig. 11.3). And in mink at least 18 months old that are homozygous for the Aleutian (gunmetal) gene, its severity is greatly diminished, usually to a mere modicum (17). Probably this outcome is another of the pleiotropic effects of the recessive mutant gene that causes lysosomal defects.

The spongiform change may be the main criterion for identifying the encephalopathy microscopically. For example, in mule deer, and to a lesser extent in elk, affected with chronic wasting disease, neuropil sponginess in the olfactory tubercle and gyri, hypothalamus, and parasympathetic nucleus of the vagus nerve is such that examining these structures is adequate to diagnose the disease (6). Of course in all three diseases, spongiform change in gray matter, as opposed to that in white matter, is a clue to their identity, as it has been for all recently described examples of the encephalopathy in animals (1).

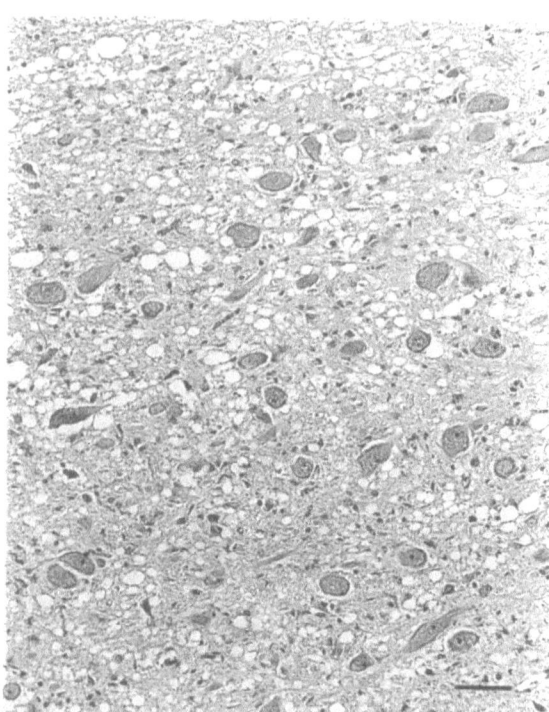

FIGURE 11.2. Chronic
wasting disease:
Spongiform change with
astrocytosis in the
parasympathetic nucleus
of the vagus nerve of a
mule deer. H&E.
Bar = 60 μm.

FIGURE 11.3.
Transmissible mink
encephalopathy: Severe
spongiform change with
astrocytosis in the rostral
midbrain tegmentum of a
pastel mink, a non-
Aleutian color phase.
H&E. Bar = 60 μm.

Dark, shrunken, angular neuronal cell bodies, which must be distinguished from artifact, make up the common form of neuronal degeneration in all three diseases. Variously designated simple atrophy, cell sclerosis, or chronic cell disease (18), it often occurs without the spongiform change. Usually accompanied by some astrocytic response, this expression of neuronal injury is especially prominent in scrapie (9, 10, 14, 19); in both sheep and goats it stands out when spongiform change is minimal or absent (Fig. 11.4). It may not seem as generally evident in the other two diseases, in part because the spongiform change is usually more readily appreciated in them and tends to overshadow the degenerate neuronal cell bodies. Uncommon in all three diseases are acute changes in neurons, such as variable chromatolysis and necrosis. In each disease, neuronal degeneration leads to loss of nerve cells, often best detected in the thalamus (10, 15, 16). In scrapie it can be conspicuous also in the granule cell layer of the cerebellar cortex (8, 14, 20). But in none of these animal diseases is the loss in any site as severe as that in the related human diseases (2, 8). This disparity may have to do with host-specific differences in the pathologic process between man and animals. Yet no doubt it is partly explained by the way sick animals are treated; they are usually denied the great effort and extensive measures used to prolong human life (21). Perhaps in many instances, then, deteriorating neurons in animals are given too little time to disappear before necropsy supervenes.

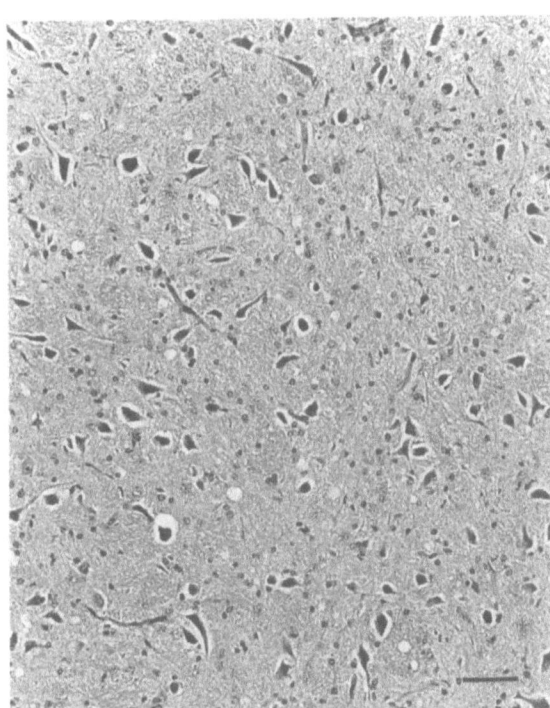

FIGURE 11.4. Scrapie: Diffuse neuronal degeneration with astrocytosis and token spongiform change in the ventral pontine nucleus of a Suffolk sheep, a British breed. H&E. Bar = 60 μm.

Of more variable prominence are neurons with large, single or multiple, usually optically empty vacuoles in the perikaryon—the histologic hallmark of scrapie (22). Such vacuolated neurons may be found as far rostral as the septal nuclei, but typically they occur mostly in the caudal brain stem, particularly in nuclei composed of large neurons (Fig. 11.5). Although present in all three diseases, nerve cells so altered are far less common in transmissible mink encephalopathy (TME) than in the other two (16). Vacuolated neurons have received far more attention in scrapie than in either of the other diseases in which the spongiform change is much better represented. But even in scrapie, such nerve cells may be scarce and hard to find (19, 23), casting doubt on their use as the sole microscopic criterion in diagnosing the disease (24).

Besides these expressions of neuronal injury, astrocytosis (hypertrophy and hyperplasia of astrocytes in gray matter) is another ever-present component of the degenerative lesion. In all three diseases, it is readily recognized in paraffin sections stained with hematoxylin and eosin. Even so, special stains, such as Cajal's gold sublimate impregnation and an immunocytochemical technique [glial fibrillary acidic protein (GFAP)], give a much more vivid picture of the astrocytic response (Fig. 11.6). In scrapie this is especially so in the cerebellar cortex, where apart from proliferation of Bergmann glia, the extent of the astrocytosis there is not made evident by

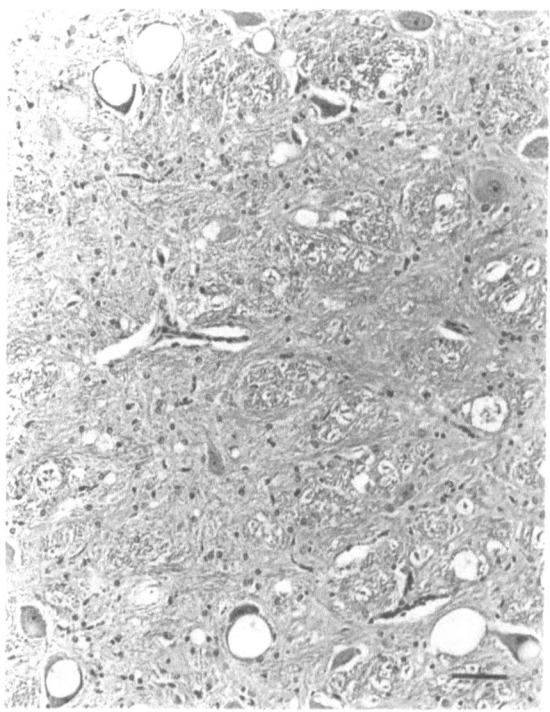

FIGURE 11.5. Scrapie: Vacuolated neurons in the lateral caudal nucleus of the medulla oblongata of a Saanen goat. H&E. Bar = 25 µm.

FIGURE 11.6. Scrapie:
Pronounced astrocytosis
in the lateral geniculate
nucleus of a Nubian goat.
Cajal's gold sublimate
impregnation.
Bar = 40 μm.

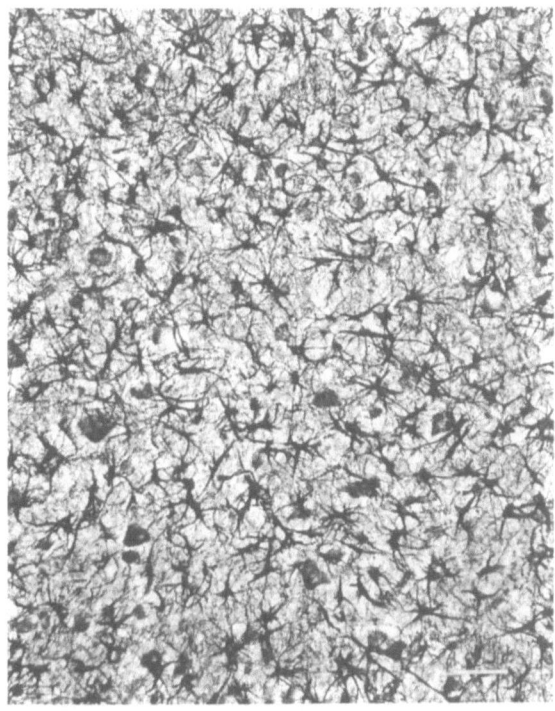

routine staining (Fig. 11.7). Most reactive astrocytes appear as naked nuclei
(18), which are often enlarged, vesicular, and oddly shaped. In TME those
in the molecular layer of the cerebral cortex, notably at depths of sulci,
typically have wisps of cytoplasm (16, 25, 26). But full-bodied gemistocytes
are unusual in all three diseases. In none does the astrocytosis result in an
appreciable increase in glial filaments forming a dense meshwork
(astrogliosis). Sometimes astrocytosis is far more striking than any other
change. This happens in scrapie (12, 14). It occurs as well in TME, most
notably in the cerebral cortex when the spongiform change is not all that
impressive (Fig. 11.8) (16, 17, 25, 26).

When the astrocytosis is disproportionately more intense than the neuronal degeneration, as is often so in scrapie, some observers have considered it the primary response in the disease (12, 14). Yet others have
dismissed it in scrapie as a nonspecific response of limited diagnostic value
(23, 24). However, as a prominent component of the degenerative lesion in
that disease, astrocytosis can be helpful in diagnosis when spongiform
change and neuronal degeneration are minimal (12). In all three diseases,
the glial response includes sparce rod cells, the only obvious examples of
activated microglia with routine staining (Fig. 11.9).

The relative prominence of the several components of the degenerative
lesion varies not only from one disease to another, as already mentioned,

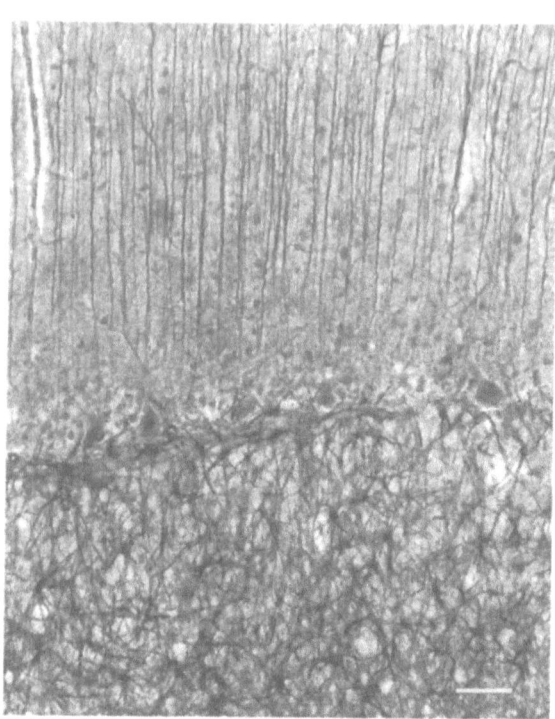

FIGURE 11.7. Scrapie: Astrocytosis in the cerebellar cortex of a Suffolk sheep. Cajal's gold sublimate impregnation. Bar = 40 μm.

FIGURE 11.8. Transmissible mink encephalopathy: Astrocytosis with little spongiform change in the frontal cortex of a sapphire mink, a color phase homozygous for the Aleutian gene. Cajal's gold sublimate impregnation. Bar = 60 μm.

FIGURE 11.9. Chronic
wasting disease:
Astrocytosis with rod
cells and moderate
spongiform change in the
caudal colliculus of a
mule deer. H&E.
Bar = 60 μm.

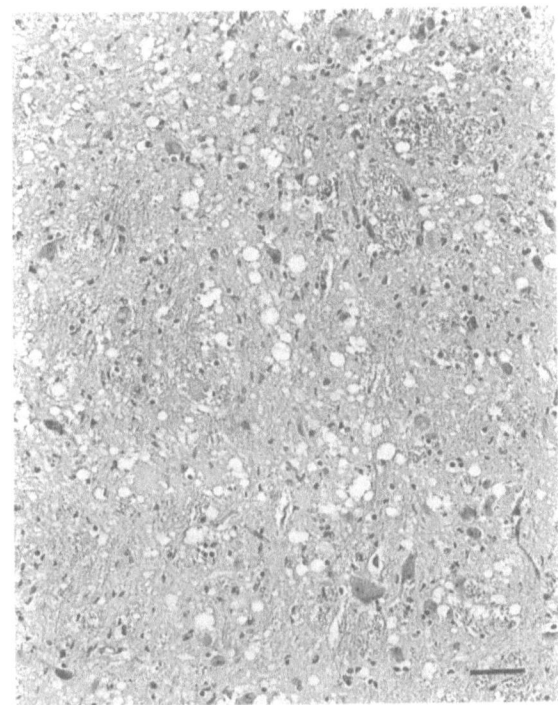

but also from one animal to another and from one part of the brain to another in each disease. Individual animal variation is more evident in scrapie, especially in sheep, than in either of the other two diseases. Some of this variation is attributable to the varied responses of the different breeds of sheep. In general, the severity of the lesion in any one of the diseases bears little relation to their clinical duration (10, 15, 16). Most likely this disparity is a reflection of unexplained wide variations in the tempo of the pathologic process from one animal to another. In scrapie-affected sheep, the neurohistologic changes may be slight or insufficient to confirm the diagnosis (19, 23). Whether full-blown encephalopathy accompanied by hardly any light microscopic changes ever occurs in the other two diseases is uncertain.

Along with the essential lesion, amyloid plaques may be seen (Fig. 11.10). These spherical structures, which are periodic acid-Schiff positive, Congophilic, and birefringent in polarized light, are present most often and most abundantly in mule deer affected with chronic wasting disease (27). They are less easily seen in elk. Most plaques occur in the cerebral cortex and diencephalon. In scrapie they are rare in both sheep (28) and goats (13). Cerebrovascular amyloid is far more common, at least in sheep (29). Plaques have not been detected in TME (16). Because they are not readily detectable by routine staining, they have little diagnostic usefulness, even in

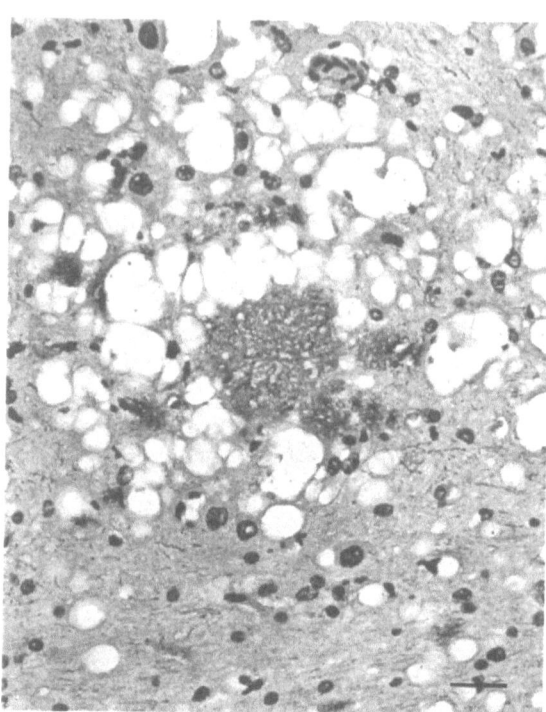

FIGURE 11.10. Chronic wasting disease: An amyloid plaque amid spongiform change in the cerebral cortex of a mule deer. Bodian stain. Bar = 25 μm. Reproduced with permission from Williams and Young (15).

mule deer. Presumably, as in mice infected with the scrapie agent (30), strain of the causative agent and genetic makeup of the host have a bearing on the occurrence of amyloid plaques in the natural diseases.

The three diseases are further distinguished by the topographic distribution of the degenerative changes. Like other examples of transmissible spongiform encephalopathy in animals (1), all three are largely diseases of subcortical gray structures extending from the septal nuclei to the caudal brain stem. In TME, the corpus striatum is consistently and severely affected (16, 25), whereas in the other two diseases it is less so (10, 13, 15). The diencephalon is a common site of the lesion in all three. Some thalamic nuclei, such as the medial geniculate, invariably undergo degeneration. The hypothalamic supraoptic and paraventricular nuclei are also affected in each disease (Fig. 11.11). In all three, notably degenerate midbrain structures are the caudal colliculus and periaqueductal gray matter. Nuclei in the pons and medulla oblongata suffer less damage in TME (16, 25) than in the other two diseases (15, 19). This is one of the distinguishing features of its neurohistologic image.

In scrapie, the cerebellar cortex is also a clinically relevant site of the degenerative lesion in most sheep and goats (8–10, 20). It is affected as well in chronic wasting disease, but probably not as consistently or as severely as in scrapie (15). In striking contrast, except for minor changes, the cerebellar cortex is spared in TME (16, 25, 31). This is another feature of the disease in mink that sets it apart from the other two.

So, too, does the widespread degeneration of the cerebral neocortex in affected mink (Fig. 11.12). The changes are most severe in the frontal lobe, particularly in dorsomedial gyri where the earliest changes occur (16, 32). Caudally, they gradually become less severe in parallel with those in the brain stem. In chronic wasting disease, changes in the neocortex are mild, mostly as scattered spongy foci in the middle and deep laminae (15). Several observers have long thought that the neocortex is free of changes in scrapie (8–10, 19). Recently, however, focal and pseudolaminar spongiform change with some neuronal degeneration and astrocytosis was found often in the frontal cortex, especially in precallosal gyri, in both sheep and goats (Fig. 11.13) (11, 13). Over the years, that far rostral level of the brain has seldom been examined in scrapie-affected animals.

To some extent the allocortex is involved as well in all three diseases. As already mentioned, the constant occurrence of spongiform change in the olfactory tubercle and gyri is notable in chronic wasting disease. In the other two diseases, these structures are less often and usually less severely affected (14, 16). The hippocampus is always affected in TME, but is much less often in the other two diseases, and then not as severely (Fig. 11.14). In all three, the amygdala is usually the site of degenerative changes. Other

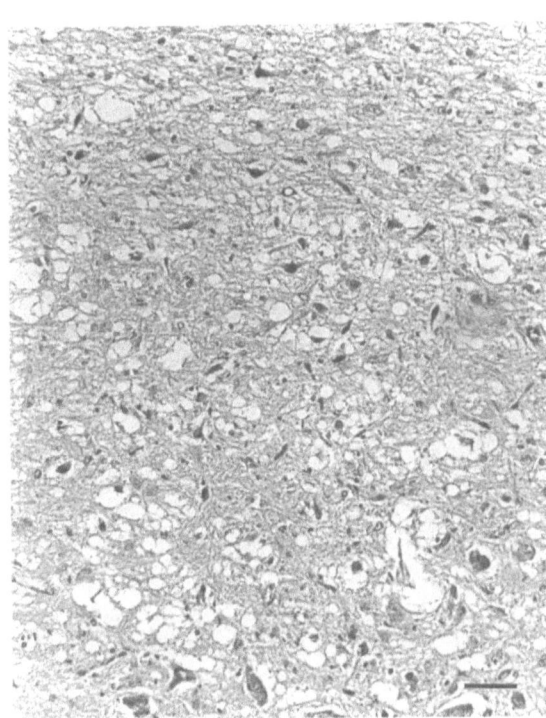

FIGURE 11.11. Chronic wasting disease: Spongiform change and neuronal degeneration in the paraventricular nucleus of the hypothalamus of a mule deer. H&E. Bar = 60 μm.

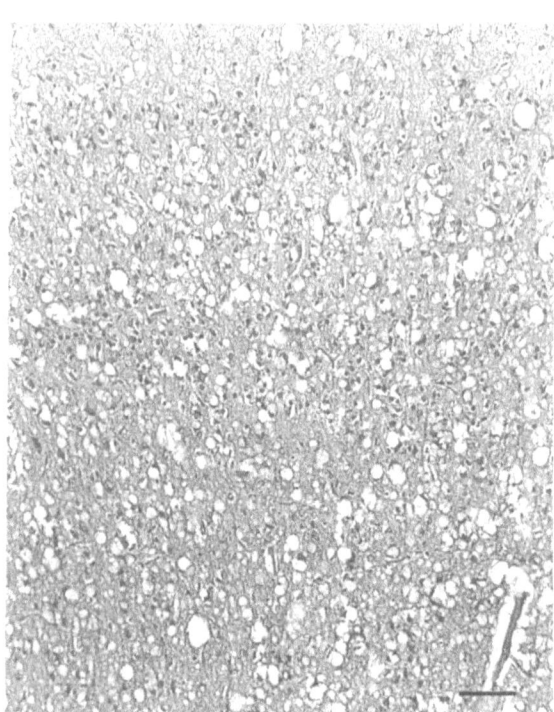

FIGURE 11.12.
Transmissible mink
encephalopathy:
Severe spongiform
degeneration of the
parietal cortex of a pastel
mink. H&E.
Bar = 60 μm.

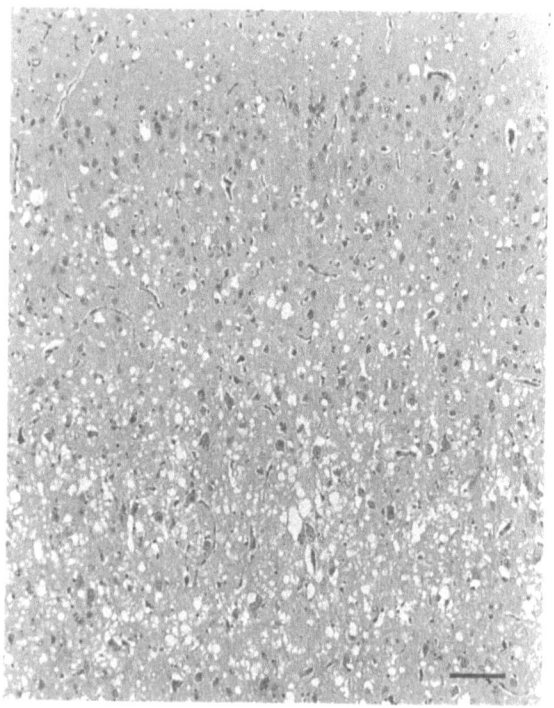

FIGURE 11.13. Scrapie:
Spongiform change in the
precallosal frontal cortex
of a Texel sheep, a Dutch
breed. H&E.
Bar = 100 μm.

FIGURE 11.14.
Transmissible mink
encephalopathy: Diffuse
spongiform change in the
hippocampus of a pastel
mink. H&E.
Bar = 100 μm.

than minor, inconstant changes, mainly vacuolated neurons most notable in scrapie and chronic wasting disease, the spinal cord is spared in all three diseases (10, 15, 16, 31). Degeneration of some long fiber tracts has been reported in scrapie-affected sheep (33, 34).

Given these differences in the topographic distribution of the neurohistologic changes, the level of the neuraxis bearing the brunt of the damage necessarily varies. In TME it is more rostral (forebrain) (16), whereas in scrapie it is more caudal, including the cerebellum (10). The neuraxial pattern in chronic wasting disease resembles that of scrapie, although it seems less restricted rostrally and places less emphasis on the cerebellum (15). Because the neurohistologic changes have a characteristic pattern of distribution in each of the three diseases, they can be looked upon as distinct neuropathologic entities. Identifying them in this way simplifies not only their postmortem diagnosis, but also their comparison with other examples of transmissible spongiform encephalopathy, whether in animals or in man.

Morphologically, all examples of the encephalopathy in animals (1) are merely variants of an underlying theme exemplified by scrapie, the prototypic disease. This sameness brings them together and points to their undoubted relatedness in cause and pathogenesis. Yet as illustrated here with three of the diseases, differences in their neurohistologic images exist; they need not only reckoning in diagnosis but also explaining. Some of the differences no doubt arise from the way the central nervous tissue in each host species responds to injury, specifically the kind brought about in these diseases. But in most animals, even the domestic ones, little is known about the responses peculiar to any of them; neuroanatomic peculiarities have gotten the attention instead (35). And in the different species the varied results of special neurohistologic techniques, nearly all originally prescribed for human nervous tissue (18), are not understood as well as needed for critical comparative studies. These concerns receive some comment in a recent textbook on veterinary neuropathology (36). Other differences in the neurohistologic images of the three diseases almost certainly reflect variations in the biologic properties of the causative agents, which probably are strains of a common one that has been modified by passage in different hosts (37).

Depending upon where it occurs in the world, scrapie is caused by various extant strains of the infectious agent. This partly explains why its morphologic expression varies more than that of the other two diseases (38). Then, too, domestic sheep and goats around the world are more varied genetically, in the sense of having many widely different fixed breed struc- tures (39), than either ranch-raised mink or captive populations of mule deer and Rocky Mountain elk. Indeed, striking differences in the morphologic expression of scrapie occur between some British breeds of sheep and several from continental Europe. As a consequence of a dead- end infection, TME arises anew each time it occurs (32). This could account for variations in its morphologic expression from one outbreak to the next, whatever the source of the causative agent might be. Even so, I have seen only one pattern of expression in four outbreaks of the disease in North America. Chronic wasting disease of cervids most likely is caused by a single strain of the pathogen responsible for its endemic occurrence in the captive herds (6). This etiologic oneness is reflected in the fairly uni- form topographic pattern of the neurohistologic changes in each cervine species (15).

Summary

In this chapter I have pointed out the obvious similarities and differences in the neurohistologic images of three examples of transmissible spongiform encephalopathy in animals. Although customarily treated as distinct nosologic entities, these diseases are after all a lot alike. Yet despite much overlapping, their neurohistologic images still differ. The extent of the differences would be made clearer by further study, even with ordinary histologic techniques, to define the full range and limitations of the neuropathologic changes in each disease. Beyond simply allowing a more comprehensive comparison of the neurohistologic images, such information would indicate how essentially a single infectious agent behaves in a variety of natural hosts (35, 40). That might provide a better understanding of the genesis, evolution, and ultimate expression of the encephalopathy than a detailed study in a single host, natural or experimental.

Acknowledgments. I am grateful to Robert Evans and M. Donald McGavin for preparing the photomicrographs. Elizabeth Williams kindly provided microscopic sections of brain and spinal cord from mule deer affected with chronic wasting disease.

References

1. Wells GAH, McGill IS. Recently described scrapie-like encephalopathies of animals: case definitions. Res Vet Sci 1992;53:1–10.
2. Bastian FO. Neuropathology. In: Bastian FO, ed. Creutzfeldt-Jakob disease and other transmissible spongiform encephalopathies. St. Louis: Mosby-Year Book, 1991:65–96.
3. Gibbs CJ Jr, Bolis L, Asher DM, et al. Recommendations of the International Roundtable Workshop on Bovine Spongiform Encephalopathy. J Am Vet Med Assoc 1992;200:164–7.
4. Detwiler LA. Scrapie. Rev Sci Tech Off Int Epiz 1992;11:491–537.
5. Marsh RF, Hadlow WJ. Transmissible mink encephalopathy. Rev Sci Tech Off Int Epiz 1992;11:539–50.
6. Williams ES, Young S. Spongiform encephalopathies in Cervidae. Rev Sci Tech Off Int Epiz 1992;11:551–67.
7. Adornato B, Lampert P. Status spongiosus of nervous tissue. Electron microscopic studies. Acta Neuropathol 1971;19:271–89.
8. Beck E, Daniel PM. Neuropathology of transmissible spongiform encephalopathies. In: Prusiner SB, McKinley MP, eds. Prions: novel infectious pathogens causing scrapie and Creutzfeldt-Jakob disease. San Diego: Academic Press, 1987:331–85.
9. Hadlow WJ, Kennedy RC, Race RE. Natural infection of Suffolk sheep with scrapie virus. J Infect Dis 1982;146:657–64.
10. Hadlow WJ, Kennedy RC, Race RE, Eklund CM. Virologic and neurohistologic findings in dairy goats affected with natural scrapie. Vet Pathol 1980;17:187–99.
11. Toumazos P. Scrapie in Cyprus. Br Vet J 1991;147:147–54.
12. Georgsson G, Gisladóttir E, Arnadóttir S. Quantitative assessment of the astrocyte response in natural scrapie of sheep. J Comp Pathol 1993;108:229–40.
13. Wood JLN, Done SH. Natural scrapie in goats: neuropathology. Vet Rec 1992;131:93–6.
14. Hadlow WJ. The pathology of experimental scrapie in the dairy goat. Res Vet Sci 1961;2:289–314.
15. Williams ES, Young S. Neuropathology of chronic wasting disease of mule deer (*Odocoileus hemionus*) and elk (*Cervus elaphus nelsoni*). Vet Pathol 1993; 30:36–45.
16. Eckroade RJ, ZuRhein GM, Hanson RP. Experimental transmissible mink encephalopathy: brain lesions and their sequential development. In: Prusiner

SB, Hadlow WJ, eds. Slow transmissible diseases of the nervous system, vol 1. New York: Academic Press, 1979:409–49.

17. Marsh RF, Sipe JC, Morse SS, Hanson RP. Transmissible mink encephalopathy: reduced spongiform degeneration in aged mink of the Chediak-Higashi genotype. Lab Invest 1976;34:381–6.

18. Okazaki H. Fundamentals of neuropathology. New York, Tokyo: Igaku-Shoin, 1983:1–23.

19. Zlotnik I. The pathology of scrapie: a comparative study of lesions in the brain of sheep and goats. Acta Neuropathol 1962;(suppl 1):61–70.

20. van Bogaert L, Dewulf A, Palsson PA. Rida of sheep. Pathological and clinical aspects. Acta Neuropathol 1978;41:201–6.

21. Dayan AD. Comparative neuropathology of aging: studies on the brains of 47 species of vertebrates. Brain 1971;94:31–42.

22. Palmer AC. Distribution of vacuolated neurons in brains of sheep affected with scrapie. J Neuropathol Exp Neurol 1960;19:102–10.

23. Fraser H. The pathology of natural and experimental scrapie. In: Kimberlin RH, ed. Slow virus diseases of animals and man. Amsterdam: North-Holland, 1976:267–305.

24. Kimberlin RH. Scrapie as a model slow virus disease: problems, progress, and diagnosis. In: Kurstak E, Kurstak C, eds. Comparative diagnosis of viral diseases, vol 3. New York, London: Academic Press, 1981:349–90.

25. Hadlow WJ, Karstad L. Transmissible encephalopathy of mink in Ontario. Can Vet J 1968;9:193–6.

26. Marsh RF, Hanson RP. Transmissible mink encephalopathy: neuroglial response. Am J Vet Res 1969;30:1643–54.

27. Guiroy DC, Williams ES, Yanagihara R, Gajdusek, DC. Topographic distribution of scrapie amyloid—immunoreactive plaques in chronic wasting disease of captive mule deer (*Odocoileus hemionus*). Acta Neuropathol 1991;81:475–8.

28. Beck E, Daniel PM, Parry HB. Degeneration of cerebellar and hypothalamo-neurohypophysial systems in sheep with scrapie; and its relationship to human system degenerations. Brain 1964;87:153–76.

29. Gilmour JS, Bruce ME, Mackeller A. Cerebrovascular amyloidosis in scrapie-affected sheep. Neuropathol Appl Neurobiol 1986;12:173–83.

30. Bruce ME, Dickinson AG, Fraser H. Cerebral amyloidosis in scrapie in the mouse: effect of agent strain and mouse genotype. Neuropathol Appl Neurobiol 1976;2:471–8.

31. Johannsen U, Hartung J. Infektiöse Enzephalopathie beim Nerz. 2 Mitteilung: pathologisch-morphologische Untersuchungen. Monatsschr Veterinaermed 1970;25:389–95.

32. Hadlow WJ, Race RE, Kennedy RC. Temporal distribution of transmissible mink encephalopathy virus in mink inoculated subcutaneously. J Virol 1987; 61:3235–40.

33. Wight PAL. The histopathology of the spinal cord in scrapie disease of sheep. J Comp Pathol 1960;70:70–83.

34. Palmer AC. Wallerian type degeneration in sheep scrapie. Vet Rec 1968; 82:729–31.

35. Koestner A. An introduction to comparative neuropathology. Methods Achiev Exp Pathol 1967;3:55–85.

36. Summers BA, Cummings JF, deLahunta A. Veterinary neuropathology. St. Louis: Mosby, 1995.
37. Gajdusek DC. Subacute spongiform encephalopathies: transmissible cerebral amyloidoses caused by unconventional viruses. In: Fields BN, Knipe DM, Chanock RM, et al., eds. Virology, vol 2. 2nd ed. New York: Raven Press, 1990:2289–324.
38. Hadlow WJ. To a better understanding of natural scrapie. In: Bradley R, Savey M, Marchant B, eds. Sub-acute spongiform encephalopathies. Dordrecht, Boston, London: Kluwer Academic, 1991:117–30.
39. Mason IL, Crawford RD. Global status of livestock and poultry species. In: National Research Council. Managing global genetic resources: livestock. Washington, DC: National Academy Press, 1993:141–69.
40. Gresham GA, Jennings AR. An introduction to comparative pathology. A consideration of some reactions of human and animal tissues to injurious agents. London, New York: Academic Press, 1962.

12

Analysis of Risk Factors and Active Surveillance for BSE in Argentina

A.A. Schudel, B.J. Carrillo, E.L. Weber, J. Blanco Viera,
E.J. Gimeno, C. Van Gelderen, E. Ulloa, A. Nader, B.G. Cané,
and R.H. Kimberlin

Bovine spongiform encephalopathy (BSE) is a new member of the scrapie-family of diseases that was first recognized in the United Kingdom in November 1986 (1) and subsequently developed into a major epidemic. BSE was caused by the contamination of cattle feed with a scrapie-like agent (2). The vehicle of infection was meat and bone meal (MBM), which is one of the two products of the rendering industry, and was commonly used as a protein-rich supplement to concentrated feedstuffs. Tallow, the other product of rendering, is also used in cattle feeds, but is not implicated as a source of infection (2). The association between BSE and the feeding of concentrates is the reason why the great majority of BSE cases have occurred in dairy cows and in beef suckler cows that originated from dairy herds (3). Although some cows with BSE were infected as adults, most were infected during calfhood (4).

The exposure of cattle to infection started abruptly during the winter of 1981–82 (2). A major reason for this was the sudden cessation of solvent extraction in rendering processes, which resulted in the loss of two partial scrapie-inactivation steps and allowed some infectivity to survive in MBM (5).

Scrapie is believed to have been the original cause of the BSE epidemic because sheep are the only recognized natural reservoir of scrapie-like agents in Britain and cattle are known to be susceptible to scrapie when administered parenterally (6, 7). Once food-borne infection had been established in cattle, the epidemic was amplified by the recycling of BSE infection from cattle to cattle via the same type of feeds that were contaminated with scrapie (8). However, even at the peak of the epidemic, the average contamination of MBM was low, causing the disease to occur at a low within-herd incidence. It has been estimated that the average contamination of MBM in BSE affected herds could have been as low as 10 oral LD_{50} units of cattle-adapted BSE per ton of feed (9).

The feeding of ruminant-derived protein to all ruminants was banned in the United Kingdom in July 1988 to stop new infections (10). However, since the median incubation period of BSE is around 4 to 6 years (5, 11) the epidemic did not reach a peak until 1992. The subsequent decline provides strong supporting evidence that MBM was the major vehicle of infection (12). Although it is too soon to exclude the possibility of maternal or lateral transmission of BSE, as happens with sheep scrapie (13, 14), there is still no clear evidence that either occurs (15, 16). Any natural transmission of BSE would be insufficient to sustain the epidemic in cattle (11, 17).

The origin and development of the United Kingdom epidemic resulted from the simultaneous presence of four main factors (5, 18):

1. A large sheep population relative to that of cattle. A wide geographic distribution of sheep with scrapie and, subsequently, cattle with BSE is necessary to account for the more or less simultaneous appearance of the earliest clinical cases of BSE throughout England and Wales (18) and the dramatic nationwide increase in the number of cases, due to the recycling of BSE infection, via feed.
2. A sufficiently high incidence of endemic scrapie infection in sheep.
3. Conditions of rendering that allowed the survival of small but significant amounts of scrapie and BSE infectivity.
4. The use of substantial quantities of ovine and bovine MBM in cattle feeds.

Factors 1, 2, and 4 existed in Britain for many years prior to BSE and the third factor was limiting until the conditions of rendering were changed around 1981 to 1982 (2, 5). Any one or more of these factors could be limiting in other countries, two of which, Argentina (19) and the United States (20), have published detailed risk assessments based particularly on factors 1 and 4, which are the easiest to quantify. The evidence from these and unpublished risk assessments indicates that no other countries possess the same combination of risk factors due to scrapie, as occurred in the United Kingdom. The United Kingdom experience of low-dose exposure of cattle to bovine-adapted BSE makes comparable epidemics in other countries extremely unlikely (8, 9).

Nevertheless, by November 1994 over 225 cases of BSE had occurred in 10 other countries. In six countries—Canada, Denmark, Germany, Islas Malvinas (Falkland Islands), Italy, and the Sultanate of Oman—cases of BSE have occurred only in cattle that were exported from, and presumably infected in, the United Kingdom. Ireland and Portugal have also had cases in cattle imported from United Kingdom and some in native-born cattle. France and Switzerland have reported BSE cases only in native cattle, at least some of which are likely to have been the result of feeding MBM imported directly, or indirectly from the United Kingdom (9).

Since the occurrence of the epidemic in the United Kingdom and following the recommendations of the international organizations—particularly the Office International des Epizooties—the Argentinian Animal Health Services implemented the following activities:

Aug. 1990: Ban on the importation of bovines or bovine by-products from countries with BSE (21).
Nov. 1991: Analysis of BSE risk factors (22).
Nov. 1992: BSE surveillance programme.
Dec. 1993: BSE surveillance programme: First report (23).
Dec. 1994: Analysis of BSE risk factors (updated). BSE surveillance programme: annual report (24).

This chapter reassesses the risk of undetected BSE being present in Argentina (19). Guided by the experience of BSE in the United Kingdom and elsewhere, consideration is given to the risks from scrapie (even though this has never been reported in Argentina), from imported cattle, and from imported MBM. The risk from imported embryos is also considered because of the evidence that, under experimental conditions, scrapie infection can be transmitted by the transfer of unwashed embryos (25), and because of a degree of uncertainty about bovine embryos as a source of BSE infection. It is concluded that the past importation of live cattle into Argentina constitutes the only potential source of risk of BSE, albeit an extremely small one. This possible risk is addressed in the second part of this chapter which reports the results of an active surveillance program of three populations of cattle, conducted over the period November 1992 to December 1993 (23).

The first two populations revealed themselves either as suspected field cases of neurologic disease or as animals that were found to be in poor conditions on inspection prior to slaughter. The third and largest group consisted of clinically normal dairy cows, which are the only group of cattle in Argentina to have a theoretical risk of feed-borne exposure to BSE infection. The cows were selected at random in abattoirs to represent a high proportion of dairy farms and provinces in Argentina (23).

With all three populations, mainly animals of 5 years of age or greater were examined, so as to maximize the chances of detecting subclinical infection and clinical disease. All cattle were examined for spongiform encephalopathy by brain histology (26–28), and some were also examined for the presence of the disease-specific, modified form of the PrP protein (29–34). The results of the surveillance are consistent with the conclusion of the risk assessment that Argentina has a high probability of being free from BSE.

BSE Risk Assessment

The risk assessment conducted in Argentina analyzed the four main factors involved:

Livestock inventory and geographical distribution.
Veterinary services and animal health conditions.
The slaughter of cattle and the disposal of animal waste.
Feed industry.

Livestock Inventory and Geographical Distribution

Argentina has very large populations of both sheep (19.8 million) and cattle (54.5 million), but, in contrast to the United Kingdom, these are geographically separate to a large extent (19, 22). Nearly half of the sheep population is in Patagonia, south of the Colorado River, which includes the provinces of Chubut, Neuquen, Río Negro, Santa Cruz, and Tierra del Fuego (Fig. 12.1). In these areas, sheep outnumber cattle by 50 to 1. Most of the cattle population is found in nine provinces further north and to the east. Dairy cattle that form 4.2% of the total are even more restricted, and about 95% are found in three provinces, Buenos Aires, Córdoba, and Santa Fé (Fig. 12.1). Significant numbers of both sheep and cattle are found only in Buenos Aires, Corrientes, and Entre Ríos (19, 22). Because cattle need better quality pasture than sheep, there is virtually no common pasturing even when both species are kept on the same farm.

Animal Health

The second aspect to consider is the animal health system, in which the animal health control programs for foot-and-mouth-disease and sheep scab guarantee the inspection for vaccination or dipping of all cattle and sheep population at least twice a year for each disease. The program of rabies and the recently imposed notification of all cases with neurologic symptoms in bovines give the opportunity for detection of suspected clinical cases. It should also be noted that there is the ante- and postmortem inspection at all slaughterhouses in Argentina (19, 22).

Argentina has never reported cases of scrapie. There have been very few opportunities for the introduction of scrapie to Argentina; since 1980, only 18 sheep have been imported from countries (Germany and Spain) where scrapie has occurred subsequently (19).

Slaughter of Cattle and Disposal of Animal Waste

The separation of cattle and sheep continues at slaughter; 564 abattoirs in Argentina specialize in the slaughter of cattle, and 423 slaughter sheep. Only two large abattoirs in the city of Buenos Aires slaughter both species in significant numbers (19). This geographic and functional separation of slaughterhouses ensures that ovine and bovine wastes are rarely mixed, just

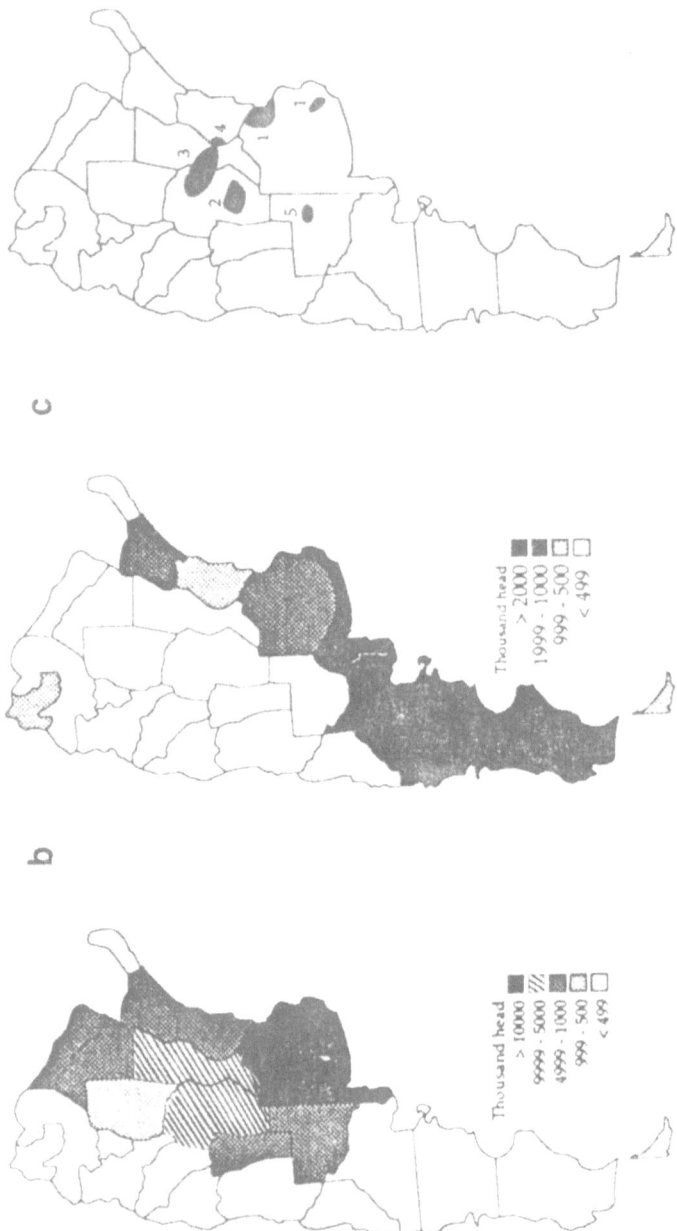

FIGURE 12.1. Livestock inventory and geographical distribution in Argentina. (a) Cattle population; (b) sheep population; (c) dairy cattle.

the opposite to the situation in the United Kingdom. With one exception, all sheep wastes are disposed of by incineration, sterilization (3 hours under pressure, at temperatures over 120°C) or by burial on farms. Even in the largest Federal abattoirs (whence meat can be exported), the amount of sheep waste produced each day is too small for economic rendering (19). The exception is the two abattoirs in Buenos Aires that slaughter both sheep and cattle. In 1993 it was estimated that no more than 120 tons of MBM, out of a total production 183,000 tons (Fig. 12.2), would have come from sheep. The only substantial production of ruminant-derived MBM is from cattle slaughtered in Federal abattoirs. In Argentina, rendering is still carried out in traditional batch cookers heated directly with steam under pressure (19, 22). However continuous-flow systems, similar to those used in the United Kingdom, are being introduced; they use indirect heating at atmospheric pressure. Tallow has a variety of industrial use in Argentina but none goes into animal feeds.

Feed Industry

The raising of sheep and cattle in Argentina employs highly extensive production systems that make maximum use of all-the-year-round pasture as a low-cost food resource. Concentrated feeds are not used for sheep and

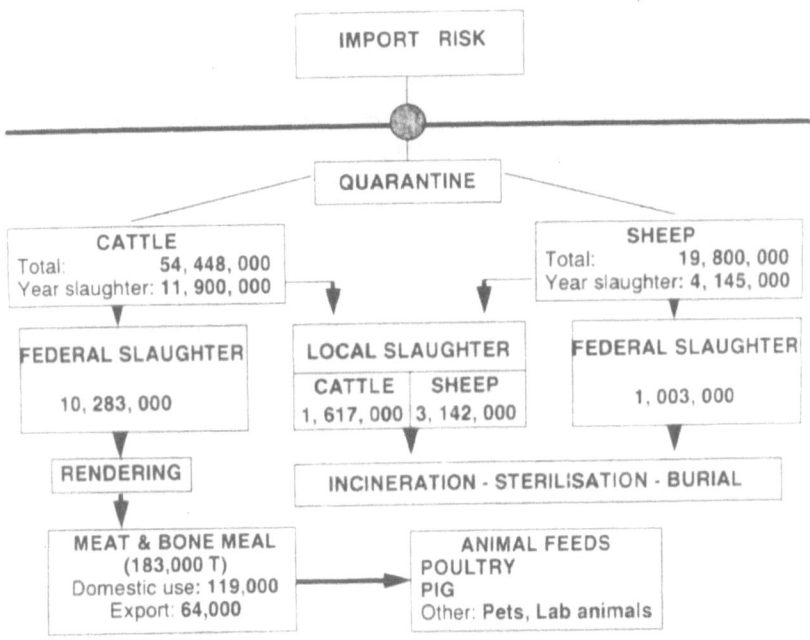

FIGURE 12.2. System model of BSE epidemiology in Argentina.

very rarely for beef cattle. Until recently small quantities of concentrated feeds were fed to dairy cattle seasonally, when there was insufficient pasture (19). The last few years have seen an increase in the number of dairy cattle and also in milk productivity, from an average of 2,400 liters/cow in 1988 to nearly 3,000 liters/cow in 1993. The latter change has been associated with a greater feeding of concentrates to dairy cows. However, in marked contrast to the United Kingdom and many other countries, MBM is not used in cattle feeds because it is too expensive, more so even than bran, which is the most common plant source of protein.

Argentina seems never to have imported MBM, and currently exports about 35% of the total production (Fig. 12.2). The remainder is fed to nonruminant species, mainly chickens and pigs.

The only ingredient of animal origin that is fed to cattle (but not to sheep) is bone ash. Since this is produced by incineration, it would be of extremely low risk even if it came from infected animals. Given the major importance of MBM as the vehicle of scrapie/BSE infection in the United Kingdom, the absence of any significant feed links within and between the sheep and cattle populations makes the occurrence of BSE in Argentine cattle exceedingly unlikely. In addition, the animal health authorities have recently imposed a complete ban on the use of MBM from ruminant origin in cattle feeds.

There is a potential threat of isolated cases of BSE occurring in imported live cattle that were infected in the country of origin. Table 12.1 shows that 19 cattle were introduced to Argentina from England and Scotland during 1980 to 1982. But as this coincides with the time of origin of the United Kingdom epidemic, the risks of the imported animals being infected are very low. Cattle have also been imported from four other countries with cases of BSE. Of these, the largest number of imports, and also the most recent, have come from Canada, which itself has reported a single case of imported BSE (35). Canada has also been a source of embryos and semen,

TABLE 12.1. Cattle imported from countries that have reported cases of BSE (1980–1993).[a]

Country	Number
Canada	386
England	18
Germany	43
Italy	83
Scotland	1
Switzerland	15

[a] Embryos and semen were imported from Canada from 1991 until October 1994. Semen was imported from Germany in 1993 and from Italy in 1992.

TABLE 12.2. Cattle imported from countries that have not reported cases of BSE (1980–1993).[a]

Country	Number
Austria	20
Brazil	2,668
Chile	390
Netherlands	3
New Zealand	16
Paraguay	108
Spain	59
United States	1,054
Uruguay	75,578[b]

[a] Embryos and semen have been imported from Australia, Belgium, Spain, New Zealand, Sweden, and the United States.
[b] Animals sent for slaughter.

some imported as recently as October 1994, but there is no clear evidence that either of these tissues can transmit BSE, least of all semen (36, 37). All the animals imported from countries with BSE, which are still alive, are under surveillance.

Among source countries that have not reported cases of BSE, and excepting the special case of Uruguay (Table 12.2), the largest number of imported cattle have come from Brazil and the United States. However, Brazil probably uses little or no MBM in cattle feeds, and the United States has a low risk of BSE occurring from endemic scrapie (20).

In conclusion, the risks of imported BSE in Argentina are low but they cannot be eliminated entirely. This uncertainty increases as the passage of time reveals cases of BSE in more and more of the source countries; for example, Italy reported a case of BSE as recently as November 1994 (37). This situation emphasizes the need for the ongoing surveillance program.

Surveillance Program

The main activities performed by the surveillance program since 1992 included:

Training staff and equipping laboratories for BSE diagnosis (Central Veterinary Laboratory [CVL], United Kingdom).
Informing farmers and veterinarians.
Notification of neurologic disease and surveillance network.
Examination of cases with CNS disease symptoms.

A surveillance system was established with technical personnel from the Servicio Nacional de Sanidad Animal (SENASA), the Instituto Nacional de Tecnología Agropecuaria (INTA), universities, and from a private diagnostic laboratory. The system started operating in the field and in slaughterhouses in November 1992, and the data obtained by December 1994 are presented in this chapter.

The risk assessment considered the possibility that BSE may have already entered Argentina through the importation of live cattle, embryos, and semen. Although this risk is low, and because MBM is not fed to cattle, it could not possibly be the cause of more than a very few isolated cases in the population at higher risk, namely old dairy cows.

For the surveillance, bovine brains were collected from dairy cattle, most of which were more than 5 years old and belonged to one of the following categories: 40 brains from animals suspected of neurologic disease that were sent from the field to the laboratory; 91 brains from animals in poor condition detected antemortem in slaughterhouses; and 888 brains from healthy animals selected at random in slaughterhouses on the population at higher risk (dairy cattle over 5 years old).

Eighty-five percent of the country's bovine population, and 100% of dairy cattle reside in the geographic areas covered by the network. The total number of brains examined was 1,019. This exceeds by five the number required to detect an incidence of histologic disease of greater than 2.95 per thousand with 95% statistical confidence (38).

The materials collected for examination were as follows. For histology: brain stem from thalamus to medulla oblongata, and one cerebral hemisphere were stored in 10% formol saline (26, 28, 39). For the detection of modified PrP: the posterior third of the medulla oblongata, bulb, part of the other cerebral hemisphere, and cerebellum were immediately frozen at $-20°C$. In addition, blood samples and nasal swabs were taken from the animals with suspected neurologic disease for biochemical, bacteriologic, toxicologic, and virologic examination, as was appropriate for differential diagnosis. The samples were labeled according to a previously established code, and farm of origin, breed, sex, and age (determined by dentition) were recorded. Specimens were sent to the laboratory within 24 hours.

The histologic procedures employed were those used in previously published studies on BSE (26, 28, 39). Extreme care was taken to avoid the introduction of artifacts during the fixation. Coronal sections of fixed pieces of brain were cut 5 mm apart. The selected areas were medulla at the obex and through the caudal cerebellar peduncles, midbrain at two levels to include the superior colliculus, and the red nucleus. For samples from suspected field cases of neurologic diseases, additional sections were taken to include cerebellum, frontal cortex, and basal ganglia, temporoparietal cortex, the area rostral to the thalamus, and occipital cortex including the hippocampus.

The diagnostic criteria applied for BSE were those defined by Wells et al. (1, 26) and consisted of bilaterally symmetrical, intracytoplasmic vacuolation of neuronal perikarya in the presence of discrete, spherical vacuoles (microcystic vacuolation) in the gray matter of specific brain nuclei. These changes are distinct from edema, postmortem changes, and fixation artifacts. The brain nuclei most affected by BSE are usually the dorsal nucleus of the vagus nerve, solitary tract, spinal tract of the trigeminal nerve, reticular formation, vestibular nuclei, and the central gray matter of the mesencephalon. Vacuolation of the red nucleus (and sometimes the oculomotor nucleus) was regarded as nonspecific, according to reports of these being incidental findings that are not related to CNS pathology (1, 14). The results were grouped in three categories: positive, negative, and inconclusive.

In some of the cases of suspected neurologic disease, other lesions were seen that established the etiology of the disease. In the remaining cases, complementary examinations (bacteriologic virologic, toxicologic, and biochemical analyses of blood and nasal secretions) were carried out to make a diagnosis.

The proteinase K–resistant form of the prion protein (PrP) was extracted from pooled samples of brain following the method of Hilmert and Diringer (40), modified according to the protocol of the Central Veterinary Laboratory, Weybridge, United Kingdom (30, 41). The extracted proteins were separated in a 15% polyacrylamide gel by the method of Laemmli (42). The proteins were transferred onto a nitrocellulose membrane following the method of Towbin et al. (43).

A Western blot assay was conducted, employing an antiserum raised in rabbits against purified PrP protein from hamsters infected with the 263K strain of scrapie (kindly provided by Dr. Pocchiari, Istitutio Superiore di Sanitá, Rome). The PrP-antiserum (Sigma Catalog No. A-8025), and Naphtol-As-Mix-Phosphate/Fast Red TR salt was used as the color reagent (Sigma Catalog Nos. N-5000/F-8764).

A sample was considered negative for PrP 27-30 when a band with the electrophoretic mobility of this protein was not detected. Purified hamster PrP 27-30 was included as standard. A 1:1,000 dilution of the antiserum detected 0.09 µg of hamster PrP in the assay employed. The amount of each sample analyzed was the equivalent of approximately 1 g of brain per lane of the polyacrylamide gel (range 0.8–2.7 g).

Forty cases of suspected neurologic disease in dairy cattle were reported from 7 of the 10 provinces surveyed (Table 12.2). Although their ages ranged from 0 to 1 to >9 years (Fig. 12.3), the average (3.4 years) was less than the median age of onset of BSE (4–6 years). All the brains were examined histologically and were scored as negative for spongiform encephalopathy. Because of the difficulty of identifying minimal spongiform lesions against a background of other histologic changes, samples of all 40 brains were tested for the presence of modified PrP: all were negative.

Fourteen of the 40 brains had a variety of histologic lesions including nonsuppurative and suppurative encephalitis (9/40), nonsuppurative necrotizing encephalitis (2/40), and suppurative meningitis (2/40). In some of these cases, further laboratory studies identified the causal infection [bovine herpes virus-1 (BHV-1), *Listeria*, and *Haemophilus sommus*].

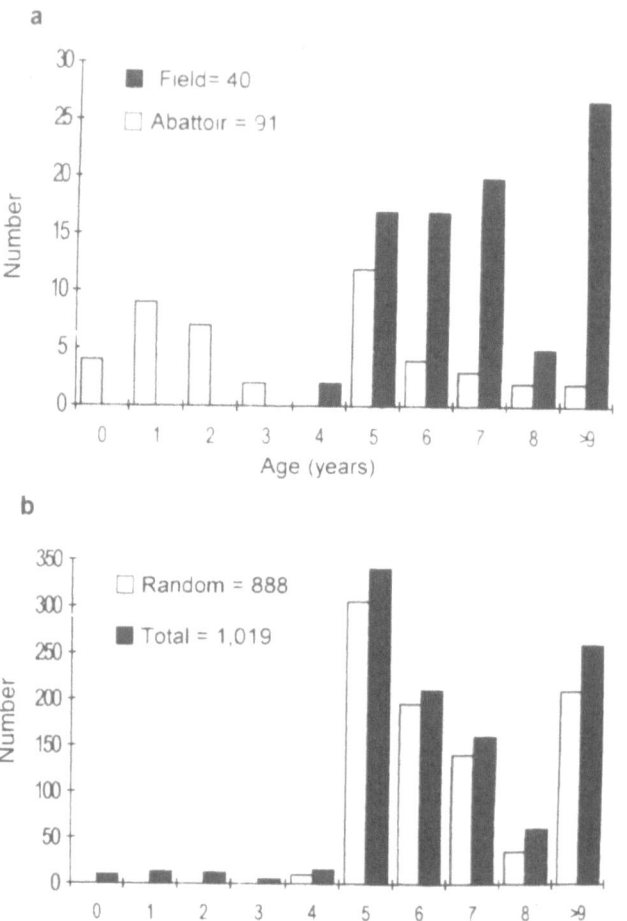

FIGURE 12.3. Age distribution and total number of cattle in each surveillance category. (a) Field = cases with signs of CNS disease; abattoir = cases in poor condition at the abattoir; (b) random = samples from apparently healthy animals from the high-risk group.

Among the remaining 26 cases, which had no significant lesions in the brain, there were 16 cases of hypocalcemia.

To increase the sample size, this part of the surveillance was extended until December 1994 when 31 more "field cases" were collected. These were examined by brain histology and for the presence of modified PrP. None satisfied the diagnostic criteria for BSE by either method. Also included were 36 cases reported in 1992, the year before the start of the surveillance program. These brains came only from the province of Buenos Aires, and there was no evidence of spongiform encephalopathy. The additional samples collected in 1992 and 1994 displayed a similar range of histopathologic lesions and alternative diagnoses as was seen in the 1993 series.

The second approach to surveillance was to look among older dairy animals that were found to be in poor condition at *antemortem* inspection in the abattoirs. This group contained 91 animals (Table 12.3). The average age was 7.1 years, and over one-quarter of the animals were older than 9 years (Fig. 12.3). All of the brain samples were negative for BSE when examined by histology and for the presence of modified PrP.

The third group consisted of 888 healthy animals that were randomly selected at slaughter to represent a large number of dairy farms (about one animal/farm) in districts of seven provinces that together contain about 85% of all dairy cattle in Argentina (Table 12.3). About 25% of the animals in this group were over 9 years of age and the average was 6.6 years. Brain samples were examined histologically for evidence of spongiform encephalopathy: 882 were negative and 6 were inconclusive. Although these six brains came from four different farms, it is significant that they were all collected at the same abattoir. The brains exhibited a generalized vacuolation of gray and white matter in the fields examined, accompanied by marked perineuronal spaces due to contraction of the neuron cell body. These findings are consistent with artifacts due to *postmortem* changes and/ or poor fixation. Frozen brain samples from these six animals were examined for the presence of modified PrP: all were negative.

It is important not to underestimate the difficulties of designing and executing a surveillance programme for BSE. Without a laboratory diagnostic test for infection, the survey had to be based on neurologic disease and brain lesions, which obviously limits the sensitivity of detection of subclinical cases. Because the experimental transmission of scrapie to cattle is associated with ill-defined clinical disease and minimal brain lesions (6, 7), the study aimed at detecting cases of BSE from a bovine source (for example, imported animals), not a sheep scrapie source. The study focused on elderly animals on the reasonable assumption that, since genetic variation in cattle is not a significant factor in the occurrence of BSE (44), BSE-infected animals would develop a clinical and/or histologic disease if they lived long enough. However, a difficult problem, which has been experienced by others (45–47), was how to further define a suitable target population for the surveillance. Given the very large cattle population in Argentina, the most practical approach was to focus on the comparatively small population of dairy cows, even though this group of animals is the only potential risk of accidental exposure to feed-borne BSE infection.

It was anticipated that restricting the surveillance to animals in the field with suspected neurologic disease would give too small a sample, particularly as this group included several relatively young animals (Fig. 12.3). The additional selection at the abattoir of cattle in poor condition allowed some exceptionally old cows to be examined. But statistically, the surveillance depended largely on the random selection of 888 healthy cows. With a total

TABLE 12.3. Distribution of samples (1993).

Province	No. of farms	No. of cattle	Breed		Sex		Category		
			Holstein	Others	Male	Female	Field	Abbatoir	Random
Buenos Aires	342	396	351	45	2	394	19	44	333
Catamarca	2	2	2	0	0	2	2	0	0
Córdoba	195	203	192	11	2	201	5	12	186
Corrientes	6	7	7	0	0	7	0	1	6
Entre Ríos	74	96	94	2	0	96	0	10	86
La Pampa	7	7	6	1	0	7	1	0	6
Río Negro	1	1	0	1	1	0	1	0	0
San Luis	9	11	6	5	0	11	5	1	5
Santa Fé	265	285	272	13	4	281	7	23	255
Liniers	11	11	11	0	0	11	0	0	11
Subtotal			941	78	9	1,010	40	91	888
Total	901	1,019	1,019		1,019		1,019		

population of 1,019 animals in the surveillance program during 1992–93, the annual prevalence of histologic disease would, if present, be no greater than 2.95 per thousand (with 95% confidence).

Against the background of the risk assessment, this figure should be regarded as an overestimate. Some idea of the degree of overestimation can be obtained as follows. The great majority of animals in the surveillance program were Holstein dairy cows (Table 12.2) over the age of 5 years (Fig. 12.3). Dairy cows constitute 4.2% of the total cattle population in Argentina, and animals older than 5 years make up about 20% of each dairy herd. If these animals are regarded as a sentinel group that represents, in terms of the risk of BSE infection, the whole cattle population, then the annual prevalence of undetected histologic disease would, at most, be about 2.5 per 100,000 of all cattle.

The results of the risk assessment and of the active surveillance program support the conclusion that Argentina has a very low risk of BSE infection being present and, therefore, of the disease occurring. A similar conclusion is likely to apply to other countries with little or no scrapie, and which use extensive systems of bovine meat and milk production that do not include the feeding of ruminant protein to cattle.

Conclusion

For all practical purposes, Argentina can be considered BSE-free. The importation of infected live cattle represents the only potential risk of introducing BSE into Argentina in the future. The continuity of the surveillance program will secure early detection of BSE and will ensure permanent maintenance of the present animal health status.

Acknowledgments. Secretarial assistance of Tessie Mac Cormack and Betty García is gratefully acknowledged.

References

1. Wells GAH, Scott AC, Johnson CT, et al. A novel progressive spongiform encephalopathy in cattle. Vet Rec 1987;121:419–20.
2. Wilesmith JW, Wells GAH, Cranwell MP, Ryan JBM. Bovine spongiform encephalopathy: epidemiological studies. Vet Rec 1988;123:638–44.
3. Wilesmith JW, Ryan JBM, Hueston WD, Hoinville LJ. Bovine spongiform encephalopathy: epidemiological features 1985–1990. Vet Rec 1992;130:90–4.
4. Wilesmith JW, Ryan JMB, Hueston WD. Bovine spongiform encephalopathy: case-control studies of calf feeding practices and meat and bone meal inclusion in proprietary concentrates. Res Vet Sci 1992;52:325–31.
5. Wilesmith JW, Ryan JBM, Atkinson MJ. Bovine spongiform encephalopathy: epidemiological studies on the origin. Vet Rec 1991;128:199–203.

6. Gibbs CJ, Safar J, Seroni M, DiMartino A, Clark WW, Hourrigan JL. Experimental transmission of scrapie to cattle. Lancet 1990;335:1275.

7. Cutlip RC, Miller JM, Race RE, et al. Intracerebral transmission of Scrapie to cattle. J Infect Dis 1994;169:814–20.

8. Wilesmith JW. Bovine spongiform encephalopathy: epidemiological factors associated with the emergence of an important new animal pathogen in Great Britain. Semin Virol 1994;5:179–87.

9. Kimberlin RH, Wilesmith JW. Bovine spongiform encephalopathy epidemiology, low dose exposure and risks. Ann NY Acad Sci 1994;724:210–20.

10. The Bovine Spongiform Encephalopathy Order 1988. Statutory instrument 1988, N 1039. HMSO, London.

11. Wilesmith JW, Wells GAH. Bovine spongiform encephalopathy. In: Chesebro BW, ed. Current topics in microbiology and immunology. New York: Springer, 1991;172:21–38.

12. Hoinville LJ. Decline in the incidence of BSE in cattle born after the introduction of the "feed ban." Vet Rec 1994;134:274–5.

13. Kimberlin RH. Bovine spongiform encephalopathy. In: Bradley R, Mathews D, eds. Transmissible spongiform encephalopathies of animals. Scientific and technical reviews. Paris: International Office of Epizootics, 1992;11:347–90 (English), 391–439 (French), 441–89 (Spanish).

14. Davis AJ, Jenny AL, Miller LD. Diagnostic characteristics of bovine spongiform encephalopathy. J Vet Diagn Invest 1991;3:266–71.

15. Wilesmith JW, Ryan JMB. Bovine spongiform encephalopathy: recent observations on the age specific incidences. Vet Rec 1992;130:491–2.

16. Wilesmith JW, Ryan JMB. Bovine spongiform encephalopathy: recent observation on the incidence during 1992. Vet Rec 1993;132:300–1.

17. Kimberlin RH. Bovine spongiform encephalopathy: an appraisal of the current epidemic in the U.K. Intervirology 1993;35:208–18.

18. Wilesmith JW. The epidemiology of bovine spongiform encephalopathy. Sem Virol 1991;2:239–45.

19. Cané BG, Gimeno EJ, Manetti JC, Van Gelderen C, Ulloa E, Schudel AA. Analysis of BSE risk factors in Argentina. 1991; Secretaría de Agricultura, Ganadería y Pesca de la Nación, ISBN 950-9853, 21-6, Capital Federal, Argentina, 11—1992, Centro Gráfico S.A., Buenos Aires, Argentina.

20. Walker KD, Hueston WD, Hurd HS, Wilesmith JW. Comparison of bovine spongiform encephalopathy risks factors in the United States and Great Britain. J Am Vet Med Assoc 1991;199:1554–61.

21. Resolution 429/90, Servicio Nacional de Sanidad Animal (SENASA) 1990; SAGyP, Argentina.

22. Cané BG, Gimeno EJ, Manetti JC, Van Gelderen C, Ulloa E, Schudel AA. Análisis de los factores de riesgo asociados a la encefalopatía espongiforme bovina en Argentina. In: Risk analysis, animal health and trade. Rev Sci Tech Off Int Epiz 1993;12:1203–34.

23. Schudel AA, Carrillo BJ, Gimeno EJ, et al. Bovine spongiform encephalopathy surveillance in Argentina. Rev Sci Tech Off Int Epiz 1994;13:801–18.

24. Schudel AA, Carrillo BJ, Weber EL, et al. Risk assessment and surveillance for bovine spongiform encephalopathy (BSE) in Argentina. Prev Vet Med 1996; 25:271–84.

25. Foster JD, McKelvey WAC, Mylne MJA, et al. Studies on maternal transmission of scrapie in sheep by embryo transfer. Vet Rec 1992;130:341–3.

26. Wells GAH, Wilesmith JW, McGill IS. Bovine spongiform encephalopathy: a neuropathological perspective. Brain Pathol 1991;1:69–78.
27. Jeffrey M. A neurological survey of brains submitted under the Bovine Spongiform Encephalopathy Orders in Scotland. Vet Rec 1992;131:332–7.
28. Miller L, Davis AJ, Jenny AL. Surveillance for lesions of bovine spongiform encephalopathy in U.S. cattle. J Vet Diagn Invest 1992;4:338–9.
29. Farquar CF, Somerville RA, Ritchie LA. Post-mortem immunodiagnosis of scrapie and bovine spongiform encephalopathy. J Virol Meth 1989;24:215–22.
30. Scott AC, Wells GAH, Stack MJ, White H, Dawson M. Bovine spongiform encephalopathy detection and quantitation of fibrils, fibril protein (PrP) and vacuolation in brain. Vet Microbiol 1990;23:295–304.
31. Ikegami Y, Ito M, Isomura H, et al. Pre-clinical and clinical diagnosis of scrapie by detection of PrP protein in tissues of sheep. Vet Rec 1991;128:271–5.
32. Katz JB, Pedersen JC, Jenny AL, Taylor WD. Assessment of Western immunoblotting for the confirmatory diagnosis of ovine scrapie and bovine spongiform encephalopathy (BSE). J Vet Diagn Invest 1992;4:447–9.
33. Mohri S, Farquar CF, Somerville RA, Jeffrey M, Foster J, Hope J. Immunodetection of a disease specific PrP fraction in scrapie-affected sheep and BSE-affected cattle. Vet Rec 1992;131:537–9.
34. Race R, Ernst D, Jenny A, Taylor W, Sutton D, Caughey B. Diagnostic implications of detection of proteinase K–resistant protein in spleen, lymph nodes, and brain of sheep. Am J Vet Res 1992;53:883–9.
35. OIE. Bovine spongiform encephalopathy in Canada. Office International des Epizooties. Disease Information 1993;6:178.
36. OIE. Office International des Epizooties, International Animal Health Code Commission. Bovine Spongiform Encephalopathy (BSE), 1992, Chapter 3.2.13.
37. OIE. Bovine spongiform encephalopathy in Italy. Office International des Epizooties. Disease Information 1994;7(42):183.
38. Spiegel MR. Theory and exercise on statistics. New York: McGraw-Hill, 1967.
39. Wells GAH, Hancock RD, Cooley WA, Richards MS, Higgins RJ, David GP. Bovine spongiform encephalopathy: diagnostic significance of vacuolar changes in selected nuclei of the medulla oblongata. Vet Rec 1989;125:521–4.
40. Hilmert H, Diringer H. A rapid and efficient method to enrich SAF-protein from scrapie brains of hamsters. Biosci Rep 1984;4:165–70.
41. Hope J, Morton LJD, Farquar CF, Multhaup G, Beyreuther K, Kimberlin RH. The major polypeptide of scrapie associated fibrils (SAF) has the same size, charge distribution and N-terminal protein sequence as predicted for the normal brain protein (PrP). EMBO J 1986;5:2591–7.
42. Laemmli UK. Cleavage of structural proteins during the assembly of the head of the bacteriophage T4. Nature (Lond) 1970;227:680–5.
43. Towbin H, Staehelin T, Gordon J. Electrophoretic transfer of proteins from polyacrylamide gels to nitrocellulose sheets: procedure and some applications. Proc Natl Acad Sci USA 1979;76:4350–4.
44. Kimberlin RH. A scientific evaluation of research into bovine spongiform encephalopathy (BSE). In: Bovine spongiform encephalopathy in Great Britain: a progress report. March 1994. Ministry of Agriculture, Fisheries and Food, U.K.
45. Hueston WD, Bleem AM, Walker KD. Bovine spongiform encephalopathy (BSE): risk assessment and surveillance in the Americas. Animal Health Insight. Fort Collins, CO: USDA-APHIS: Veterinary Services, Fall 1992:1–7.

46. USDA-APHIS-VS. Bovine Spongiform Encephalopathy (BSE): implications for the United States. Fort Collins, CO: USDA-APHIS: Veterinary Services, July 1993:1–25.
47. Bleem AM, Randall LC, Bruce Francy D, et al. Risk factors and surveillance for bovine spongiform encephalopathy in the United States. J Am Vet Med Assoc 1994;204:644–51.

13

Speculations on the Origin of BSE and the Epidemiology of CJD

Richard H. Kimberlin

The occurrence of bovine spongiform encephalopathy (BSE) has dramatically increased public awareness of the transmissible spongiform encephalopathies (TSEs) and stimulated a widespread interest in the current state of knowledge.

There is good evidence that the recycling of BSE infection from cattle to cattle, via meat and bone meal (MBM) in concentrated feeds, was the key factor causing a major epidemic in the United Kingdom. But the origin is less certain.

Because sheep are the only firmly established reservoir of TSE infection in animals, it is generally assumed that the origin of BSE was the feed-borne transmission of scrapie to cattle. The epidemiology of BSE is entirely consistent with this hypothesis. Furthermore, laboratory studies showing the differential selection of preexisting and mutant scrapie strains when transmitted from one species to another can explain why the BSE epidemic is associated with a strain of agent that is different from known scrapie strains. Nevertheless, in the absence of further supporting experimental evidence (there has been no attempt to transmit United Kingdom scrapie to cattle by oral exposure), it is difficult to ignore an alternative hypothesis.

A plausible alternative springs from many studies of the pathogenesis of scrapie in mice and hamsters showing that lifetime persistence of infection in certain tissues of the lymphoreticular system (LRS) is common to all experimental models of the disease. The observed differences between models in incubation period are mainly due to different rates of replication and spread of infection within the peripheral and central nervous systems (CNS). However, one model of scrapie is notable in that neuroinvasion fails to occur, or occurs so late that disease does not develop.

It is possible, therefore, that the origin of BSE was a naturally endemic, essentially avirulent (nonneuroinvasive or nonneuropathogenic) infection of the bovine LRS that caused an epidemic only when changes in the methods of rendering favored the positive selection of a pathogenic mutant by food-borne transmission. A consequence of this hypothesis is that eradication of BSE would not eliminate the avirulent infection from the United Kingdom cattle population. In addition, the hypothesis radically alters the risk assessment for BSE in other countries by taking sheep and scrapie out of consideration. Many countries that might also have an endemic, avirulent infection of cattle use low-temperature rendering to produce MBM, and then feed it to cattle. The fact that a comparable BSE epidemic has not occurred outside the United Kingdom is the strongest evidence against a bovine origin of BSE.

However, a degree of uncertainty remains about inapparent TSE infection in cattle and, possibly, in other species of domesticated food animals. Whether the origin of BSE was sheep or cattle, a most important lesson from the United Kingdom epidemic is the potential danger of feeding MBM derived from one species to animals of the same species. Recent legislation in the European Union prevents this happening in ruminants. It may be wise to consider applying similar measures to protect ruminants in some other countries and, possibly, other species of food animals.

There are also uncertainties about the epidemiology of Creutzfeldt-Jakob disease (CJD). Current opinion seems to favor an infectious cause of iatrogenic CJD; a genetic origin for familial cases, which are strongly associated with certain mutations in the human *PrP* gene; and somatic mutations of the *PrP* gene (or chance conversion of normal PrP protein to pathogenic forms) as the cause of sporadic CJD, which accounts for about 85% of all cases. However, the shape of the age-specific incidence curve of sporadic CJD is like that of BSE; incidence does not increase with age, as would be expected if somatic mutations or other random events were the primary cause.

It is conceivable that the different manifestations of CJD are an epidemiologic illusion of the patterns of disease occurrence not necessarily reflecting the spread of infection. A unified hypothesis is suggested in which the origin of all forms of CJD is a high-incidence endemic infection of man with a nonneuroinvasive or nonpathogenic strain of agent that is quite harmless in people carrying the normal *PrP* gene. Germline mutations of this gene might allow the common infection to be pathogenic, leading to the occurrence of disease with a genetic pattern. Mutations in the agent, followed by positive selection of the mutant in CNS tissue, would produce sporadic cases of CJD at a frequency that reflected the mutation rate of the agent.

This hypothesis makes predictions that are amenable to the techniques of agent strain typing that have been so successful with scrapie and BSE. But disproving it could be difficult in the absence of a diagnostic test for the genome of the infectious CJD agent. However, the hypothesis has very few implications for the occurrence of iatrogenic CJD. The dominant risks would arise from exposure to mutant strains of agent in the CNS (or closely associated tissues) from incipient or clinical cases of sporadic CJD, not from the use of nonneural tissues, such as blood.

Background to BSE and Key Findings

The story of the BSE epidemic has come a long way since the disease was first recognized in the United Kingdom in November 1986 (1), and it is sufficiently familiar that only a few main points need be mentioned here.

Prior to BSE, a major epidemic of human kuru (in the 1950s) and several outbreaks of transmissible mink encephalopathy (TME) in ranch-reared mink (in the 1960s) provided important precedents for the food-associated transmission of scrapie-like infections (2, 3). But whereas TME and kuru were associated with essentially uncooked material, the vehicle of BSE infection of cattle was MBM obtained by the rendering ("cooking") of waste animal tissues from slaughterhouses and used as a source of protein in cattle feeds (4). The most recent supporting evidence is the dramatic decline in the UK epidemic following the ban, introduced in July 1988 (5), on the feeding of ruminant-derived protein to ruminants (see Chapter 4).

BSE and TME are believed to have originated with the feed-borne transmission of scrapie, across the species barrier, from sheep to cattle and mink, respectively (see below). However, the large increase in the BSE epidemic, before the feed ban took effect, was due to cattle-to-cattle recycling of a bovine-adapted agent (6); in other words, there was no species barrier during much of the epidemic. Kuru developed into a major epidemic for the same reason (2), but outbreaks of TME have always been comparatively short-lived because of the absence of any significant recycling of infection in mink via feed (3).

A majority, but not all, of BSE cases were a consequence of calfhood infection leading to clinical disease after a median incubation period of about 5 years (7, 8). One of the most remarkable features of BSE is that the increase in the epidemic was mainly due to more and more new herds becoming affected, rather than to an increase in incidence within herds. Indeed, the half-yearly within-herd incidences have remained remarkably low, increasing only from about 1.5% to 3% of adults during the course of the epidemic (9). Since variations in the *PrP* and other bovine genes are not major factors in the occurrence of BSE (10), this low-incidence pattern indicates that the average contamination of MBM-containing feeds used in affected herds may have been as small as about 10 oral LD_{50} units of a cattle-adapted BSE agent per ton of feed (11).

The concept of low-dose exposure explains why comparable epidemics of BSE are not expected in any other countries, even though about 10, to date, have experienced some cases of BSE (12). Many of these cases were due to the importation, from the United Kingdom, of infected live animals or of MBM before control measures were introduced (11). It seems that only the

United Kingdom had just the right combination of BSE risk factors involving sheep, endemic scrapie, the use of ovine and bovine MBM in ruminant feeds, and changes in the conditions of rendering around 1981 to 1982 that allowed some scrapie/BSE infectivity to survive in MBM and cause an epidemic (13).

The decline in the BSE epidemic has followed a pattern reminiscent of kuru after ritual cannibalism of dead relatives ceased in the late 1950s. With both diseases, the youngest cases disappeared first and the decline was accelerated when the number of cases in the most commonly affected age groups was reduced (2, 9). Kuru is now a very rare disease and, because there appear to be no natural routes for the transmission of infection, it seems destined for extinction (14).

Although it is a little premature to make the same prediction for BSE, there is no clear evidence of transmission other than via feed. Any maternal and/or horizontal transmission that might occur is unlikely to be sufficient to sustain the BSE epidemic indefinitely (9), as occurs with sheep scrapie (15, 16). The eradication of BSE is foreseeable provided the current feed ban is maintained in full force.

The occurrence of BSE has had a considerable impact on world trade in live cattle, embryos, food, and both human and veterinary medicinal products. It has raised public awareness of the scrapie-like diseases to unprecedented levels, and the legacy of BSE will probably be a change in attitudes about the acceptability and wisdom of feeding animal protein to domesticated food animals of the same or different species. This leads to a consideration of the origin of BSE, which is the least understood part of the story.

Possible Origins of BSE in the United Kingdom

Ovine Origin of BSE

There are two reasons why the most plausible origin of the BSE epidemic was the food-borne infection of cattle with sheep scrapie. First, sheep are the only firmly established natural reservoir of TSE infection in animals (15, 16). However, there is evidence that chronic wasting disease in mule deer and elk may also occur as a natural endemic infection but this disease seems to be restricted to parts of North America (17). The second reason is that analysis of the BSE risk factors based on sheep and scrapie is consistent with the unique occurrence of epidemic BSE in the United Kingdom, and not in other countries that have scrapie (9, 11).

Laboratory studies of biologically characterized strains of scrapie in mice and hamsters show that crossing the species barrier can exert strong selective pressures favoring some strains of agent and excluding others (10). Agent-typing studies in mice suggest that the BSE agent has properties quite different from those of any contemporary or previous isolates of scrapie from United Kingdom sheep (18). Therefore, if scrapie was the origin of BSE, there was almost certainly a species barrier to the food-borne transmission of infection from sheep to cattle, and this conclusion is supported by the difficulties experienced in transmitting two sources of scrapie to cattle exposed parenterally (19).

There are two possible ways in which a unique BSE agent could have arisen from scrapie (10). One is that a spontaneous mutation of the scrapie agent occurred in sheep to produce a novel strain that was positively selected on transmission to cattle (4). But to account for the fact that the earliest BSE cases were widely distributed in England and Wales (6), this mutation must not only have occurred frequently, but the quantitatively minor mutant strain was selected across the sheep-to-cattle species barrier. The latter seems unlikely.

A more plausible hypothesis is that BSE arose from a single strain of scrapie agent that was relatively common in the United Kingdom sheep population, or a number of different strains that behaved similarly when crossing the sheep-to-cattle species barrier (10). However, for the cattle-adapted strain to be different from scrapie, the "sheep" agent must have been the origin of a common mutant strain that arose in, and was strongly selected by, cattle, probably during the first sheep-to-cattle transmission. The absence of genetic variation in cattle that influences the occurrence of BSE would naturally lead to a uniform selective pressure (10) with the result that the same (or very similar) strain of agent has been found in all seven geographically separate isolates of BSE that have been characterized so far (18).

Possible evidence against this hypothesis comes from two successful transmissions, by injection, of United States sources of scrapie to cattle, neither of which resulted in a disease that closely resembled BSE, clinically or histologically. Indeed proof of transmission depended largely on demonstrating the presence of the protease-resistant form of the PrP protein (19, 20).

However, since most of the observed BSE epidemic in the United Kingdom was a consequence of cattle-to-cattle recycling, with no species barrier, the primary transmission of scrapie to cattle may well appear to be different from cattle-adapted BSE in terms of clinical disease, pathology, and incubation period (21). It could also be significant that scrapie in the United States occurs mostly in black-faced breeds, particularly Suffolks (16), and scrapie strains that are typical of these breeds may not be representative of the major strains of United Kingdom scrapie. It is therefore important to attempt a transmission of naturally occurring strains of United Kingdom scrapie to cattle, by the oral route, to support an ovine origin of BSE.

In summary, the ovine-origin hypothesis is consistent with the epidemiologic findings of BSE (4) and with the current understanding of the behavior of scrapie strains when crossing the species barrier (10). But this evidence alone is insufficient as proof, and the hypothesis that the origin was bovine fits the data nearly as well.

Requirements for a Bovine Origin of BSE

The first major requirement for this hypothesis to be consistent with the start and subsequent development of the BSE epidemic is that a geographically widespread reservoir of infection was naturally maintained in the United Kingdom cattle population. Second, this infection was essentially nonpathogenic within the commercial life span of cattle although rare sporadic disease probably occurred undetected. The third requirement is that the changes in rendering conditions that led to sufficient feed-borne exposure of cattle to initiate the epidemic converted the avirulent infection into one that produced clinical disease in a large number of animals. To understand how this could have happened requires a brief discussion of the pathogenesis of the TSEs.

Pathogenesis of Scrapie and Related TSEs

The occurrence of infected carriers that do not develop disease has long been suspected to play a role in natural scrapie (15).They are an inevitable consequence of the median age of onset of scrapie (3 to 5 years) being close to the commercial life span of some breeds of sheep and competing with other sheep diseases of late onset. Carriers are also to be expected with infections like scrapie because the occurrence and incubation period of the clinical disease depends on two independent but closely interacting factors: the strain of agent and the genotype of the host.

The chemical nature of the scrapie agent is not at issue here. The important point is that, biologically, it behaves as though it has an independent genome (be it nucleic acid, protein, or whatever) that exhibits the familiar microbial properties of strain variation, fidelity of replication, and mutation (22). The major gene controlling natural and experimental scrapie in sheep is *Sip*, and the *Sinc* gene is its murine homologue for experimental scrapie (23, 24). Both genes are now generally accepted as being the same as the mammalian *PrP* gene, which almost certainly plays a major role in all the natural and experimental TSEs (25, 26).

Many studies have shown that, following peripheral infection of the host, early replication of the agent takes place in the lymphoreticular system (LRS). Disease only develops if and when the agent enters and then replicates in the central nervous system (CNS). Much evidence points to a neuroinvasive pathway via the visceral autonomic nerves, particularly those of the sympathetic nervous system (15). Most of the well-characterized experimental models of scrapie in mice and hamsters were derived by serial intracerebral passage of agent from clinically affected brains. Not surprisingly, therefore, most of the scrapie strains isolated are neuroinvasive and neuropathogenic. Nevertheless, there are major differences in incubation period depending on the strain of the agent and the *PrP* genotype of the host (22).

Comparisons of scrapie models of widely different incubation periods reveal only very small differences in the times of onset of agent replication in the LRS. Furthermore, there is no undue restriction on the initiation of neuroinvasion at the cellular interface between the LRS and the nervous system. Even with some of the very long incubation period models, neuroinvasion is initiated within a few weeks of infection (27). The differences lie in the subsequent stages of replication and cell-to-cell spread of infection within the peripheral and central nervous systems, the overall speed of which correlates strikingly with incubation period. Because of the limits on the sensitivity of the bioassays in these studies, the onset of detectable replication of agent in the brain was not until about halfway through the incubation period, however long or short this was (27).

However, there is one scrapie model (the 87V strain of scrapie in *Sinc* p7p7 mice) that seems to be different (28, 29). When relatively low doses of agent are injected intraperitoneally into mice, it is possible to obtain early infection and replication in the LRS, similar to that seen with other scrapie models. But a year and a half later, there was no detectable replication of the agent in the CNS, and the mice did not develop clinical scrapie within their normal life span of about 2 years. This is not because the agent cannot replicate in the CNS; direct injection into the brain readily produced disease in all the mice. The problem with intraperitoneal infection is that the 87V strain of agent has difficulty in spreading from the LRS to the CNS (15).

How an Avirulent Infection Could Cause an Epidemic of BSE

Against this background, it is easy to see how cattle might be endemically infected with a strain of agent that is essentially avirulent because it is nonneuroinvasive or nonneuropathogenic. Replication in the LRS would sustain a reservoir from which infection was spread, naturally, to other cattle, perhaps in the same ways that scrapie spreads in sheep (15, 16). Such a stable carrier state could be broken in either of two ways.

A germline mutation in the *PrP* gene could allow the avirulent strain of agent to become neuroinvasive or neuropathogenic. This seems the less likely option because the bovine *PrP* gene exhibits remarkably few polymorphisms, compared with the *PrP* genes of sheep (30), mice (31), and man (32, 33), and none is associated with the occurrence of BSE (34). Nor is there any evidence that other bovine genes might be important (10, 35). This is not to say that bovine *PrP* gene plays no role in the pathogenesis of BSE, simply that it is not a variable factor.

Therefore, the more likely event is a mutation of the avirulent agent to produce a virulent strain. The geographically widespread distribution of the earliest BSE cases (6) requires that such a mutation occur quite often in cattle. This is a reasonable assumption. For one thing the incidence of the parental, avirulent infection of cattle would need to be relatively high for it to be endemic by natural routes of transmission; and transmission studies with scrapie (36, 37) and TME (38) show that these agents mutate quite commonly.

It is suggested that the mutant strain has no selective advantage in the LRS where it would remain a minor component and not be spread naturally to any significant extent. But if the mutant strain was able to replicate efficiently in nervous tissues, it could have a great selective advantage over the avirulent strain. Depending on how rapidly the virulent strain could spread and replicate in the CNS, it might even be the cause of sporadic disease in cattle at an incidence that escaped detection in the past, i.e., 1 case in 100,000 adults per annum, which was the incidence of inapparent BSE diagnosed retrospectively in the United Kingdom, between April 1985 and November 1986 (4, 39).

However, this alone would not cause a BSE epidemic. There would have to be a rapid, positive selection of the virulent mutant coupled with widespread exposure of cattle. This could have occurred when changes in the rendering of animal wastes allowed feed-borne infection of cattle and, subsequently, cattle-to-cattle recycling of infected CNS material rapidly created a reservoir of infection of the virulent strain.

To cause disease, the feed-borne virulent strain would need to gain access to the CNS, presumably via Peyer's patches and the peripheral nervous system, as has been shown with orally infected mice (40). At the same time, it would appear that its presence in the LRS does not lead to detectable (so far) natural transmission of BSE (see Chapter 4). TME is not naturally transmitted either (except by cannibalism) and it has been suggested that the reason is that the agent does not replicate to high titers in the LRS and other extraneural tissues (41), as the scrapie agent does in sheep (42).

In this respect, BSE may be more like TME than scrapie because no infectivity has been detected anywhere outside the CNS in confirmed cases of BSE, using mouse bioassay methods (43). Some components of the LRS may still be important in the pathogenesis of BSE in initiating neuroinvasion, but without much accummulation of infectivity or spread to tissues from which it can be shed. Studies of intragastric infection of mice with scrapie suggest that after agent replication in Peyer's patches, pathogenesis could be entirely neural (40).

In conclusion, it is not difficult to imagine how a major epidemic of clinical BSE could have developed from a bovine origin once infection via feed led to the selection of a more neuroinvasive or neuropathogenic strain of agent. On this scenario, the food-borne virulent strain not only would be different from scrapie, but its spread via feed would be independent of, and superimposed on, the natural spread of the avirulent infection. One of the consequences of this is that the eradication of BSE would not eliminate the avirulent strain, which would remain endemic in the cattle population. However, making the ruminant protein feed ban (5) permanent would prevent this situation, if it exists, from causing future epidemics in the United Kingdom.

Evidence Against a Natural, Avirulent TSE Infection of Cattle

There is one major observation that argues against a bovine origin of BSE. Given the considerable international trade in breeding cattle over past decades, there is no reason to suppose that an avirulent endemic infection of cattle would not exist in many other countries as well as the United Kingdom. However, this would change the whole basis of the assessment of risk of BSE. Once sheep and scrapie are removed from consideration, there would be a risk in many countries where bovine MBM is made by processes similar to those used in the United Kingdom and is fed to cattle in large quantities. The absence of a United Kingdom-like epidemic in any other country, notably the United States, which has an active surveillance program (44), is the only strong evidence against a bovine origin of BSE.

However, the situation is clouded by an outbreak of TME in the United States in 1985. Because the ranch owner commonly fed "downer" cattle to his mink, it was suggested that the origin of infection in this, and previous American outbreaks of TME, was cattle, not sheep (45). The difficulty is that many mink ranchers feed "downer" cattle to mink (46) and the 1985 outbreak is the only one in the United States since 1963. Therefore, TME would appear to be a rare consequence of feeding bovine material to mink, perhaps material infected with a mutant bovine agent that, in contrast to the parental strain, happened to be pathogenic for mink across the species barrier.

But it remains equally plausible that rare occurrences of TME are due to scrapie, which has existed in the United States for several decades (16). An outbreak could result from a single, unnoticed or forgotten occasion when just one scrapie infected sheep was fed to mink. Given that scrapie can be the cause of death in sheep without any of the usual preceding clinical signs (47), an outbreak of TME could arise when a seemingly "healthy" dead sheep was used to make a batch of mink feed. Yet another possibility is that cattle may occasionally become infected with scrapie and that one such animal was the cause of the 1985 outbreak of TME.

The rarity of TME makes it very difficult to be certain where the infection comes from and the available evidence is neither for nor against the existence of a natural TSE infection in cattle. But without an appropriate laboratory test, it is difficult to be sure that cattle are not endemically infected with an avirulent agent. And in theory, the same could be true of other species, including man.

Epidemiology of Creutzfeldt-Jakob Disease (CJD)

Background and Problems

The inability to test for infection has bedeviled epidemiologic studies of some of the other scrapie-like diseases. The problem is that, depending on the degree of variation in the strains of agent and host *PrP* gene and the precise interaction between the two, the patterns of disease occurrence may or may not be an accurate reflection of the spread of the infectious agent.

BSE lies at the simplest end of the spectrum. The epidemic seems to have involved, essentially, a single (virulent) strain of agent and a single host *PrP* phenotype so that most infected calves develop the disease if they are kept beyond the peak age of onset of BSE, about 4 to 6 years (Fig. 13.1). The same is broadly true of kuru and TME in that there is no evidence that variation in the *PrP* gene (or any other host gene) obscures the spread of infection. Consequently the epidemiology of all three diseases is relatively simple and an infectious cause has never been in doubt.

Not so with scrapie, where the pronounced influence of host genetic factors made it difficult to establish that scrapie is naturally contagious (48).

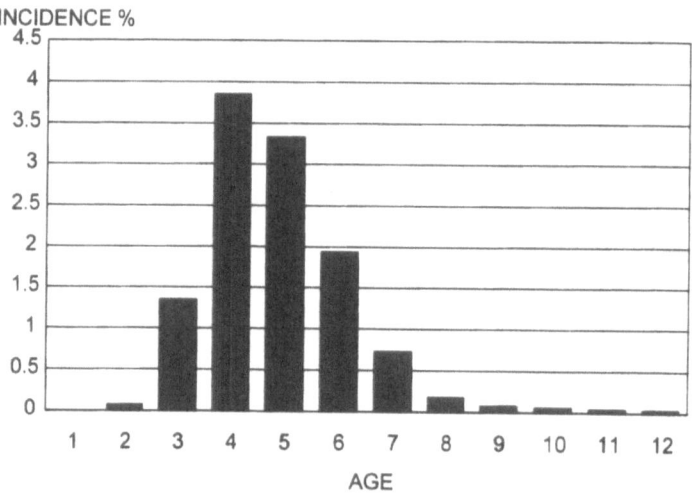

FIGURE 13.1. Age-specific incidence rates (percent) of 5,802 confirmed cases of BSE in herds affected during 1990. The data have not been corrected for the effects of culling, but this does not substantially alter the general pattern of distribution of BSE cases in each age group. (The figure was kindly provided by Mr. J.W. Wilesmith.)

With the recent discovery that the occurrence of disease is associated with certain polymorphisms of the *PrP* gene, it is now possible to interpret the epidemiology of scrapie more precisely in terms of the spread of infection and the specific *PrP* genotypes of individual sheep (30).

In contrast, Gerstmann-Sträussler-Scheinker disease (GSS) and familial CJD show such strong associations with specific mutations of the human *PrP* gene that, were it not for the fact they are experimentally transmissible (49, 50), they would be universally regarded as diseases of a solely genetic origin (32). Sporadic CJD is different again; it seems to lie in an epidemiologic no-man's-land in appearing to be neither infectious nor genetic. The growing opinion is that the different epidemiologic patterns of the TSEs indicate different etiologies—infectious, genetic, or spontaneous—as is illustrated by the three classic epidemiologic manifestations of CJD (25, 32).

Clearly iatrogenic CJD is caused by the parenteral exposure of patients to infected material during the course of treatment, notably with cadaveric human growth hormone (51). Fortunately, iatrogenic CJD accounts for fewer than 1% of all cases.

About 10% to 15% of CJD cases (including GSS) are familial and have a very strong association with one or other mutations in the *PrP* gene (33, 52, 53). These mutations, it is argued, predispose the normal PrP protein to adopt a posttranslational conformation that behaves like a pathogenic, transmissible agent. This idea of a "pseudoinfectious" agent receives support from studies of transgenic mice which express a modification of the *PrP* gene that is the equivalent of the human codon 102 mutation associated with GSS (52). With time, many of these mice spontaneously develop a neurologic disease characterized by spongiform encephalopathy. Small amounts of the abnormal form of PrP protein are also found in the CNS, and transmission studies indicate the presence of low levels of infectivity (54). The seemingly de novo appearance of a transmissible agent would be the key piece of evidence that GSS is truly a genetic disease and, given its fundamental importance, it is reasonable to reserve judgment until further studies can prove that this, and not contamination, is the only tenable interpretation.

However, the majority of CJD cases (about 85%) occur sporadically in many countries throughout the world, at an annual incidence of about 1 case per million population (55, 56). No *PrP* gene mutations are known to be necessarily associated with sporadic CJD (57), but, in terms of the prion hypothesis, it could be caused by somatic mutations in the *PrP* gene or some other chance event that leads to a spontaneous conversion of normal PrP protein into a pathogenic form (25). The problem with this hypothesis is that one might expect the incidence of sporadic CJD to increase with age (58). In fact, the age-specific incidence of CJD is clearly lower in people over 80 years of age than in the 60- to 69-year age group (Fig. 13.2). The shapes of the age-specific incidence curves for BSE (Fig. 13.1) and sporadic

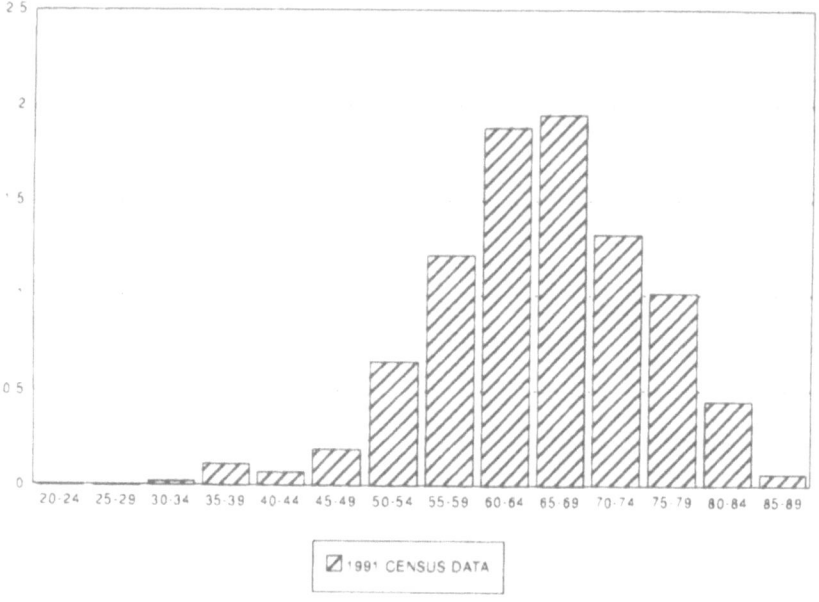

FIGURE 13.2. Age-specific incidence rates of 573 definite and probable cases of sporadic CJD in Great Britain during 1970 to 1994, a time period of 25 years (N = 573). Incidence rates are expressed per million and are based on census data for 1991. Almost identical incidence rates were obtained using census data for 1981. (The figure was kindly provided by Dr. R.G. Will.)

CJD (Fig. 13.2) are sufficiently similar to suggest that sporadic CJD, like BSE, has an infectious etiology.

Indeed, there is no logical necessity to interpret the genetic and sporadic patterns of CJD in terms of fundamentally different etiologies. Since experimental transmissibility is a major unifying attribute of all three forms of CJD (49), it is possible that the different epidemiologic patterns reflect differences in the extent to which the *PrP* gene is the limiting factor in the development of disease that is primarily caused by an infectious agent. The key question concerns the natural reservoir of CJD infection.

Unified Hypothesis for the Epidemiology of CJD

A great deal of evidence shows that the worldwide distribution of sporadic CJD differs from those of scrapie and the consumption of sheep products (55, 56), and there is no evidence that clearly implicates any other zoonotic source of infection. The simplest alternative is that a reservoir of CJD infection is naturally endemic in man, exactly analogous to the hypothesis discussed above for the bovine origin of BSE.

Stated briefly, the hypothesis is that a substantial proportion of the world's human population is infected with a common strain(s) of the CJD agent that is completely harmless in people carrying the normal *PrP* gene because it cannot invade the CNS or it is nonpathogenic. This equilibrium between the strain of the agent and the *PrP* genotype could be upset by one of two events, each causing disease to appear with a totally different epidemiologic pattern.

Specific germline mutations of the human *PrP* gene might allow the otherwise harmless infection to become pathogenic. If the incidence of infection in the general population is high enough, the pattern of the disease would be genetic, and only experimental transmissibility would betray the existence of an infectious agent. This corresponds precisely to the situation with GSS and familial CJD.

A mutation of the agent to produce a strain that, in contrast to the parental strain, was neuroinvasive and neuropathogenic would lead to a completely random pattern of CJD occurrence at an incidence that reflected the rate at which mutations appeared. One main requirement of this proposal is that the pathogenic strain should not contribute significantly to the epidemiology of the infection, otherwise clusters of cases would occur. This requirement can be met if it is assumed that the person-to-person spread of infection depends on the reservoir of infection in the LRS, and that pathogenic mutants are only selected in the nervous system.

Another requirement is suggested by the shape of the age-specific incidence curve (Fig. 13.2), which implies that infection with the common strain occurs in childhood or adolescence, and that the median incubation period is 40 to 50 years. This is not improbable. Some of the growth hormone cases of CJD are associated with incubation periods of up to at least 25 years (51), and the experience with the most recent, albeit very rare, cases of kuru is consistent with incubation times of 30 years or more (14).

Medical virology is familiar with the concept of common infections being the cause of rare diseases; for example, subacute sclerosing panencephalitis (SSPE) is a highly infrequent consequence of infection with measles virus (59). Perhaps a more appropriate example is progressive multifocal leukoencephalopathy (PML), an extremely rare disease affecting the CNS that peaks in the sixth decade of life. PML is primarily associated with a human polyoma virus (JCV) that occurs as a very common infection during childhood. However, the disease is almost never seen in children and its late onset appears to be due to one of a number of disorders involving immunosuppression (60).

In contrast to SSPE and PML, an obvious limitation of the unified hypothesis for CJD is the difficulty of testing for the presence of a common, nonpathogenic TSE infection of man by current bioassay methods, unless the agent happens to be pathogenic for the animal species into which it is injected. Studies of the transmission of a CJD-like agent to hamsters by the injection of normal human buffy coat cells have produced totally contradictory results (61, 62). One way forward might be to use transgenic mice that express different or chimeric *PrP* genes in high copy number (63). Ultimately, what may be needed is a nonbiologic, laboratory test for the agent, but a specific test may not be possible until the genome of the agent has been identified.

However, the unified hypothesis makes at least two predictions that are currently testable. First, CJD isolates from the brains of familial cases should be different (the common strain) from sporadic CJD isolates (mutant strains). No systematic strain-typing of CJD isolates has been carried out, but it may be worth noting that transmissions from cases of sporadic CJD to primates or mice have been rather more successful than from familial cases (49, 50). This is reminiscent of the difference between the uniform transmissibility of BSE (single strain) to mice compared with transmissions of sheep scrapie (multiple strains), not all of which have been successful (18).

The second prediction concerns the isolates made from the LRS (and other extraneural tissues) compared with those from the CNS. With familial CJD (and GSS), isolates from the two types of tissue would tend to be similar because the common CJD strain would be present in both. However, a different pattern would be expected with cases of sporadic CJD because the common strain would be in the LRS, but the CNS would contain the mutant strain. There is tantalizing evidence that this might be true. Successful transmissions to squirrel monkeys from nonneural tissues of patients with sporadic CJD are very much rarer than transmissions from the CNS (49). Unfortunately, this may not be a reflection of differences in agent strain, but of titer, as was suggested earlier for TME and BSE (see above).

The standard biologic methods of agent strain typing, which have been used so successfully with scrapie and BSE (18), are sufficiently discriminating to address these predictions about familial and sporadic CJD. Several isolates would be required to establish such differences and much would depend on agent strain differences in man being reflected by different biologic properties in whichever species (and *PrP* genotype) of laboratory animal that was used. But there is good evidence that the genomic identity of scrapie strains is preserved on transmission from one species to another (37).

Consequences and Conclusions

The striking associations between specific mutations of the human *PrP* gene and the occurrence of GSS, familial CJD, and fatal familial insomnia (FFI) (64) have favored the idea that these mutations are the cause of CJD and that the experimentally transmissible agent is simply a consequence. But the uncertainties surrounding the etiology of sporadic CJD, which is much more common, leave room for alternative hypotheses. An hypothesis that unifies the etiology, not just of the different types of CJD but all the TSEs, has some aesthetic and intellectual attractions. It also focuses attention on the risks of iatrogenic transmission of infection between, and within, species.

Reservoir of CJD Infection in Man

At first sight, the idea of a common CJD infection of man seems very disturbing, given that a substantial number of products of human origin are used in human medicine when there is no protective species barrier. But it should be recognized that exposure to the common CJD agent, by whatever means, would be of little consequence to people carrying the normal *PrP* gene. Similarly, people with a mutant *PrP* gene might only be at a marginally greater risk of disease from iatrogenic exposure to the common CJD agent if they were not already infected. The risks of iatrogenic transmission would be dominated by potential exposure to the mutant CJD strain in neural tissues derived from incipient or clinical cases of sporadic CJD.

Unfortunately, this risk is already too familiar (51). All iatrogenic cases of CJD have been associated with contamination arising from the CNS (implanted EEG electrodes, neurosurgical instruments) or from tissues in close proximity to it (corneal transplants, dura mater transplants, human growth hormone, and other pituitary hormones). By contrast, one of the most common procedures involving nonneural tissue, namely blood transfusion, does not appear to be a major risk factor for CJD (65).

Reservoir of TSE Infection in Cattle

Compared with the above situation with iatrogenic CJD, the consequences to man of the bovine origin hypothesis are unlikely to be very serious because any risks from food and pharmaceuticals would be considerably reduced by the cattle-to-man species barrier, and there would be no recycling of infection in man as there was with kuru. It is also emphasized that the existence of an avirulent endemic infection of cattle would predate the BSE epidemic by many years, even decades. Moreover, a similar historical exposure of man would have occurred in all other countries that have the natural bovine infection, irrespective of the occurrence of BSE cases. Of course, it could be argued that avirulent TSE infection of cattle is the origin of sporadic cases of CJD, but it is unlikely that a bovine reservoir of infection would be better able than an ovine reservoir to account for the uniformly rare occurrence of sporadic CJD throughout the world.

The greater public health concern focuses on the United Kingdom, where the emergence of a strain of BSE that is pathogenic for cattle required preemptive measures to minimize the exposure of people, simply because this strain appears to be novel and it arose comparatively recently. One of the most important of these measures was the exclusion of the, so-called, specified bovine offals from the human food chain (66). In retrospect, the cattle-adapted BSE agent existed several years before the disease was first recognized in November 1986, and possibly as early as 1982. The evidence is that four of the seven cattle from whose clinically affected brains this agent was subsequently isolated were in the national herd that year (see Chapter 3). However, it should be noted that CJD has occurred in the United Kingdom at the same rate as in other countries with few or no cases of BSE, and this surveillance will continue (67; see Chapter 27).

For the reasons given above, the main consequences of a reservoir of avirulent TSE infection in cattle would be to animal health. The results of strain-typing studies (18) leave few doubts about the food-borne transmission of (virulent) BSE agent, across the species barrier, to cause TSE in domestic cats and some species of zoo animals in, or from, the United Kingdom (68). However, it is reasonable to assume that these cases of TSE arose from primary infection before measures had been taken, in 1990, to minimize the exposure to BSE, notably the exclusion of the specified bovine offals from all animal feeds (69).

The unique lesson of BSE is the potential danger of intraspecies recycling of a primary infection from a different species, or of a mutant strain of agent from the same species. These risks are greatest for food animals when the rendering and feed industries are linked by the production and use of MBM.

In theory, the risks can be reduced by improving the conditions of rendering. To this end, the European Union (EU) has outlawed processes that are demonstrably unable to inactivate the BSE agent and has set minimum standards for the other processes (70). But it is much more difficult to determine, with sufficient confidence, the rendering conditions that would adequately disinfect the TSE agents (21). The surest way to prevent the recycling of virulent or avirulent infection in sheep, cattle, and other ruminants is not to feed them with ruminant-derived protein or any animal protein, as was recently adopted throughout the EU (71).

Although there is no evidence that TSEs occur naturally in the other main food species in the United Kingdom, pigs are experimentally susceptible to high parenteral doses of BSE (35) and poultry might be because they carry a *PrP*-like gene (72). Oral exposure of these species to the BSE agent in the UK is now greatly reduced by the extension of the specified bovine offals ban (69). And exposure to scrapie continues in many countries, as in the past, without any discernible ill effects.

Nevertheless, because within-species recycling of pig and poultry protein is common, there is at least a theoretical risk to these nonruminants if primary food-borne infection was introduced from another species (e.g., sheep) or if inapparent reservoirs of TSE infection exist in pigs or poultry. The absence, in either species, of TSE epidemics concurrent with the epidemic of BSE indicates that such risks are low and, for this reason, there is no justification for precipitate action. But perhaps over time, the precautions for ruminants might be extended so that no species of food-animal is fed protein from its own species.

Acknowledgments. The author is extremely grateful to Mr. J.W. Wilesmith and to Dr. R.G. Will who provided Figure 13.1 and Figure 13.2, respectively, and who were also kind enough to comment on the manuscript, the preparation of which owes much to the diligence of Sara Kimberlin.

References

1. Wells GAH, Scott AC, Johnson CT, et al. A novel progressive spongiform encephalopathy in cattle. Vet Rec 1987;121:419–20.
2. Gajdusek DC. Unconventional viruses and the origin and disappearance of Kuru. Science 1977;197:943–60.
3. Marsh RF, Hanson RP. On the origin of transmissible mink encephalopathy. In: Prusiner SB, Hadlow WJ, eds. Slow transmissible diseases of the nervous system, vol 1. New York. Academic Press, 1979:451–60.
4. Wilesmith JW, Wells GAH, Cranwell MP, Ryan JMB. Bovine spongiform encephalopathy: epidemiological studies. Vet Rec 1988;123:638–44.
5. The Bovine Spongiform Encephalopathy Order 1988. Statutory Instrument No. 1039. London: Her Majesty's Stationary Office, 1988.
6. Wilesmith JW. The epidemiology of bovine spongiform encephalopathy. Semin Virol 1991;2:239–45.
7. Wilesmith JW, Ryan JBM, Hueston WD, Hoinville LJ. Bovine spongiform encephalopathy: epidemiological features 1985–1990. Vet Rec 1992;130:90–4.
8. Wilesmith JW, Ryan JBM, Hueston WD. Bovine spongiform encephalopathy: case-control studies of calf feeding practices and meat and bone meal inclusion in proprietary concentrates. Res Vet Sci 1992;52:325–31.
9. Wilesmith JW. Bovine spongiform encephalopathy: epidemiological factors associated with the emergence of an important new animal pathogen in Great Britain. Semin Virol 1994;5:179–87.
10. Kimberlin RH. Bovine spongiform encephalopathy: an appraisal of the current epidemic in the U.K. Intervirology 1993;35:208–18.
11. Kimberlin RH, Wilesmith JW. Bovine spongiform encephalopathy (BSE): epidemiology, low dose exposure and risks. Ann NY Acad Sci 1994;724:210–20.
12. Bradley R, Marchant B, eds. Transmissible spongiform encephalopathies. Proceedings of a consultation on BSE with the Scientific Veterinary Committee of the Commission of the European Communities, Brussels; 1993 Sept 14–15; European Commission Document VI/4131/94-EN. Brussels, 1994.
13. Wilesmith JW, Ryan JBM, Atkinson MJ. Bovine spongiform encephalopathy: epidemiological studies on the origin. Vet Rec 1991;128:199–203.
14. Klitzman RL, Alpers MP, Gajdusek DC. The natural incubation period of Kuru and the episodes of transmission in three clusters of patients. Neuroepidemiology 1984;3:3–20.

15. Kimberlin RH. Unconventional "slow" viruses. In: Collier LH, Timbury MC, eds. Topley and Wilson's principles of bacteriology, virology and immunity, vol 4. 8th ed. London: Edward Arnold, 1990:671–93.

16. Detwiler LA. Scrapie. Sci Tech Rev Int Off Epiz 1992;11:491–537.

17. Williams ES, Young S. Spongiform encephalopathies in Cervidae. Sci Tech Rev Int Off Epiz 1992;11:551–67.

18. Bruce ME, Chree A, McConnell I, Foster J, Pearson G, Fraser H. Transmission of bovine spongiform encephalopathy and scrapie to mice: strain variation and the species barrier. Philos Trans R Soc Lond [B] 1994;343:405–11.

19. Gibbs CJ Jr, Safar J, Ceroni M, Di Martino A, Clark WW, Hourrigan JL. Experimental transmission of scrapie to cattle. Lancet 1990;335:1275.

20. Cutlip RC, Miller JM, Race RE, et al. Intracerebral transmission of scrapie to cattle. J Infect Dis 1994;169:814–20.

21. Kimberlin RH. A scientific evaluation of research into bovine spongiform encephalopathy (BSE). In: Bradley R, Marchant B, eds. Transmissible spongiform encephalopathies. Proceedings of a consultation on BSE with the Scientific Veterinary Committee of the Commission of the European Communities, Brussels; 1993 Sept 14–15; European Commission Document VI/4131/94-EN: Brussels, 1994:455–77.

22. Bruce ME, McConnell I, Fraser H, Dickinson AG. The disease characteristics of different strains of scrapie in *Sinc* congenic mouse lines: implication for the nature of the agent and host control of pathogenesis. J Gen Virol 1991;72: 595–603.

23. Dickinson AG, Outram GW. Pathogenesis of experimental scrapie. In: Bock G, Marsh J, eds. Novel Infectious Agents and the Central Nervous System. Ciba Foundation Symposium No. 135. Chichester: Wiley, 1988;63–83.

24. Foster JD, Dickinson AG. Genetic control of scrapie in Cheviot and Suffolk sheep. Vet Rec 1988;123:159.

25. Prusiner SB. Molecular biology and genetics of prion diseases. Philos Trans R Soc Lond [B] 1994;343:447–63.

26. Weissmann C, Bueler H, Fischer M, Sauer A, Aguet M. Susceptibility to scrapie in mice is dependent on PrPc. Philos Trans R Soc Lond [B] 1994;343: 431–3.

27. Kimberlin RH, Walker CA. Incubation periods in six models of intraperitoneally injected scrapie depend mainly on the dynamics of agent replication within the nervous system and not the lymphoreticular system. J Gen Virol 1988;69:2953–60.

28. Bruce ME. Agent replication dynamics in a long incubation period model of mouse scrapie. J Gen Virol 1985;66:2517–22.

29. Collis SC, Kimberlin RH. Long term persistence of scrapie infection in mouse spleens in the absence of clinical disease. FEMS Microbiol Lett 1985;29:111–4.

30. Goldmann G, Hunter N, Smith G, Foster J, Hope J. PrP genotype and agent effects in scrapie: change in allelic interaction with different isolates of agent in sheep, a natural host of scrapie. J Gen Virol 1994;75:989–95.

31. Westaway D, Cooper C, Turner S, Da Costa M, Carlson GA, Prusiner SB. Structure and polymorphism of the mouse prion protein gene. Proc Natl Acad Sci USA 1994;91:6418–22.

32. Collinge J, Palmer M. Molecular genetics of human prion diseases. Philos Trans R Soc Lond [B] 1994;343:371–8.

33. Goldfarb LG, Brown P, Cervenakova L, Gajdusek DC. Genetic analysis of Creutzfeldt-Jakob disease and related disorders. Philos Trans R Soc Lond [B] 1994;343:379–84.

34. Hunter N, Goldmann W, Smith G, Hope J. Frequencies of *PrP* gene variants in healthy cattle and cattle with BSE in Scotland. Vet Rec 1994;135:400–3.

35. Dawson M, Wells GAH, Parker BNJ, et al. Transmission studies of BSE in cattle, pigs and domestic fowl. In: Bradley R, Marchant B, eds. Transmissible spongiform encephalopathies. Proceedings of a Consultation on BSE with the Scientific Veterinary Committee of the Commission of the European Communities, Brussels; 1993 Sept 14–15; European Commission Document VI/4131/94-EN: Brussels, 1994:161–7.

36. Bruce ME, Dickinson AG. Biological evidence that scrapie agent has an independent genome. J Gen Virol 1987;68:79–89.

37. Kimberlin RH, Walker CA, Fraser H. The genomic identity of different strains of mouse scrapie is expressed in hamsters and preserved on reisolation in mice. J Gen Virol 1989;70:2017–25.

38. Kimberlin RH, Cole S, Walker CA. Transmissible mink encephalopathy (TME) in Chinese hamsters: identification of two strains of TME and comparisons with scrapie. Neuropathol Appl Neurobiol 1986;12:197–206.

39. Kimberlin RH. Bovine spongiform encephalopathy. Sci Tech Rev Int Off Epiz 1992,11:347–90 (English), 391–439 (French), 441–89 (Spanish).

40. Kimberlin RH, Walker CA. Pathogenesis of scrapie in mice after intragastric infection. Virus Res 1989;12:213–20.

41. Hadlow WJ, Race RE, Kennedy RC. Temporal distribution of transmissible mink encephalopathy virus in mink inoculated subcutaneously. J Virol 1987; 61:3235–40.

42. Hadlow WJ, Kennedy RC, Race RE. Natural infection of Suffolk sheep with scrapie virus. J Infect Dis 1982;146:657–64.

43. Fraser H, Foster JD. Transmission to mice, sheep and goats and bioassay of bovine tissues. In: Bradley R, Marchant B, eds. Transmissible spongiform encephalopathies. Proceedings of a Consultation on BSE with the Scientific Veterinary Committee of the Commission of the European Communities, Brussels; 1993 Sept 14–15; European Commission Document VI/4131/94-EN: Brussels, 1994:145–59.

44. Bleem AM, Crom RL, Francy DB, Hueston WD, Kopral C, Walker K. Risk factors and surveillance for bovine spongiform encephalopathy in the United States. J Am Vet Med Assoc 1994;204:644–51.

45. Marsh RF, Bessen RA, Lehmann S, Hartsough GR. Epidemiological and experimental studies on a new incident of transmissible mink encephalopathy. J Gen Virol 1991;72:589–94.

46. Robinson MM. Transmissible encephalopathy research in the United States. In: Bradley R, Marchant B, eds. Transmissible spongiform encephalopathies. Proceedings of a Consultation on BSE with the Scientific Veterinary Committee of the Commission of the European Communities, Brussels; 1993, Sept 14–15; European Commission Document VI/4131/94-EN: Brussels, 1994: 261–7.

47. Clark AM, Moar JAE. Scrapie: a clinical assessment. Vet Rec 1992;130:377–8.

48. Kimberlin RH. The aetiology and genetic control of natural scrapie. Nature (Lond) 1979;278:303–4.

49. Brown P, Gibbs CJ Jr, Rodgers-Johnson P, et al. Human spongiform encephalopathy: The National Institutes of Health series of 300 cases of experimentally transmitted disease. Ann Neurol 1994;35:513–29.
50. Muramoto T, Kitamoto T, Tateishi J, Goto I. Successful transmission of Creutzfeldt-Jakob disease from human to mouse verified by prion protein accumulation in mouse brains. Brain Res 1992;599:309–16.
51. Brown P, Preece MA, Will RG. "Friendly fire" in medicine: hormones, homografts, and Creutzfeldt-Jakob disease. Lancet 1992;340:24–7.
52. Hsiao K, Baker HF, Crow TJ, et al. Linkage of a prion protein missense variant to Gerstmann-Straussler syndrome. Nature (Lond) 1989;338:342–5.
53. Gabizon R, Rosenman H, Meiner Z, et al. Mutation in codon 129 of the prion protein gene in Libyan Jews with Creutzfeldt-Jakob disease. Philos Trans R Soc Lond [B] 1994;343:385–90.
54. Hsiao KK, Groth D, Scott M, et al. Serial transmission in rodents of neurodegeneration from transgenic mice expressing mutant prion protein. Proc Natl Acad Sci USA 1994;91:9126–30.
55. Brown P, Cathala F, Raubertas RF, Gajdusek DC, Castaigne P. The epidemiology of Creutzfeldt-Jakob disease: conclusion of a 15-year investigation in France and review of the world literature. Neurology 1987;37:895–904.
56. Will RG, Matthews WB, Smith PG, Hudson C. A retrospective study of Creutzfeldt-Jakob disease in England and Wales 1970–1979 II: epidemiology. J Neurol Neurosurg Psychiatry 1986;49:749–55.
57. Palmer MS, Collinge J. Mutations and polymorphisms in the prion protein gene. Human Mut 1993;2:168–73.
58. Wood JLN, Hall AJ. Creutzfeldt-Jakob disease. Lancet 1991;338:182–3.
59. Bellman MH, Dick G. Surveillance of subacute sclerosing panencephalitis. Br Med J 1980; 281:393–4.
60. Stoner GL, Walker DL, Webster H deF. Age distribution of progressive multifocal leukoencephalopathy. Acta Neurol Scand 1988;78:307–12.
61. Manuelidis EE, Manuelidis L. A transmissible Creutzfeldt-Jakob disease-like agent is prevalent in the human population. Proc Natl Acad Sci USA 1993;90:7724–28.
62. Godec MS, Asher DM, Kozachuk WE, et al. Blood buffy coat from Alzheimer's disease patients and their relatives does not transmit spongiform encephalopathy to hamsters. Neurology 1994;44:1111–5.
63. Telling GC, Scott M, Hsiao KK, et al. Transmission of Creutzfeldt-Jakob disease from humans to transgenic mice expressing chimeric human-mouse prion protein. Proc Natl Acad Sci USA 1994;91:9936–40.
64. Monari L, Chen SG, Brown P, et al. Fatal familial insomnia and familial Creutzfeldt-Jakob disease: different prion proteins determined by a DNA polymorphism. Proc Natl Acad Sci USA 1994;91:2839–42.
65. Esmonde TFG, Will RG, Slattery JM, et al. Creutzfeldt-Jakob disease and blood transfusion. Lancet 1993;341:205–7.
66. The Bovine Offal (Prohibition) Regulations 1989. Statutory Instrument No. 2061. London: HMSO, 1989.
67. Alperovitch A, Brown P, Weber T, Pocchiari M, Hofman A, Will R. Incidence of Creutzfeldt-Jakob disease in Europe in 1993. Lancet 1994;343: 918.

68. Kirkwood JK, Cunningham AA. Epidemiological observations on spongiform encephalopathies in captive wild animals in the British Isles. Vet Rec 1994;135:296–303.
69. The Bovine Spongiform Encephalopathy (No. 2) (Amendment) Order 1990. Statutory Instrument No. 1930. London: Her Majesty's Stationary Office, 1990.
70. European Commission Decision 94/382 of 27 June 1994.
71. European Commission Decision 94/381 of 27 June 1994.
72. Harris DA, Falls DL, Johnson FA, Fischbach GD. A prion-like protein from chicken brain copurifies with an acetylcholine receptor-inducing activity. Proc Natl Acad Sci USA 1991;88:7664–8.

Part III

Biochemistry: Protein Chemistry, Molecular Biology, and Molecular Genetics of the Spongiform Encephalopathies

14

Structure and Biologic Characteristics of Prion Protein (Scrapie Amyloid): Implications for the Safety of Naturally Derived Biologics

Jiri Safar

Bovine spongiform encephalopathy (BSE) of cattle; scrapie of sheep and goat; encephalopathy of mink; chronic wasting disease of mule deer and elk; and kuru, Creutzfeldt-Jakob disease (CJD), Gerstmann-Sträussler-Scheinker syndrome (GSS) and fatal familial insomnia (FFI) of humans are all members of the family of transmissible, degenerative diseases of the central nervous system (CNS) (1). Due to their unusually long incubation time and unique properties, the causative agents were originally classified as "unconventional" or slow viruses. These degenerative diseases are characterized neuropathologically by (a) progressive vacuolization in the dendritic and axonal processes and neuron cell bodies, and to a lesser extent, in astrocytes and oligodendrocytes; (b) reactive gliosis; and (c) variable extracellular amyloid deposits, formed by the scrapie (CJD) isoform of prion protein, PrP^{Sc}, and/or prion protein (scrapie amyloid), PrP27-30 (1, 2).

Although all forms of the above-mentioned human diseases are very rare, with an overall annual incidence of around 1 case per million in the general population, there is a higher incidence in rural Slovakia, Chile, and among Sephardic Jews. Growing interest in this group of diseases (3) is in part due to the recent outbreak of BSE in Britain with $\geq$400 cows diagnosed per week, resulting in a threat of transmission to other mammalian species, and the fear of possible human transmission (4). Moreover, the biologically unique transmissible agent and its pathogenesis have important implications for understanding neurodegenerative disorders, aging, and possibly Alzheimer's disease (1, 5). There is a clinical and pathologic analogy between spongiform encephalopathy and Alzheimer's disease in that both are characterized by deposits of brain amyloid, and the affected patients develop progressive dementia. However, the underlying mechanisms differ significantly. Kuru, CJD, and GSS compose the transmissible brain amyloidoses, with amyloid deposits formed by the PrP protein, whereas Alzheimer's disease is not transmissible (6), and extracellular amyloid is formed by β/A4 peptide (1, 7).

Recent molecular biologic and genetic evidence has demonstrated that the abnormal isoform of the host cell-derived prion protein plays a central role in the transmission and development of spongiform encephalopathies (prion disease) (1, 2, 8). The prion protein undergoes a three-stage transition: *cellular isoform of prion protein* (PrP^C) → *scrapie (CJD) isoform of prion protein* (PrP$^{Sc/CJD}$) → *prion protein* (PrP27-30) (9). Determining the molecular mechanism of prion protein transformation is critical to understanding the transmission and pathogenesis of spongiform encephalopathies (prion diseases) (2), as well as amyloid formation (1) and the resulting neuronal degeneration (10).

In an attempt to interpret the available data, several hypotheses exist: (a) the virino hypothesis, with a conventional nucleic acid–containing agent triggering secondary amyloid formation by PrP27-30 (11); (b) the transmissible amyloidoses hypothesis, with autopatterning of the tertiary structure and nucleated polymerization of scrapie amyloid (prion protein, PrP27-30) (1); (c) the prion hypothesis, in which the normal isoform of prion protein (PrP^C) is converted into a scrapie or CJD isoform (PrP$^{Sc/CJD}$) by a change in the tertiary structure and/or different posttranslational modification and/or by a ligand, and able to replicate itself with no nucleic acid involvement (2); and (d) the "unifying hypothesis," where PrP^{Sc} would trigger conversion of PrP^C into PrP^{Sc} with the latter having an affinity for an intracellular ligand. This intracellular ligand, possibly a nucleic acid, would then be amplified and could propagate strain-specific characteristics in the next host (12).

The Proteins and the Agent

The prion protein (scrapic amyloid, PrP27-30), is formed from an abnormal isoform of prion protein, PrP^{Sc}, by limited N-terminal proteolytic cleavage. After translation, the full-length protein undergoes complex posttranslational modifications: the signal peptide is cleaved; a single disulfidic bridge between Cys179 and Cys214 is formed; two N-linked glycosylation sites at Asn181 and Asn197 are occupied and processed into highly complex sugar chains with a terminal neuraminic acid (13, 14); and the C-terminus is replaced by a glycolipid (15). The normal cellular isoform of prion protein, PrP^C, is a glycolipid-anchored, protease-sensitive membrane glycoprotein with an apparently identical molecular mass and charge, found predominantly in synaptosomal-microsomal membrane fractions of neurons in normal brain and on the surface of the hamster neuronal cells in explant culture (2).

During experimental scrapie and CJD infection, PrP^C is gradually converted into the protease-resistant PrP$^{Sc/CJD}$, which has a different membrane interaction (16, 17), physicochemical behavior (18), and potential for amyloid formation. Furthermore, PrP^{Sc}, even as a monomer, is inseparable from infectivity (14). The close association of this host-coded PrP$^{Sc/CJD}$ with infectivity is the unique and, until recently, controversial feature in the molecular biology of spongiform encephalopathies. However, there is considerable evidence for the conclusion that the prion protein (scrapie/CJD amyloid) is indeed an integral component of the scrapie or CJD infectious unit, and possibly the only one: (a) PrP^{Sc} and infectivity copurify in immunoaffinity and size-exclusion chromatography (14, 19); (b) in preparative isoelectric focusing, PrP^{Sc} has the same charge as the infectious agent (20); (c) PrP27-30 and the infectious agent have similar physicochemical behavior and subcellular distribution (18); (d) PrP27-30 is proportionate to the agent titer and may be partially neutralized by PrP-specific antibodies (14, 19); (e) denaturation, hydrolysis, selective modification and proteolysis of PrP27-30 also diminish the biologic activity (9); (f) the infectious agent replicates only in those cell clones producing PrP^{Sc} and not in PrP^{Sc}-negative cells (21); (g) the procedures modifying nucleic acids do not diminish infectivity (9); (h) several point mutations, deletions, and insertions in the human PrP gene are linked with the development of GSS, human familial forms of CJD, and fatal familial insomnia (2, 3, 8, 22); (i) the primary structure and point mutations in hamster and mouse PrP genes influence the incubation time and also partially affect the agent's biologic behavior (2); (j) the point mutation analogous to a human GSS codon 102 mutation induces in a transgenic mice disorder pathologically indistinguishable from human GSS (23), and the disease is then transmissible to nontransgenic animal lines (24); and (k) transgenic mice homozygous for the disrupted PrP^C gene are resistant to scrapie infection (25).

Molecular Biology

The gene encoding the normal cellular isoform of the prion protein, PrP^C, is found on the short arm of human chromosome 20 and chromosome 2 in mice; the second of its two exons contains the entire transcribed 253 codon region of the gene (5, 8). The gene is highly conserved and the greatest levels of expression are found in neurons where PrP messenger RNA (mRNA) is expressed about 50 times more than in glia; lower levels are found in most tissues. The PrP gene is developmentally regulated and maintains steady expression during the life span of the host (2). The precise physiologic function of the normal isoform PrP^C is unknown, and transgenic mice homozygous for the disrupted PrP^C gene develop and behave normally (26). However, hippocampal slices from transgenic mice homozygous for the disrupted PrP gene have weakened γ-aminobutyric acid type A

[GABA(A)] receptor-mediated fast inhibition and impaired long-term potentiation (27). This finding suggests that PrP^C may play a role in synaptic transmission, and the loss of normal function may be involved in the epileptiform activity seen in CJD or may contribute to the early synaptic loss and neuronal degeneration found in these diseases (27).

The serine residue at codon 231, exposed in rough endoplasmic reticulum after cleavage of the C-terminal hydrophobic peptide, is covalently modified by glycolipid (15); this moiety then solely anchors the normal PrP^C in the membranes. The protein is transported toward the cell surface through the cis- and trans-Golgi apparatus where sugar chains are processed into complex, branched types with terminal sialic acid, fucose, and galactose (13, 14); the $T_{\frac{1}{2}}$ of surface-expressed murine PrP^C in N2a glioblastoma cells is ~5.5 hours. In cloned scrapie-infected cells, the surface-expressed PrP^C is internalized, converted into PrP^{Sc}, and then probably accumulates in endosome-like vesicles with $T_{\frac{1}{2}}$ being ~15 hours (21, 28, 29).

Structure of Model Proteins and Peptides in the Solid State: Clues to Scrapie Amyloid Formation

Amyloid formation may be described in the simplest terms as a transition from a monomeric solution into the solid state of aggregates, with the morphology of typical fibrils, amorphous aggregates, or films (30). Because of the lack of data on solution → solid-state protein conformational transitions, we recently tested the possibility of whether or not aggregation by itself induced the β-sheet secondary structure. As model systems, we used three synthetic peptides corresponding to the helical, turn, or random protein structures, and three structurally well-described proteins: all-α helical myoglobin, mixed α-helical/β-sheet RNase A, and all-β-sheet structured concanavalin A. Using circular dichroism (CD) and infrared (IR) spectroscopy, together with high-resolution transmission electron microscopy (TEM), we directly compared the solution and aggregate conformations, and followed the stability of different secondary structure elements in transition from solution to aggregates in thin films (30).

The experiments demonstrate that the solid state of aggregated proteins is noncrystalline and amorphous. Surprisingly, the transition from monomer in solution into the aggregated solid state induces no or only minor changes in the protein or peptide secondary structure. However, exposure of protein films to high temperature changes the various types of secondary structure, such as turns, α-helices, and random elements into uniform β-sheets. Once the films are β-sheet structured, they are very stable, even at temperatures above 100°C. The changes are irreversible, and after cooling to room temperature the proteins retain the new β-sheet secondary structure (30).

The data provide evidence that aggregation itself does not necessarily change the secondary structure of model proteins or peptides. However, at higher temperatures, proteins and peptides in the solid state show a general conformational pathway with different elements of their secondary structure converting into β-sheets (30). Thus, any protein or peptide is potentially amyloidogenic and may be irreversibly converted into a β-sheet–structured solid state. However, in most proteins, such conformational transitions are separated by a large energy barrier, thereby preventing amyloid formation (30).

Conformation of Prion Protein (Scrapie Amyloid)

The intrinsic insolubility of PrP27-30 and PrP^{Sc} makes it very difficult to directly examine the secondary and tertiary structure of these proteins and the mechanism of their association. The solvents or detergents able to solubilize aggregated PrP^{Sc} and PrP27-30 are denaturing as judged by loss of infectivity, protease resistance, or both (14, 31, 32). The only efficient method to disperse aggregated PrP27-30 proteins, maintain protease resistance and infectivity, and prevent aggregation is to reconstitute the proteins into liposomes or detergent-lipid complexes (31). Although the existence of some lower oligomers cannot be completely dismissed, there is considerable evidence that PrP27-30 is present in liposomes as a monomer or dimer (31, 33, 34).

The CD spectrum of PrP^{Sc} incorporated in liposomes is unusual and has the characteristics of an aromatically shifted type II β-sheet band (35). The secondary structure analysis of CD spectra of PrP^{Sc} yield 34% β-sheet with less than 20% α-helix; the rest of the structure is random or in β-turns. When PrP^{Sc} protein is allowed to fully aggregate into the solid state in thin films, the perturbation in the CD spectrum indicates the presence of a secondary structure with less α-helix and more β-sheets (35).

The amino-terminal region of PrP^{Sc}, aggregated or dispersed in liposomes, is accessible and sensitive to limited proteinase K cleavage with no change in infectivity. The remaining proteinase K-resistant carboxyl-terminal core, PrP27-30, when incorporated into liposomes, is an all β-structure (43%) with no α-helix, and with 57% of the remainder of the molecule present in β-turns or random structure (35). The β-sheet content agrees well with the IR spectroscopy (47% and 54%) data obtained for PrP27-30 (36, 37). After complete aggregation in thin films, the amplitude of the CD spectrum of PrP27-30 reproducibly increases two- to threefold when compared to liposomes, indicating a higher β-sheet content (35). This can be only the result of a conversion of secondary structure from random or β-turn elements into β-sheets, or an intermolecular alignment of external β-sheet in thin films, or both.

Our data provide direct in vitro experimental evidence for the β-sheet assembly of PrP^{Sc} and PrP27-30 into amyloid occurring in vivo in plaques and for the importance of the N-terminal cleavage for correct alignment into amyloid. Surprisingly, the polymerization does not change infectivity, and despite difficulties with dispersion of aggregates, most is recovered in a biologically active form (35). In contrast, the increase in infectivity of PrP27-30 incorporated into liposomes probably reflects the efficient dissociation of PrP27-30 aggregates; a similar observation was previously made with different phospholipids (19, 31, 33).

The secondary structure of PrP^{Sc} deduced from CD spectra (35) has a higher β-sheet and lower α-helix content compared with PrP^{C} (38). Therefore, it is possible that amyloid formation and disease transmission result solely from a conformational transition in the PrP monomer, and it is essential to understand the initial step of this structural transformation. The resulting polymerization and proteolytic processing into mature amyloid is a consequence of that first critical step, analogous to the unexplained "seeding" in the nucleated polymerization hypotheses (1, 39). The transition may be apparently triggered endogenously by point mutations or insertions in the PrP protein sequence, or exogenously by PrP^{Sc} or PrP27-30. The phenomenon of prion protein (scrapie amyloid) transformation adds an additional level of complexity to the general problem of determining the mechanism by which a specific amino acid sequence directs the folding into an unique conformation (40, 41).

The Folding Intermediate Concept of Prion Protein (Scrapie Amyloid) Formation

Our recent spectroscopic conformational data on dissociation and unfolding of PrP^{Sc}, and in more detail on its proteinase-resistant core, PrP27-30, demonstrated that both proteins fold through monomeric conformational intermediates (35, 42), and established the folding intermediate (framework) concept of PrP protein folding (43). The conformational intermediates of PrP27-30 and PrP^{Sc} are critical because (a) they are the direct thermodynamic and conformational precursors of the scrapie amyloid aggregate; and (b) comparison of the monomeric intermediates and PrP27-30 aggregate provides essential information about the stability, mechanism of association, and accompanying conformational transitions (41, 42).

Advances in fluorescence, CD, and nuclear magnetic resonance (NMR) spectroscopy, with the capability of monitoring changes in protein conformation occurring in vitro in a millisecond (ms) time scale and during equilibrium unfolding, have dramatically affected our understanding of how proteins fold and unfold (40, 41). Current data available from a variety of proteins demonstrate that they do not fold in a single step, but rather through distinct conformational intermediates with nonnative, nondenatured secondary and tertiary structures [vectorial "framework" concept, for review see (40, 41)]. During the first step, in less than 5 ms, protein molecules in many different unfolded states collapse into more compact structures, forming a few discrete populations of molecules with similar elements of secondary structure and nonpolar surfaces. The next phase, lasting 5 to 1,000 ms, involves formation of a single population of molecules with a native-like secondary structure, but still consisting of the unstable tertiary structure [molten globule intermediate (41, 44)]. Finally, in a phase that may last hours, the concerted formation of many noncovalent interactions throughout the protein molecules establish the native tertiary and secondary structure (41).

Kinetic and thermodynamic laws require PrP^C and PrP^{Sc} to undergo an analogous stepwise conformational transition in vivo (40, 41). Therefore, the critical problems of scrapie amyloid (prion) protein formation are (a) determining the intermediate steps in the folding pathway of PrP^C; (b) specifying the intermediate steps of the alternative folding pathway of PrP^{Sc}; and finally (c) explaining how both pathways are conformationally and thermodynamically linked to establish the critical crossroad (42, 43). Such detailed knowledge is essential to better understand the pathogenesis of the disease and will help to introduce new concepts for rational pharmacologic and gene therapy (41, 45).

The conformation of the prion protein folding intermediate, dissociated from aggregates at low concentrations of guanidine hydrochloride (Gdn HCl) or by decreasing pH, meets the definition of a molten globule folding intermediate (35, 42). This state is structurally and thermodynamically distinct from both the native and denatured states of the protein. It is defined as a compact, partly ordered conformation with nonspecific tertiary structure, high secondary structure content, significant structural flexibility, and an intrinsic tendency to aggregate as a result of a more exposed hydrophobic surface (41, 46).

The molten globule state of any protein is significantly less stable than its native form (46). This lower stability of the molten globule conformation could provide a thermodynamic "tunnel" for the transition from a largely α-helical PrP^C into a more β-structured PrP^{Sc} and aggregates. This phenomenon may be an important mechanism to facilitate PrP^C → PrP^{Sc} conversion because such a process is separated by a large energy barrier in model proteins and peptides (35, 43). The hypothetical ligand or binding with chaperonins could stabilize PrP^{Sc} in the molten globule form, thereby further favorably diminishing the energy barrier.

The complexity and more than two-state equilibrium dissociation and unfolding of PrP^{Sc} and PrP27-30 precluded precise calculation of the stability (ΔG^{app}) of the proteins (35, 42). However, the midpoints of the conformational transitions of each intermediate are proportionate to their stability. By analyzing the transition midpoints, it is apparent that the PrP^{Sc} or PrP27-30 aggregates are significantly more stable than the molten globule conformation (35), and therefore may represent a kinetic trap for normally monomeric and metastable conformational intermediates. Additional thermodynamic stabilization with a concurrent increase in β-sheet content in the secondary structure is achieved by association and proteolytic cleavage of the N-terminal regions of PrP^{Sc} with realignment of PrP27-30 into mature amyloid (35). Amino terminal cleavage is an important additional irreversible step with the resultant aggregates being more stable than those of PrP^{Sc} (35). Thermodynamic stability of the protein is closely related to the turnover rate (40), and the high stability of PrP^{Sc} and PrP27-30 aggregates alone may account for their accumulation.

The conformations, defined as molten globule intermediates, exist either as kinetic protein folding intermediates (41) or equilibrium intermediates (46). The compact early kinetic intermediates of protein folding pathways occur during the early folding phases (47), and are probably the universal step in protein folding. The conformationally similar metastable protein forms appear in equilibrium in some apoproteins, or are stabilized at equilibrium by low pH (A-forms) or by interaction with chaperonins (42). Which mechanism is involved in formation of the monomeric molten globule intermediate of prion protein and result in aggregation and amyloid formation in vivo is the critical question in the pathogenesis of the disease.

Conformational Links Between Prion Protein and Infectivity

Heat Denaturation

An important characteristic of agents causing spongiform encephalopathies (prion disease) is their exceptional thermal resistance (1, 48–51). This thermoresistance is very difficult to explain, and, in the case of the "protein only" prion model (2), it implies reversibility of the protein thermal unfolding or an absence of unfolding at high temperatures. Our experiments with purified PrP27-30 in thin films exposed to dry heat for 30 min indicate no change of infectivity at 100°C and only a minor drop at 132°C (52). At the same time, the secondary structure of the PrP27-30 is heat resistant in the solid state with no significant changes after exposure to 132°C (52). The secondary structure of PrP27-30 in the solid state apparently has the necessary thermal stability to be a carrier of infectivity. The heat stability of the β-sheet–structured prion protein (scrapie amyloid) may be explained by the higher thermodynamic stability of the β-sheets and absence of free water in the solid state (53).

Our experiments have elucidated a major source of controversy found in many scrapie or CJD heat inactivation studies. It is apparent that the level of hydration and aggregation of prion protein is critical for heat inactivation, and small differences in the experimental conditions related to the physical state of the sample, hydration, and dissociation will vastly affect the results.

Effect of Solvents

The conformation of any protein is not static, but rather is controlled by the dynamic relationship of enthalpy and entropy of the protein's amino acid sequences and the solvent (53). As a result, various solvents may induce conformations different from that found in water. When a protein is exposed to a different solvent in the solid state, it retains the new conformation even after eliminating the solvent in a vacuum (solvent "memory" effect) (30). This enables the secondary structure of the protein to change into a new and stable one. Because PrP27-30 reconstituted from dry films is infectious, we used this phenomenon to determine the impact of organic solvents on the structure of PrP27-30 in the solid state, and to correlate the resultant various stable conformations with the film's infectivity (52).

Solvent replacement studies of the PrP27-30 secondary structure in the solid state followed by CD spectroscopy indicate uniform conversion of native β-pleated sheet-like secondary and tertiary structure into α-helices (52). The strongest α-helix inducer was formic acid, a known inactivator of scrapie infectivity (54). The structural effect of formic acid may be explained by formation of temporary hydrogen bonds in the protein's secondary structure. During formic acid evaporation and in the absence of water, the formic acid hydrogen bonds are replaced by intrachain hydrogen bonds forming an α-helix in a zipper-like manner (37). Simulaneously, the PrP27-30 film loses infectivity. Trifluoroacetic acid and fluorinated alcohols include a smaller degree of β-to-α secondary and tertiary structure transitions of PrP27-30, and proportionally diminish the film's residual infectivity (52). Surprisingly, inactivation of infectivity in the solid state is not associated with an anticipated denatured state, but rather with transformation of β-sheet into α-helices (52).

Effect of Chemical Denaturants

When PrP^{Sc} or PrP27-30 is exposed to different concentrations of chemical denaturants, such as urea (32) or guanidine hydrochloride (Gdn HCl) (32, 35, 42), the effects strictly depend on the final concentration of the denaturant (35). At a low concentration of Gdn HCl, the aggregates of both proteins dissociate into partially folded, "molten globule" monomers (see above); the secondary structure of monomers partially unfold at an intermediate Gdn HCl concentration, and fully unfold at high Gdn HCl concentrations. The response of the prion protein to the increasing concentration of Gdn HCl is best described as $A \leftrightarrow I_1 \leftrightarrow I_2 \leftrightarrow U$, where A is an aggregate, I_1 and I_2 equilibrium intermediates, and U the completely unfolded conformation of the protein (35, 42). The scrapie amyloid dissociated intermediate I_1 has a substantial portion of ordered secondary structure and a tertiary structure close to the unfolded form (see above) (35, 42). The dissociation of PrP27-30 is fully reversible from up to 2.5 M Gdn HCl, as judged by return of native-like CD spectrum and rapid association. The complete reversibility of dissociation and simultaneous conformational changes of PrP27-30 coincide with full recovery of infectivity following dilution of the Gdn HCl (35, 42). In contrast, the irreversible loss of infectivity at ≥2.5 M Gdn HCl correlates with irreversible unfolding of PrP27-30 with rapid formation of amorphous misfolded aggregates upon dilution (35).

A similar dependency of infectivity on native conformation of prion proteins is observed by combining the effect of low concentrations of Gdn HCl and heat (42). When the infectious preparations of PrP27-30 in 2 M Gdn HCl are gradually heated up to 90°C, the protein undergoes noncooperative unfolding. The process is irreversible, and after cooling the protein solution back to room temperature, diluting Gdn HCl, or both, the protein forms misfolded aggregates with nonnative conformation (42). Simultaneously, the preparations of PrP27-30 lose all infectivity (42).

Our experiments provide strong evidence for a structure-function relationship between the native, β-pleated sheet-like secondary and tertiary structure of the prion protein, and infectivity. Another interpretation is that the observed changes in prion protein conformation parallel the conformational changes of a still undetected subfraction of prion protein or scrapie agent responsible for scrapie transmission. However, this is only speculative, and it is therefore possible that the β-pleated sheet secondary and specific tertiary structure of the PrP protein is solely responsible for its propagation and amplification. An equally plausible explanation is that the native, β-sheet–structured conformation of PrP27-30 is critical for ligand binding, and only the entire complex can propagate, amplify, and form amyloid. The proposed ligand does not diminish the key role of the prion protein structure, because any alterations in PrP27-30 conformation result in modified ligand binding (55), and therefore in different infectivity. The biologic data on various "strains" of infectivity isolated from animals with an identical primary sequence of PrP protein seem to support the ligand concept (12), or alternatively, different posttranslational modification and targeting (2, 12). The "protein-only" model is difficult to reconcile with current models of protein folding and thermodynamics (41) because it infers the existence of multiple stable conformations of the protein with the same primary sequence under the same conditions, able to specifically replicate from passage-to-passage in vivo.

Acknowledgment. The author thanks Devera G. Schoenberg, MS, for editing the manuscript.

References

1. Gajdusek DC. Transmissible and nontransmissible dementias: distinction between primary cause and pathogenetic mechanism in Alzheimer's disease and aging. Mt Sinai J Med 1988;55:3–5.
2. Prusiner SB. Molecular biology and genetics of prion diseases. Philos Trans R Soc Lond [Biol] 1994;343:447–63.
3. Brown P, Goldfarb LG, Gajdusek DC. The new biology of spongiform encephalopathy: infectious amyloidoses with a genetic twist. Lancet 1991; 337:1019–22.
4. Bradley R, Wilesmith JW. Epidemiology and control of bovine spongiform encephalopathy (BSE). Br Med Bull 1993;49:932–59.
5. Prusiner SB. Genetic and infectious prion diseases. Arch Neurol 1993;50: 1129–53.
6. Godec MS, Asher DM, Kozachuk WE, et al. Blood buffy coat from Alzheimer's disease patients and their relatives does not transmit spongiform encephalopathy to hamsters. Neurology 1994;44:1111–5.
7. Glenner GG, Murphy MA. Amyloidosis of the nervous system. J Neurol Sci 1989;94:1–28.
8. Prusiner SB, Hsiao KK. Human prion diseases. Ann Neurol 1994;35:385–95.
9. Prusiner SB. Chemistry and biology of prions. Biochemistry 1992;31:12277–88.
10. Safar J. Infectious amyloid, prions, unconventional viruses, and disease. Neurobiol Aging 1994;15:279–81.
11. Czub M, Braig HR, Diringer H. Pathogenesis of scrapie: study of the temporal development of clinical symptoms, of infectivity titers and scrapie-associated fibrils in brains of hamsters infected intraperitoneally. J Gen Virol 1986;67:2005–9.
12. Weissmann C. A "unified theory" of prion propagation. Nature 1991;352: 679–83.
13. Endo T, Groth D, Prusiner SB, Kobata A. Diversity of oligosaccharide structures linked to asparagines of the scrapie prion protein. Biochemistry 1989;28:8380–8.
14. Safar J, Wang W, Padgett MP, et al. Molecular mass, biochemical composition, and physicochemical behavior of the infectious form of the scrapie precursor protein monomer. Proc Natl Acad Sci USA 1990;87:6373–7.
15. Stahl N, Borchelt DR, Hsiao K, Prusiner SB. Scrapie prion protein contains a phosphatidylinositol glycolipid. Cell 1987;51:229–40.
16. Safar J, Ceroni M, Gajdusek DC, Gibbs CJ Jr. Differences in the membrane interaction of scrapie amyloid precursor proteins in normal and scrapie- or Creutzfeldt-Jakob disease-infected brains. J Infect Dis 1991;163:488–94.
17. Stahl N, Borchelt DR, Prusiner SB. Differential release of cellular and scrapie prion proteins from cellular membranes by phosphatidylinositol-specific phospholipase C. Biochemistry 1990;29:5405–12.
18. Safar J, Ceroni M, Piccardo P, et al. Subcellular distribution and physicochemical properties of scrapie-associated precursor protein and relationship with scrapie agent. Neurology 1990;40:503–8.
19. Gabizon R, McKinley MP, Groth D, Prusiner SB. Immunoaffinity purification and neutralization of scrapie prions. Proc Natl Acad Sci USA 1988;85:6617–21.

20. Ceroni M, Piccardo P, Safar J, Gajdusek DC, Gibbs Jr CJ. Scrapie infectivity and prion protein are distributed in the same pH range in agarose isoelectric focusing. Neurology 1990;40:508–13.

21. Taraboulos A, Raeber AJ, Borchelt DR, Serban D, Prusiner SB. Synthesis and trafficking of prion proteins in cultured cells. Mol Biol Cell 1992;3:851–63.

22. Prusiner SB. Inherited prion diseases. Proc Natl Acad Sci USA 1994;91:4611–4.

23. Hsiao KK, Scott M, Foster D, Groth DF, DeArmond SJ, Prusiner SB. Spontaneous neurodegeneration in transgenic mice with mutant prion protein. Science 1990;250:1587–90.

24. Hsiao K, Groth D, Scott M, et al. Serial transmission in rodents of neurodegeneration from transgenic mice expressing mutant prion protein. Proc Natl Acad Sci USA 1994;91:9126–30.

25. Bueler H, Aguzzi A, Sailer A, et al. Mice devoid of PrP are resistant to scrapie. Cell 1993;73:1339–47.

26. Bueler H, Fischer M, Lang Y, et al. Normal development and behaviour of mice lacking the neuronal cell-surface PrP protein. Nature 1992;356:577–82.

27. Collinge J, Whittington MA, Sidle KCL, et al. Prion protein is necessary for normal synaptic function. Nature 1994;370:295–7.

28. Caughey B, Race RE, Ernst D, Buchmeier MJ, Chesebro B. Prion protein biosynthesis in scrapie-infected and uninfected neuroblastoma cells. J Virol 1989;63:175–81.

29. Caughey B, Neary K, Buller R, et al. Normal and scrapie-associated forms of prion protein differ in their sensitivities to phospholipase and proteases in intact neuroblastoma cells. J Virol 1990;64:1093–101.

30. Safar J, Roller PP, Ruben GC, Gajdusek DC, Gibbs CJ Jr. Secondary structure of proteins associated in thin films. Biopolymers 1993;33:1461–76.

31. Gabizon R, Prusiner SB. Prion liposomes. Biochem J 1990;266:1–14.

32. Prusiner SB, Groth D, Serban A, Stahl N, Gabizon R. Attempts to restore scrapie prion infectivity after exposure to protein denaturants. Proc Natl Acad Sci USA 1993;90:2793–7.

33. Gabizon R, McKinley MP, Prusiner SB. Purified prion proteins and scrapie infectivity copartition into liposomes. Proc Natl Acad Sci USA 1987;84:4017–21.

34. Bellinger KC, Kempner E, Groth D, Gabizon R, Prusiner SB. Scrapie prion liposomes and rods exhibit target sizes of 55,000 Da. Virology 1988;164:537–41.

35. Safar J, Roller PP, Gajdusek DC, Gibbs CJ Jr. Conformational transitions, dissociation, and unfolding of scrapie amyloid (prion) protein. J Biol Chem 1993;268:20276–84.

36. Caughey BW, Dong A, Bhat KS, Ernst D, Hayes SF, Caughey WS. Secondary structure analysis of the scrapie-associated protein PrP27-30 in water by infrared spectroscopy. Biochemistry 1991;30:7672–80.

37. Gasset M, Baldwin MA, Fletterick RJ, Prusiner SB. Perturbation of the secondary structure of the scrapie prion protein under conditions that alter infectivity. Proc Natl Acad Sci USA 1993;90:1–5.

38. Pan K-M, Baldwin M, Nguyen J, et al. Conversion of α-helices into β-sheets features in the formation of scrapie prion proteins. Proc Natl Acad Sci USA 1993;90:10962–6.

39. Come JH, Fraser PE, Lansbury PT. A kinetic model for amyloid formation in the prion disease—importance of seeding. Proc Natl Acad Sci USA 1993; 90:5959–63.

40. Jaenicke R. Protein folding: local structures, domains, subunits, and assemblies. Biochemistry 1991;30:3147–61.
41. Matthews CR. Pathways of protein folding. Annu Rev Biochem 1993;62:653–83.
42. Safar J, Roller PP, Gajdusek DC, Gibbs CJ Jr. Scrapie amyloid (prion) protein has the conformational characteristics of an aggregated molten globule folding intermediate. Biochemicstry 1994;33:8375–83.
43. Safar J, Roller PP, Ruben GC, Gajdusek DC, Gibbs CJ Jr. Conformational pathways of scrapie amyloid (prion) protein and the structure-function relationship with infectivity. In: Iqbal K, Mortimer JA, Winblad B, and Wisniewski HM, eds. Research advances in Alzheimer's disease and related disorders. London: John Wiley & Sons, 1995:775–81.
44. Ptitsyn OB, Pain RH, Semitsonov GV, Zerovnik E, Razgulyaev OI. Evidence for a molten globule state as a general intermediate in protein folding. FEBS Lett 1990;262:20–4.
45. Matthews BW. Structural and genetic analysis of protein stability. Annu Rev Biochem 1993;62:139–60.
46. Kuwajima K. The molten globule state as a clue for understanding the folding and cooperativity of globular-protein structure. Proteins: Struct Funct Genet 1989;6:87–103.
47. Garvey EP, Matthews CR. Effects of multiple replacements at a single position on the folding and stability of dihydrofolate reductase from *Escherichia coli*. Biochemistry 1989;28:2083–93.
48. Brown P, Rohwer RG, Gajdusek DC. Newer data on the inactivation of scrapie virus or Creutzfeldt-Jakob disease virus in brain tissue. J Infect Dis 1986; 153:1145–8.
49. Rohwer RG. Scrapie: Virus-like size and virus-like susceptibility to inactivation of the infectious agent. Nature 1984;308:658–62.
50. Rohwer RG. Scrapie shows a virus-like sensitivity to heat inactivation. Science 1984;223:600–2.
51. Rohwer RG. Estimation of scrapie nucleic acid MW from standard curves for virus sensitivity to ionizing radiation. Nature 1986;320:381.
52. Safar J, Roller PP, Gajdusek DC, Gibbs CJ Jr. Thermal stability and conformational transitions of scrapie amyloid (prion) protein correlate with infectivity. Protein Sci 1993;2:2206–16.
53. Schulz GE, Schirmer RH. Principles of protein structure. New York, Berlin, Heidelberg, Tokyo: Springer-Verlag, 1979:314.
54. Brown P, Wolff A, Gajdusek DC. A simple and effective method for inactivating virus infectivity in formalin-fixed tissue samples from patients with Creutzfeldt-Jakob disease. Neurology 1990;40:887–90.
55. Jaenicke R. Folding and association of proteins. Prog Biophys Mol Biol 1987;49:117–237.

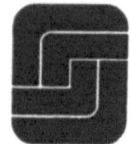

15

Prion Strains and Neuromuscular Disease in PrP Transgenic Mice

GEORGE A. CARLSON

Experimental scrapie in goats provided the first suggestion of prion strains (1). Two distinct constellations of clinical signs, labeled nervous or drowsy, bred true through subsequent intracerebral inoculations, demonstrating the existence of strains or variants of scrapie agent. Subsequent transmissions of both goat and sheep scrapie into experimental rodents led to isolation of multiple scrapie "strains" (2). Scrapie isolates that exhibited distinct and stable properties in the same inbred mouse strain provided compelling evidence for agent-specified information, thought to be encoded by a poly-nucleotide (3). The existence of microbiologic strains of agent continues to be presented as an argument against the concept of a proteinaceous infectious particle devoid of functional nucleic acid. However, the only known functional component of prions is PrP^{Sc}, a disease-specific, posttranslational derivative of the normal prion protein isoform, PrP^{C} (4). Beta-pleated sheet secondary structure predominates in PrP^{Sc}, with little, if any, found in PrP^{C}, and the two isoforms may differ only in their conformations (5, 6). PrP^{Sc} is necessary, and probably sufficient, for transmission of scrapie. No scrapie-specific nucleic acid has ever been detected in purified prion preparations, and to be present at one molecule per infectious unit, a prion-specific nucleic acid would need to have fewer than 50 nucleotides (7). Properties of viral, bacterial, and fungal strains reflect their nucleic acid genomes, with the interesting exception of prion-like epigenetic phenomena in yeast (8). Because prions are devoid of functional nucleic acid, other mechanisms must account for the existence of strains.

Incubation Time Differences Among Mouse Prion Strains Result from PrP^C:PrP^{Sc} Interactions

Prion Protein Genotype Determines Scrapie Incubation Time

Although the isolation and characterization of most mouse scrapie isolates predates the discovery of PrP (9) and its chromosomal *Prnp* gene (10, 11), incubation time profiles in mice with *Prnp^a*, *Prnp^b*, or *Prnp^a/Prnp^b* genotypes were used to define scrapie strains (12). The predominant influence of this single gene, previously known as *Sinc*, on mouse scrapie incubation time was described over 30 years ago (13, 14), but the gene's chromosomal location was unknown until demonstration of linkage with the prion protein gene (15). It is now known that *Prnp* itself, rather than a linked locus, controls scrapie incubation period. The *a* and *b* alleles of *Prnp* differ at codons 108 (Leu/Phe) and 189 (Thr/Val) (16). We borrow the term *allotype* from immunogenetic nomenclature to distinguish between the PrP-A and PrP-B proteins encoded by the *a* and *b* alleles of *Prnp*, reserving the term *isoform* to distinguish between PrP^C and PrP^{Sc}. It is important to stress that the influence of each *Prnp* allele on prion incubation time depends on the scrapie strain that is inoculated. Prolongation of incubation times for the Rocky Mountain Lab (RML) isolate by the *b* allele is dominant with *a/a* homozygotes becoming ill ≤150 days after inoculation, while *a/b* and *b/b* mice show no signs of disease until at least 200 days. In contrast, the *a* allele appears to be overdominant in prolonging 22A incubation period; *Prnp^b* homozygous mice have shorter incubation times (approximately 200 days) than *a* homozygous animals (roughly 400 days). F1 hybrids have longer 22A incubation times (~500 days) than either parent. Such isolate-specific patterns of allelic interaction with the prion incubation time gene were taken as evidence that the scrapie agent had a genome independent of the host (3, 17). Sheep also show similar scrapie strain-dependent differences in the effects of particular PrP gene alleles on incubation time (18–20).

Host-Directed Changes in Prion Properties

Passage history can have a profound influence on prion strain characteristics. As noted above, 22A has a relatively short 200-day incubation time in mice homozygous for the *b* allele of *Prnp*, while *Prnp^a* mice have intervals of ~400 days between inoculation and illness. Although this incubation time profile is stable if 22A is maintained in *Prnp^b* homozygotes, only PrP^{Sc}-B 22A prions produce long incubation times in *Prnp^a* mice. Incubation time rapidly shortens through subsequent passages through mice expressing PrP-A (3). If scrapie were due to infection by a virus, the shortening of the incubation period might reflect host selection for more rapidly replicating mutants present in the 22A isolate. An explanation invoking preferential interaction between homologous PrP allotypes is more viable.

The first suggestion of an effect of PrP allotype matching was the shorter incubation time in $Prnp^b$ homozygous mice produced by RML PrP^{Sc}-B prions (passed through mice homozygous for the b allele of $Prnp$) than by RML prions comprised of PrP^{Sc}-A (21). The interval between inoculation and illness for homologous prions was reduced by over 100 days (from more than 300 to approximately 200 days). It is important to note, however, that RML prion incubation periods in $Prnp^b$ mice never became as short as in a/a homozygotes regardless of passage history. The species barrier to scrapie transmission is a more dramatic demonstration of the effect of PrP homology between donor and host (22). Only a small percentage of mice inoculated with Syrian hamster (sHa) prions become ill, and only after 500 days or more. Expression of PrP transgenes from Syrian hamsters overcomes the species barrier with incubation times as short as 50 days in high copy number mice (23, 24). Transgenic mice expressing sHaPrP inoculated with sHa prions produce only sHa prions, while mouse-derived prions elicit production of mouse but not sHa scrapie agent. Clearly, interaction between PrP^C and PrP^{Sc} is an essential feature of prion replication.

Prion Strain Properties Reflect Interactions Between PrPC and PrP^{Sc}

Freedom from nature's constraint of three possible combinations for two alleles was provided by the availability of Tg and gene-ablated mice, and spurred new approaches to understanding prion strains and reevaluation of old concepts. For more than 30 years, the term *dominant* was used to describe the effects of the b allele of $Prnp$ in prolonging scrapie incubation time (14, 15). There also was strong genetic evidence that $Prnp$, rather than a distinct locus, controlled scrapie incubation period (16, 25, 26). For these reasons, we expected that expression of $Prnp^b$ transgene would prolong the RML prion incubation time of $Prnp^a$ homozygous mice. To our surprise, Tg($Prnp^b$) mice had shorter incubation times than their non-Tg litter mates (27). One interpretation of this result was the possibility of a distinct incubation time locus, not included within the $Prnp$ cosmid insert used to produce the mice. To test the hypothesis of an incubation time locus genetically distinct from $Prnp$, we determined the influence of an "authentic" $Prnp^b$ allele (or its tightly linked incubation time locus) on the RML isolate incubation time in Tg15 mice. Tg($Prnp^b$)15 mice have three copies of the $Prnp^b$-containing cosmid insert derived from I/LnJ mice. Tg15 mice hemizygous for the transgene array were crossed with B6.I-1 (B6.I-$Prnp^b$/Co) congenic mice and inoculated with RML isolate. B6.I mice have long RML incubation times and are homozygous for I/LnJ-derived alleles of $Prnp$ and tightly linked genes (26); other genes are derived from the short incubation time B6 strain. All offspring of this cross are *Prnp* heterozygous, with half expected to be transgene-negative and half carrying the three copy transgene array. *Prnp^a* homozygous mice with the transgene array had incubation times of 115 ± 3 d compared to 144 ± 5 d for transgene negative *Prnp^a* mice. In comparison, transgene-positive *Prnp^a*/*Prnp^b* heterozygous mice had incubation times of 166 ± 2 d, not as long as "authentic" *Prnp^a*/*Prnp^b* mice but a significantly longer than transgene positive *Prnp^a* mice.

Our recent reevaluation of the genetic control of scrapie incubation time provides an explanation for these results. Incubation time is inversely proportional to the amount of PrP^C expressed. Mice carrying one normal *Prnp* allele and one nonfunctional allele ablated through homologous recombination have very long incubation times. The apparent dominance of the *b* allele in prolonging RML incubation time in *Prnp* heterozygous mice may actually be due to the reduced amount of PrP^C produced from a single copy of *Prnp^a*. Based on these and similar results, we suggest that RML prions convert PrP^C-A into PrP^Sc-A more rapidly than they convert PrP^C-B to PrP^Sc-B. *Prnp^b* transgene expression in *Prnp^a* homozygous mice would not be expected to affect the supply of PrP^C-A, and additional PrP^C, even the less efficiently converted PrP^C-B allotype, would only shorten incubation time. To summarize, the supply of PrP^C-A rather than an active dominant effect of PrP^C-B, is the primary determinant of RML incubation time (28, 29).

The 22A scrapie isolate produces disease more rapidly in *Prnp^b* mice than in *Prnp^a* animals. In further contrast with RML and similar common isolates, *Prnp* heterozygous mice have longer incubation times than either parent as illustrated by incubation times in B6 (405 ± 2 d), B6.I (194 ± 10 d) and their F1 hybrid (508 ± 14 d). Incubation times in Tg15 mice revealed that increasing the amount of PrP-B shortened incubation time. 22A was inoculated into mice carrying one (+/0) or two (+/+) copies of the *Prnp^b* transgene array from Tg15 in addition to the two endogenous copies of the *Prnp^a* allele. Hemizygous Tg15(+/0) mice had much shorter incubation times than "authentic" *Prnp^a*/*Prnp^b* heterozygotes (395 ± 12 d vs. 508 ± 14 d), and incubation times were further shortened by transgene homozygosity (286 ± 15 d). Incubation times of Tg15 (+/+) mice with six copies of *Prnp^b* and two copies of *Prnp^a* remained longer than those of B6.I mice with two copies of *Prnp^b*, however. This indicates that prolongation of 22A incubation time is a dominant, positive result of *Prnp^a* expression.

Even though results using Tg15 mice to dissect 22A isolate properties are dramatically different from those obtained with the RML isolate, they also can be explained by allotype-preferences in PrP^C-PrP^Sc interactions. PrP^Sc in 22A prions converts PrP^C-B to PrP^Sc more efficiently than it does PrP^C-A, resulting in shorter incubation times for B6.I than B6 mice. A telling feature of the incubation time profile for 22A is the overdominance of the *a* allele of *Prnp* in prolonging incubation period. Dickinson and Outram (30) proposed a replication site hypothesis to account for overdominance, postulating that dimers of the *Sinc* gene product feature in the replication of the scrapie agent. This model assumes that PrP^C-B dimers are more readily converted to PrP^{Sc} than are PrP^C-A dimers and that PrP^C-A:PrP^C-B heterodimers would be very resistant to conversion to PrP^{Sc}. Increasing the ratio of PrP-B to PrP-A would shorten incubation times by favoring the formation of PrP^C-B homodimers. A homologous mechanism could account for the relative rarity of individuals heterozygous for the met/val polymorphism at codon 129 of the human PrP gene in spontaneous Creutzfeldt-Jakob disease (CJD) (31).

An alternative model breaks PrP^C-PrP^{Sc} interactions into two distinct aspects, binding affinity and conversion rate. A higher affinity of 22A PrP^{Sc} for PrP^C-A than PrP^C-B, with inefficient conversion of PrP^{Sc}-A could account for the exceptionally long incubation time of *Prnp* heterozygotes. Under this scenario, the reduction in the supply of 22A prions available for interaction with the PrP^C-B product of the single *Prnp^b* allele would slow disease progression. Results using the 87V prion isolate support the possibility of high affinity PrP^C-PrP^{Sc} interactions in the absence of conversion to PrP^{Sc} and favor the possibility that binding affinities and rate constants may be sufficient to explain incubation time profiles of prion strains. Only *Prnp^b* mice became uniformly ill after *Prnp^a*, *Prnp^b*, *Prnp^a/Prnp^b* and Tg15 mice were inoculated with 87V prions. Only a single *Prnp^a* mouse out of 31 inoculated developed scrapie. In contrast to the results with other scrapie isolates, expression of the three copy *Prnp^b* transgene array did not alter scrapie susceptibility. The postulate of a high-binding affinity of PrP^C-A for 87V PrP^{Sc}-B without conversion to PrP^{Sc}-A is sufficient to explain these results.

Although the amino acid sequence of host PrP^C places major constraints on disease manifestation and incubation time, additional levels of complexity, either higher order structure or cofactors, are necessary for prion diversity. For example, both PrP^{Sc}-A and PrP^{Sc}-B can "encode" the RML prion-specified property of shorter incubation times in *Prnp^a* mice than in *Prnp^b* animals. In other words, even though interactions between homologous allotypes favor shorter incubation times (21), the incubation time profile and induced pathologic changes are stable regardless of passage history. In contrast, *Prnp^a* mice appear unable to replicate the properties specified by 22A or 87V prions, suggesting that PrP-B is able to adopt conformations or bind cofactors not available to PrP-A.

Spontaneous Disease in Mice Overexpressing Wild-Type Prion Protein Transgenes

As illustrated above, Tg mice have provided crucial information on the nature of prion replication in experimental scrapie. The additional finding that uninoculated Tg with high copy numbers of wild-type (wt) PrP genes succumbed to spontaneous neurologic disease has important implications for sporadic prion disease (32). Disease occurred in mice carrying high copy numbers of SHaPrP, Mouse (Mo) PrP-B, or Sheep (She) PrP transgenes. Within each Tg line, there were a variety of clinical signs whose spectrum overlapped those seen in other lines. For example, Tg(SHaPrP+/+)7 mice (+/+ indicates homozygosity for the transgene array) developed progressive disease presenting with tremors, ataxia, head-bobbing, and an abnormal hunched posture, while the majority of Tg(Prn-p^b+/−)94 mice (+/− indicates hemizygosity for the transgene array) presented with hind-limb paresis progressing to paralysis. It is not known whether these differences reflect the transgene or the genetic background of the host.

Age at onset of disease decreases as the amount of PrP^C increases. The comparison between homozygous Tg(SHaPrP+/+)7 and hemizygous Tg(SHaPrP+/−)7 mice is particularly helpful. Homozygous Tg7 mice have ~120 copies of SHaPrP and develop disease at 468 ± 8 (SEM) days and die at 546 ± 5 days. In contrast, no hemizygous Tg7 animals with ~60 copies of the transgene array showed any sign of neurologic illness before 650 days of age. Homozygous Tg7 mice express 367 ± 34 µg of SHaPrP per gram brain protein, while hemizygous Tg7 animals had 195 ± 12 µg/g. No clinical disease was evident at similar ages in Tg(SHaPrP)71 and Tg(SHaPrP)81 mice that expressed 128 ± 12 and 56 ± 7 µg/g of the hamster transgene. A similar relationship is seen in mice expressing mouse PrP-B, with disease occurring only in high copy number lines.

Tg mice with spontaneous neurologic dysfunction exhibited changes in the CNS similar to scrapie with focal spongiform degeneration of the gray matter, localized to the stratum lacunosum moleculare of the hippocampus, the superior colliculus, and midbrain tegmentum. Mild astrogliosis accompanied the spongiform change. These changes were more focal and much less intense than those found in mice inoculated with rodent prion isolates and seemed insufficient to account for the profound clinical signs. The hind-limb paresis in Tg94 and gait abnormalities in Tg7 led to examination of the spinal cord, sciatic nerve, and muscle. Although no obvious pathology in spinal cord was noted, dramatic pathologic changes were seen in both skeletal muscle and peripheral nerve that were not a consequence of normal aging. Quadriceps muscle from affected transgenics had scattered degenerating fibers, increased numbers of fibers with central nuclei, active phagocytosis of fibers, and large variation in fiber size. Similar changes were found in all skeletal muscle groups including the diaphragm and intercostals; cardiac and smooth muscle appeared to be unaffected. It is likely that muscle degeneration resulted from overexpression of wtPrP. In addition to these primary myopathic changes, fiber type grouping indicated neurogenic rearrangement. As suggested by the grouping of type I muscle fibers that normally show an even distribution among type II fibers, there were significant abnormalities in the sciatic nerves of Tg7 and Tg94 mice. Large numbers of thinly myelinated axons accompanied a mild loss of large myelinated axons. Again, none of these changes were seen in non-Tg mice or in mice with low copy numbers of wtPrP transgenes.

Detectable levels of proteinase-resistant PrP^{Sc} were not associated with spontaneous neuromuscular disease in mice overexpressing wtPrP. Immunoreactive PrP was demonstrable in both brain and muscle of clinically ill mice but was susceptible to proteinase K digestion. Although likely due to excessive amounts of wtPrP, the mechanism of disease is unknown. PrP^{C} levels in brain and muscle are not obviously different between clinically ill and young Tg mice, suggesting that the late-onset of disease does not reflect a requirement for accumulation of PrP and perhaps indicating an accumulation of damage resulting from excess PrP^{C}. Toxicity in tissue culture has been demonstrated for a PrP synthetic peptide corresponding to residues 106–126 of human PrP (33). Some known PRNP mutations that cause familial prion disease might possibly exert their effects by decreasing the turnover of PrP^{C} (34). Similarly, prion disease pathology might reflect local elevation in concentration due to accumulation of PrP^{Sc} rather than acquisition of novel toxic properties arising from the conversion of PrP^{C} to PrP^{Sc}.

Although PrP^{Sc} is not apparent in Tg(SHaPrP+/+)7 mice, brain extracts from clinically ill mice have transmitted neurologic disease to hamsters (20). Although additional work is needed to confirm and extend these findings, transmissibility of spontaneous disease from animals without germline PrP mutations has important implications for naturally occurring prion diseases of humans and animals.

Summary

Results from experiments using transgenic mice suggest that incubation time profiles of prion strains can be explained solely by affinity of PrP^{C}-PrP^{Sc} binding and the rate constant of subsequent conversion of PrP^{C} to PrP^{Sc}. We do not suggest that our results prove that PrP^{Sc} is the sole functional component of prions. However, a single model, at least superficially based on the laws of mass action, can account for the incubation time properties of a variety of prion strains without invoking an additional macromolecule that encodes information (28, 29). A key question in prion replication and reproduction of strain specific properties is whether simple physical interactions between PrP^{C} and PrP^{Sc} are sufficient or whether cofactors or metabolic activity is required. Recent studies suggest that some radiolabeled PrP^{C} mixed with excess PrP^{Sc} becomes proteinase K–resistant (35). Different-sized proteinase-resistant PrP fragments result from infection with the hyper or drowsy strains of hamster transmissible mink encephalopathy (36) and similar fragments can result following admixture in the cell free system (35). Unfortunately, excess PrP^{Sc} in the reaction mixture precludes determination of whether propagation of strain-specific proteinase resistance properties in the cell free system reflects prion replication. Cell free acquisition of PrP^{Sc} properties by PrP^{C} appears to require denaturation of both isoforms at the start of the reaction. Although self-propagation of PrP^{Sc} multimers is a possibility, evidence from transgenic mice suggests the involvement of an additional host component in prion replication (37). Interestingly, the involvement of chaperone proteins has been demonstrated in the propagation of the yeast prion-like factor [psi$^+$] (38). Definitive determination of the biochemical basis for scrapie strain behavior will likely require understanding the nature of interaction between the cellular and scrapie isoforms of PrP, the physical differences between the two isoforms, and the participation of chaperones or other molecules in protein folding and unfolding.

Regardless of the precise mechanism for prion replication, the finding that excessive concentrations of wtPrP^{C} can lead to transmissible disease has implications for sporadic disease. Overexpression of PrP^{C} due either to heritable regulatory mutations or to environmental insults has the potential to lead to de novo synthesis of infectious prions.

Acknowledgments. Thanks are due to Drs. Stanley B. Prusiner, Steven J. DeArmond, David Westaway, Shu-Lian Yang, Michael Scott, Marilyn Torchia, Darlene Groth, Glen Telling, and many other members of the University of California, San Francisco scrapie group whose research and insight formed the basis for this chapter.

References

1. Pattison IH, Millson GC. Scrapie produced experimentally in goats with special reference to the clinical syndrome. J Comp Pathol 1961;71:101–8.
2. Zlotnik I. Slow, latent, and temperate virus infections: "Observations on the experimental transmission of scrapie of various origins to laboratory animals." Washington, DC: NINDB Monograph No. 2, 1965.
3. Bruce ME, Dickinson AG. Biological evidence that the scrapie agent has an independent genome. J Gen Virol 1987;68:79–89.
4. Prusiner SB. Molecular biology of prion diseases. Science 1991;252:1515–22.
5. Pan K-M, Baldwin M, Nguyen J, et al. Conversion of alpha-helices into beta-sheets features in the formation of the scrapie prion proteins. Proc Natl Acad Sci USA 1993;90:10962–6.
6. Safar J, Roller PP, Gajdusek DC, Gibbs CJ Jr. Conformation transitions, dissociation, and unfolding of scrapie amyloid (prion) protein. J Biol Chem 1993;269:20276–84.
7. Kellings K, Meyer N, Mirenda C, Prusiner SB, Riesner D. Further analysis of nucleic acids in purified scrapie prion preparations by improved return refocussing gel electrophoresis (RRGE). J Gen Virol 1992;73:1025–9.
8. Wickner RB. [URE3] as an altered *URE2* protein: evidence for a prion analog in *Saccharomyces cerevisiae.* Science 1994;264:566–9.
9. Bolton DC, McKinley MP, Prusiner SB. Identification of a protein that purifies with the scrapie prion. Science 1982;218:1309–11.

10. Oesch B, Westaway D, Wächli M, et al. A cellular gene encodes scrapie PrP 27–30 protein. Cell 1985;40:735–46.
11. Sparkes RS, Simon M, Cohn VH, et al. Assignment of the human and mouse prion protein genes to homologous chromosomes. Proc Natl Acad Sci USA 1986;83:7358–62.
12. Dickinson AG, Fraser H. Scrapie: pathogenesis in inbred mice: an assessment of host control and response involving many strains of agent. In: Katz M, Meuler V, eds. Slow virus infections of the CNS. New York: Springer-Verlag, 1977:3–14.
13. Dickinson AG, MacKay JMK. Genetical control of the incubation period in mice of the neurological disease, scrapie. Heredity 1964;19:279–88.
14. Dickinson AG, Meikle VMH, Fraser HG. Identification of a gene which controls the incubation period of some strains of scrapie agent in mice. J Comp Pathol 1968;78:293–9.
15. Carlson GA, Kingsbury DT, Goodman PA, et al. Linkage of prion protein and scrapie incubation time genes. Cell 1986;46:503–11.
16. Westaway D, Goodman PA, Mirenda CA, McKinley MP, Carlson GA, Prusiner SB. Distinct prion proteins in short and long scrapie incubation period mice. Cell 1987;51:651–62.
17. Dickinson AG, Meikle VMH. Host-genotype and agent effects in scrapie incubation: change in allelic interaction with different strains of agent. Mol Gen Genet 1971;112:73–9.
18. Goldman W, Hunter N, Foster JD, Salbaum JM, Beyreuther K, Hope J. Two alleles of a neural protein gene linked to scrapie in sheep. Proc Natl Acad Sci USA 1990;87:2476–80.
19. Westaway D, Zuliani V, Cooper CM, et al. Homozygosity for prion protein alleles encoding glutamine-171 renders sheep susceptible to natural scrapie. Genes Dev 1994;8:959–69.
20. Westaway D, Carlson GA, Prusiner SB. On safari with PrP: prion diseases of animals. Trends Microbiol 1995;3:141–7.
21. Carlson GA, Westway D, DeArmond SJ, Peterson-Torchia M, Prusiner SB. Primary structure of prion protein may modify scrapie isolate properties. Proc Natl Acad Sci USA 1989;86:7475–9.
22. Pattison IH. Experiments with scrapie with special reference to the nature of the agent and the pathology of the disease. In: Gajdusek DC, Gibbs CJ, Alpers M, eds. Slow, latent and temperate virus infections. NINDB Monograph No. 2. Washington, DC: U.S. Government Printing Office, 1965:249–57.
23. Scott M, Foster D, Mirenda C, et al. Transgenic mice expressing hamster prion protein produce species-specific scrapie infectivity and amyloid plaques. Cell 1989;59:847–57.
24. Prusiner SB, Scott M, Foster D, et al. Transgenetic studies implicate interactions between homologous PrP isoforms in scrapie prion replication. Cell 1990;63:673–86.
25. Carlson GA, Goodman PA, Lovett M, et al. Genetics and polymorphism of the mouse prion gene complex: control of scrapie incubation time. Mol Cell Biol 1988;8:5528–40.
26. Carlson GA, Ebeling C, Torchia M, Westaway D, Prusiner SB. Delimiting the location of the scrapie prion incubation time gene on chromosome 2 of the mouse. Genetics 1993;133:979–88.

27. Westaway D, Mirenda CA, Foster D, et al. Paradoxical shortening of scrapie incubation times by expression of prion protein transgenes derived from long incubation time mice. Neuron 1991;7:59–68.

28. Carlson GA, Ebeling C, Yang SL, et al. Prion isolate specified allotypic interactions between the cellular and scrapie prion proteins in congenic and transgenic mice. Proc Natl Acad Sci USA 1994;91:5690–4.

29. Carlson GA, DeArmond SJ, Torchia M, Westaway D, Prusiner SB. Genetics of prion diseases and prion diversity in mice. Philos Trans R Soc Lond B 1994;343:363–9.

30. Dickinson AG, Outram GW. The scrapie replication-site hypothesis and its implications for pathogenesis. In: Prusiner SB, Hadlow WJ, eds. Slow transmissible diseases of the nervous system, vol 2. New York: Academic Press, 1979: 13–31.

31. Palmer MS, Dryden AJ, Hughes JT, Collinge J. Homozygous prion protein genotype predisposes to sporadic Creutzfeldt-Jakob disease. Nature 1991; 352:340–2.

32. Westaway D, DeArmond SJ, Cayetano-Canlas J, et al. Degeneration of skeletal muscle, peripheral nerves and the central nervous system in transgenic mice overexpressing wild-type prion proteins. Cell 1994;76:117–29.

33. Forloni G, Angeretti N, Chiesa R, et al. Neurotoxicity of a prion protein fragment. Nature 1993;362:543–6.

34. Whittington MA, Sidle KCL, Gowland I, et al. Rescue of neurophysiological phenotype seen in PrP null mice by transgene encoding human prion protein. Nature Genetics 1995;9:197–201.

35. Bessen RA, Kocisko DA, Raymond GJ, Nandan S, Lansbury PT, Caughey B. Nongenetic propagation of strain-specific properties of scrapie prion protein. Nature 1995;375:698–700.

36. Bessen RA, Marsh RF. Biochemical and physical properties of the prion protein from two strains of the transmissible mink encephalopathy agent. J Virol 1992;66:2096–101.

37. Telling GC, Scott M, Hsiao KK, et al. Transmission of Creutzfeldt-Jakob disease from humans to transgenic mice expressing chimeric human-mouse prion protein. Proc Natl Acad Sci USA 1995;91:9936–40.

38. Chernoff YO, Lindquist SL, Ono B-I, Inge-Vechtomov SG, Liebman SW. Role of the chaperone protein Hsp104 in propagation of the yeast prion-like factor [psi+]. Science 1995;268:880–4.

16

Deciphering Prion Diseases with Transgenic Mice

GLENN C. TELLING, MICHAEL SCOTT, AND STANLEY B. PRUSINER

Prions cause a group of human and animal neurodegenerative diseases that are now classified together because their etiology and pathogenesis involve modification of the prion protein (PrP) (1). Prion diseases are manifest as infectious, genetic, and sporadic disorders (Table 16.1). These diseases can be transmitted among mammals by the infectious particle designated "prion" (2). Despite intensive searches over the past three decades, no nucleic acid has been found within prions (3–6), yet a modified isoform of the host-encoded PrP designated PrP^{Sc} is essential for infectivity (1, 7–10). In fact, considerable experimental data argue that prions are composed exclusively of PrP^{Sc}. Earlier terms used to describe the prion diseases include *transmissible encephalopathies*, *spongiform encephalopathies*, and *slow virus diseases* (11–13).

The quartet of human (Hu) prion diseases are frequently referred to as kuru, Creutzfeldt-Jakob disease (CJD), Gerstmann-Sträussler-Scheinker (GSS) disease, and fatal familial insomnia (FFI). Kuru was the first of the human prion diseases to be transmitted to experimental animals and it has often been suggested that kuru spread among the Fore people of Papua New Guinea by ritualistic cannibalism (12, 14). The experimental and presumed human-to-human transmission of kuru led to the belief that prion diseases are infectious disorders caused by unusual viruses similar to those causing scrapie in sheep and goats. Yet the occurrence of CJD in families, first reported almost 70 years ago (15, 16), was perplexing, to say the least. The significance of familial CJD remained unappreciated until mutations in the protein coding region of the PrP gene on the short arm of chromosome 20 were discovered (17–19). The earlier finding that brain extracts from patients who had died of familial prion diseases inoculated into experimental animals often transmit disease, posed a conundrum that was resolved with the genetic linkage of these diseases to mutations of the PrP gene (20–22).

TABLE 16.1. Human prion diseases.

Disease	Etiology
Kuru	Infection
Creutzfeldt-Jakob disease	
Iatrogenic	Infection
Sporadic	Unknown
Familial	PrP mutation
Gerstmann-Sträussler-Scheinker disease	PrP mutation
Fatal familial insomnia	PrP mutation

The most common form of prion disease is sporadic CJD. Many attempts to show that the sporadic prion diseases are caused by infection have been unsuccessful (23–26). The discovery that inherited prion diseases are caused by germline mutation of the PrP gene raised the possibility that sporadic forms of these diseases might result from a somatic mutation (21). The discovery that PrP^{Sc} is formed from the cellular isoform of the prion protein, PrP^C, by a posttranslational process (27) and that overexpression of wild-type (wt) PrP transgenes produces spongiform degeneration and infectivity de novo (28) has raised the possibility that sporadic prion diseases result from the spontaneous conversion of PrP^C into PrP^{Sc}.

CJD has a worldwide incidence of ~1 case per 10^6 population (29). Less than 1% of CJD cases are infectious and all of those appear to be iatrogenic. Between 10% and 15% of prion disease cases are inherited while the remaining cases are sporadic. Kuru was once the most common cause of death among New Guinea women in the Fore region of the Highlands (14, 30, 31) but has virtually disappeared with the cessation of ritualistic cannibalism (32). Patients with CJD frequently present with dementia but ~10% of patients exhibit cerebellar dysfunction initially. Patients with either kuru or GSS usually present with ataxia, while those with FFI manifest insomnia and autonomic dysfunction (33–35).

PrP^{CJD} has been found in the brains of most patients who died of prion disease. The term *PrP^{CJD}* is preferred by some investigators when referring to the abnormal isoform of HuPrP in human brain. Here, PrP^{Sc} is used interchangeably with PrP^{CJD}. PrP^{Sc} is always used after human CJD prions have been passaged into an experimental animal since the nascent PrP^{Sc} molecules are produced from host PrP^C and the PrP^{CJD} in the inoculum only serves to initiate the process. In the brains of some patients with inherited prion diseases as well as transgenic (Tg) mice expressing mouse (Mo) PrP with the human GSS point mutation (Pro → Leu), detection of PrP^{Sc} has been problematic despite clinical and neuropathologic hallmarks of neurodegeneration (36, 37). Presumably, neurodegeneration is due, at least in part, to the abnormal metabolism of mutant PrP (36). Of note, horizontal transmission of neurodegeneration from the brains of patients with inherited prion diseases to inoculated rodents has been less frequent than with sporadic cases (22). Whether this distinction between transmissible and nontransmissible inherited prion diseases will persist is unclear. Tg mice expressing a chimeric Hu/Mo PrP gene have been found to be highly susceptible to Hu prions from sporadic and iatrogenic CJD cases (38). These Tg(MHu2M) mice should make the use of apes and monkeys for the study of human prion diseases unnecessary and allow for tailoring the PrP^C translated from the transgene to match the sequence of the PrP^{CJD} in the inoculum. Other Tg mouse studies have demonstrated that PrP^{Sc} in the inoculum interacts preferentially with homotypic PrP^C during the propagation of prions (39, 40). PrP^C is the cellular isoform of the prion protein that has been identified in all mammals and birds examined to date; PrP^C is anchored to the external surface of cells by a glycolipid moiety and its function is unknown (41).

Scrapie is the most common natural prion disease of animals. An investigation into the etiology of scrapie followed the vaccination of sheep for looping ill virus with formalin-treated extracts of ovine lymphoid tissue unknowingly contaminated with scrapie prions (42). Two years later, more than 1,500 sheep developed scrapie from this vaccine. While the transmissibility of experimental scrapie became well established, the spread of natural scrapie within and among flocks of sheep remained puzzling. Parry (43, 44) argued that host genes were responsible for the development of scrapie in sheep. He was convinced that natural scrapie is a genetic disease that could be eradicated by proper breeding protocols. He considered its transmission by inoculation of importance primarily for laboratory studies and communicable infection of little consequence in nature. Other investigators viewed natural scrapie as an infectious disease and argued that host genetics only modulates susceptibility to an endemic infectious agent (45).

The offal of scrapied sheep in Great Britain is thought to be responsible for the current epidemic of bovine spongiform encephalopathy (BSE) or mad cow disease (46). Prions in the offal from scrapie-infected sheep appear to have survived rendering that produced meat and bone meal (MBM). After BSE was recognized, MBM produced from domestic animals was banned from further use. Since 1986 when BSE was first recognized, >160,000 cattle have died of BSE. Whether humans will develop CJD after consuming beef from cattle with BSE prions is of considerable concern.

The fundamental event in prion diseases seems to be a conformational change in PrP. All attempts to identify a posttranslational chemical modification that distinguishes PrP^{Sc} from PrP^C have been unsuccessful to date (47). PrP^C contains ~45% α-helix and is virtually devoid of β-sheet (48). Conversion to PrP^{Sc} creates a protein that contains ~30% α-helix and 45% β-sheet. The mechanism by which PrP^C is converted into PrP^{Sc} remains unknown but PrP^C appears to bind to PrP^{Sc} to form an intermediate complex during the formation of nascent PrP^{Sc}.

Prions differ from all other known infectious pathogens in several respects. First, prions do not contain a nucleic acid genome that codes for their progeny. Viruses, viroids, bacteria, fungi, and parasites all have nucleic acid genomes that code for their progeny. Second, the only known component of the prion is a modified protein that is encoded by a cellular gene. Third, the major, and possibly only, component of the prion is PrP^{Sc}, which is a pathogenic conformer of PrP^{C}.

As our knowledge of the prion diseases increases and more is learned about the molecular and genetic characteristics of prion proteins, these disorders will undoubtedly undergo modification with respect to their classification. Indeed, the discovery of the PrP and the identification of pathogenic PrP gene mutations have already forced us to view these illnesses from perspectives not previously imagined.

Transgenetics and Gene Targeting

While transgenetic studies have yielded a wealth of new knowledge about infectious, genetic, and sporadic prion diseases, the laborious production of Tg mice limits the number of studies that can be performed. The relatively long gestation period of mice coupled with the need to do microinjections of fertilized embryos prevents the creation of the very large numbers of different Tg mice that would yield the greatest amount of new information. Assays that permit screening of a multitude of possible phenotypes in genetic experiments are generally the most informative. While the limited number of mice expressing different transgenes is definitely a liability, experiments with Tg mice expressing foreign and mutant PrP molecules have been extraordinarily useful in advancing our understanding of prion biology. It is important to stress that transgenetic studies can readily yield an incomplete, and sometimes erroneous interpretation of the data if the number of lines of mice examined expressing a particular construct is inadequate. Defining an adequate number of lines is difficult, but certain comparisons of lines expressing high and low levels of a given PrP transgene have proved to be quite helpful (37, 39).

Species Barriers for Transmission of Prion Diseases

The passage of prions between species is a stochastic process characterized by prolonged incubation times (49–51). Prions synthesized de novo reflect the sequence of the host PrP gene and not that of the PrP^{Sc} molecules in the inoculum (52). On subsequent passage in a homologous host, the incubation time shortens to that recorded for all subsequent passages and it becomes a nonstochastic process. The species barrier concept is of practical

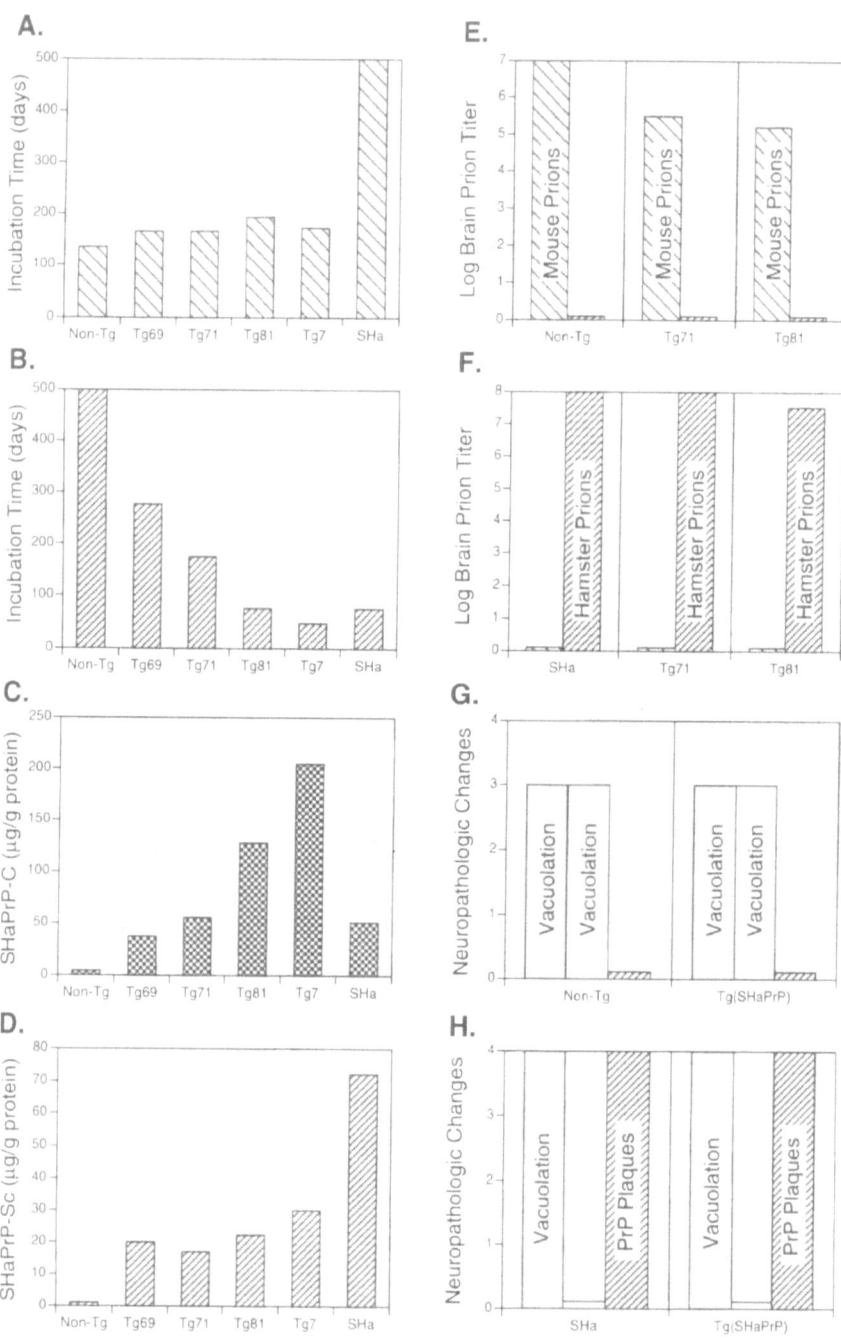

FIGURE 16.1. Transgenic (Tg) mice expressing Syrian hamster (SHa) prion protein exhibit species-specific scrapie incubation times, infectious prion synthesis and neuropathology). (A) Scrapie incubation times in nontransgenic mice (Non-Tg) and four lines of Tg mice expressing SHaPrP and Syrian hamsters inoculated

importance in assessing the risk for humans of developing CJD after consumption of scrapie-infected lamb or BSE-infected beef (46, 53–56).

To test the hypothesis that differences in PrP gene sequences might be responsible for the species barrier, Tg mice expressing SHaPrP were constructed (39, 57). The PrP genes of Syrian hamsters and mice encode proteins differing at 16 positions. Incubation times in four lines of Tg(SHaPrP) mice inoculated with Mo prions were prolonged compared with those observed for non-Tg, control mice (Fig. 16.1A). Inoculation of Tg(SHaPrP) mice with SHa prions demonstrated abrogation of the species barrier resulting in abbreviated incubation times due to a nonstochastic process (Fig. 16.1B) (39, 57). The length of the incubation time after inoculation with SHa prions was inversely proportional to the level of SHaPrP^C in the brains of Tg(SHaPrP) mice (Fig. 16.1B,C) (39). SHaPrP^{Sc} levels in the brains of clinically ill mice were similar in all four Tg(SHaPrP) lines inoculated with SHa prions (Fig. 16.1D). Bioassays of brain extracts from clinically ill Tg(SHaPrP) mice inoculated with Mo prions revealed that only Mo prions but no SHa prions were produced (Fig. 16.1E). Conversely, inoculation of Tg(SHaPrP) mice with SHa prions led to only the synthesis of SHa prions (Fig. 16.1F). Thus, the de novo synthesis of prions is species specific and reflects the genetic origin of the inoculated prions. Similarly, the neuropathology of Tg(SHaPrP) mice is determined by the genetic origin of prion inoculum. Mo prions injected into Tg(SHaPrP) mice produced a neuropathology characteristic of mice with scrapie. A moderate degree of

◄ ────────────────────────────────

Figure 16.1 *Continued*   intracerebrally with ~10^6 ID$_{50}$ units of Chandler Mo prions serially passaged in Swiss mice. The four lines of Tg mice have different numbers of transgene copies: Tg69 and 71 mice have two to four copies of the SHaPrP transgene, whereas Tg81 have 30 to 50 and Tg7 mice have >60. Incubation times are number of days from inoculation to onset of neurologic dysfunction. (B) Scrapie incubation times in mice and hamsters inoculated with ~10^7 ID$_{50}$ units of Sc237 prions serially passaged in Syrian hamsters and as described in (A). (C) Brain SHaPrP^C in Tg mice and hamsters. SHaPrP^C levels were quantitated by an enzyme-linked immunoassay. (D) Brain SHaPrP^{Sc} in Tg mice and hamsters. Animals were killed after exhibiting clinical signs of scrapie. SHaPrP^{Sc} levels were determined by immunoassay. (E) Prion titers in brains of clinically ill animals after inoculation with Mo prions. Brain extracts from Non-Tg, Tg71, and Tg81 mice were bioassayed for prions in mice (left) and hamsters (right). (F) Prion titers in brains of clinically ill animals after inoculation with SHa prions. Brain extracts from Syrian hamsters as well as Tg71 and Tg81 mice were bioassayed for prions in mice (left) and hamsters (right). (G) Neuropathology in Non-Tg mice and Tg(SHaPrP) mice with clinical signs of scrapie after inoculation with Mo prions. Vacuolation in gray (left) and white matter (center); PrP amyloid plaques (right). Vacuolation score: 0 = none, 1 = rare, 2 = modest, 3 = moderate, 4 = intense. (H) Neuropathology in Syrian hamsters and transgenic mice inoculated with SHa prions. Degree of vacuolation and frequency of PrP amyloid plaques as described in (G).

vacuolation in both the gray and white matter was found while amyloid plaques were rarely detected (Fig. 16.1G) (Table 16.2). Inoculation of Tg(SHaPrP) mice with SHa prions produced intense vacuolation of the gray matter, sparing of the white matter, and numerous SHaPrP amyloid plaques characteristic of Syrian hamsters with scrapie (Fig. 16.1H).

Overexpression of wtPrP Transgenes

During transgenetic studies, we discovered that uninoculated older mice harboring high copy numbers of wtPrP transgenes derived from Syrian hamsters, sheep, and PrP-B mice spontaneously developed truncal ataxia, hind-limb paralysis, and tremors (58). These Tg mice exhibited a profound necrotizing myopathy involving skeletal muscle, a demyelinating polyneuropathy, and focal vacuolation of the CNS. Development of disease was dependent on transgene dosage. For example, Tg(SHaPrP$^{+/+}$)7 mice homozygous for the SHaPrP transgene array regularly developed disease between 400 and 600 days of age, while hemizygous Tg(SHaPrP$^{+/0}$)7 mice also developed disease, but after >650 days.

Attempts to demonstrate PrP^{Sc} in either muscle or brain were unsuccessful but transmission of disease with brain extracts from Tg(SHaPrP$^{+/+}$)7 mice inoculated into Syrian hamsters did occur. These Syrian hamsters had PrP^{Sc} as detected by immunoblotting and spongiform degeneration (D.

TABLE 16.2. Species-specific prion inocula determine the distribution of spongiform change and deposition of PrP amyloid plaques in transgenic mice.

		SHa prions					Mo prions		
		Spongiform change[a]		PrP plaques[b]			Spongiform change[a]		PrP plaques[b]
Animal	n^c	Gray	White	Frequency	Diameter[d]	n^c	Gray	White	Frequency
Non-Tg		N.D.[e]		N.D.		10	+	+	−
Tg 69	6	+[f]	−	Numerous	6.5 ± 3.1 (389)	2	+	+	−
Tg 71	5	+	−	Numerous	8.1 ± 3.6 (345)	2	+	+	−
Tg 81	7	+	−	Numerous	8.3 ± 3.0 (439)	3	+	+	Few
Tg 7	3	+[g]	−	Numerous	14.0 ± 8.3 (19)	4	+	+	−
SHa	3	+	−	Numerous	5.7 ± 2.7 (247)		N.D.		N.D.

[a] Spongiform change evaluated in hippocampus, thalamus, cerebral cortex, and brain stem for gray matter and the deep cerebellum for white matter.
[b] Plaques in the subcallosal region were stained with SHaPrP mAb 13A5, anti-PrP rabbit antisera R073 and trichrome stain.
[c] Number of brains examined.
[d] Mean diameter of PrP plaques given in microns ± standard error with the number of observations in parentheses.
[e] N.D., not determined.
[f] +, present; −, not found.
[g] Focal: confirmed to the dorsal nucleus of the raphe.

Groth and S.B. Prusiner, unpublished data). Serial passage with brain extracts from these animals to recipients was observed. De novo synthesis of prions in Tg(SHaPrP$^{+/+}$)7 mice overexpressing wtSHaPrP^C provides support for the hypothesis that sporadic CJD does not result from infection but rather is a consequence of the spontaneous, although rare, conversion of PrP^C into PrP^{Sc}. Alternatively, a somatic mutation in which mutant SHaPrP^C is spontaneously converted into PrP^{Sc} as in the inherited prion diseases could also explain sporadic CJD. These findings as well as those described below for Tg(MoPrP-P101L) mice argue that prions are devoid of foreign nucleic acid, in accord with many earlier studies that use other experimental approaches as described above.

Ablation of the PrP Gene

Ablation of the PrP gene in Tg (Prnp$^{0/0}$) mice has, unexpectedly, not affected the development of these animals (59). In fact, they are healthy at almost 2 years of age. Prnp$^{0/0}$ mice are resistant to prions (Fig. 16.2) and do not propagate scrapie infectivity (8, 9).

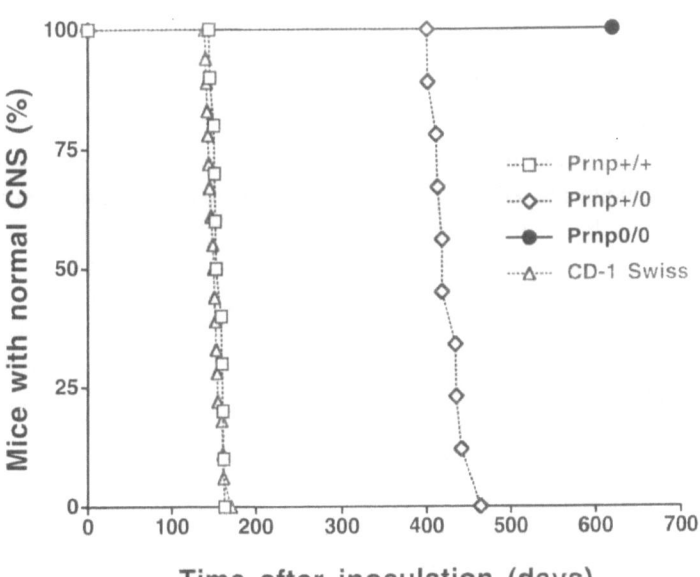

FIGURE 16.2. Incubation times in PrP gene ablated Prnp$^{0/+}$ and Prnp$^{0/0}$ mice as well as wt Prnp$^{+/+}$ and CD-1 mice inoculated with RML mouse prions. The RML prions were heated and irradiated at 254 nm prior to intracerebral inoculation into CD-1 Swiss mice (open triangles), Prnp$^{+/+}$ mice (open squares), Prnp$^{0/+}$ mice (open diamonds) or Prnp$^{0/0}$ mice (filled circle).

Prnp$^{0/0}$ mice crossed with Tg(SHaPrP) mice were rendered susceptible to SHa prions but remained resistant to Mo prions (8, 9). Since the absence of PrP^C expression does not provoke disease, it is likely that scrapie and other prion diseases are a consequence of PrP^{Sc} accumulation rather than an inhibition of PrP^C function (59).

Mice heterozygous (Prnp$^{0/+}$) for ablation of the PrP gene had prolonged incubation times when inoculated with Mo prions (Fig. 16.2) (9). The Prnp$^{0/+}$ mice developed signs of neurologic dysfunction at 400 to 460 days after inoculation. These findings are in accord with studies on Tg(SHaPrP) mice in which increased SHaPrP expression was accompanied by diminished incubation times (Fig. 16.1B) (39).

Since Prnp$^{0/0}$ mice do not express PrP^C, we reasoned that they might more readily produce α-PrP antibodies. Prnp$^{0/0}$ mice immunized with Mo or SHa prion rods produced α-PrP antisera that bound Mo, SHa, and Hu PrP (9). These findings contrast with earlier studies in which α-MoPrP antibodies could not be produced in mice presumably because the mice had been rendered tolerant by the presence of MoPrP^C (60–62). That Prnp$^{0/0}$ mice readily produce α-PrP antibodies is consistent with the hypothesis that the lack of an immune response in prion diseases is due to the fact that PrP^C and PrP^{Sc} share many epitopes. Whether Prnp$^{0/0}$ mice produce α-PrP antibodies that specifically recognize conformational dependent epitopes present on PrP^{Sc} but absent from PrP^C remains to be determined.

Modeling of GSS in Tg(MoPrP-P101L) Mice

The codon 102 point mutation found in GSS patients was introduced into the MoPrP gene and Tg(MoPrP-P101L)H mice were created expressing high (H) levels of the mutant transgene product. The two lines of Tg(MoPrP-P101L)H mice designated 174 and 87 spontaneously developed CNS degeneration, characterized by clinical signs indistinguishable from experimental murine scrapie and neuropathology consisting of widespread spongiform morphology and astrocytic gliosis (36) and PrP amyloid plaques (Fig. 16.3) (37). By inference, these results contend that PrP gene mutations cause GSS, familial CJD, and FFI.

Brain extracts prepared from spontaneously ill Tg(MoPrP-P101L)H mice transmitted CNS degeneration to Tg196 mice expressing low levels of the mutant transgene product and some Syrian hamsters (37). Many Tg196 mice and some Syrian hamsters developed CNS degeneration between 200 and 700 days after inoculation, while inoculated CD-1 Swiss mice remained well. Serial transmission of CNS degeneration in Tg196 mice required about 1 year, while serial transmission in Syrian hamsters occurred after ~75 days (37). Although brain extracts prepared from Tg(MoPrP-P101L)H mice transmitted CNS degeneration to some inoculated recipients, little or no PrP^{Sc} was detected by immunoassays after limited proteolysis. Undetect-

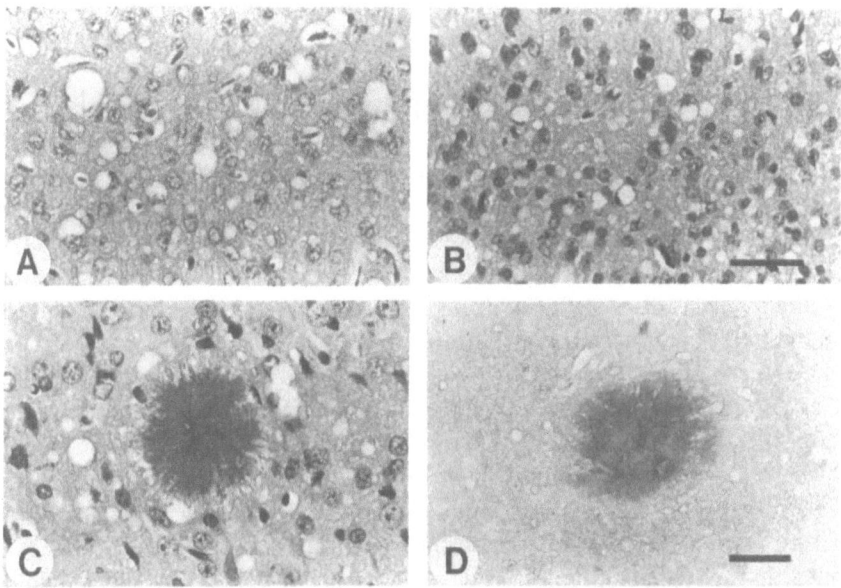

FIGURE 16.3. Neuropathology of Tg(MoPrP-P101L) mice developing neuro-degeneration spontaneously. (A) Vacuolation in cerebral cortex of a Swiss CD-1 mouse that exhibited signs of neurologic dysfunction at 138 days after intracerebral inoculation with ~10^6 ID_{50} units of RML scrapie prions. (B) Vacuolation in cerebral cortex of a Tg(MoPrP-P101L) mouse that exhibited signs of neurologic dysfunction at 252 days of age. (C) Kuru-type PrP amyloid plaque stained with periodic acid-Schiff in the caudate nucleus of a Tg(MoPrP-P101L) mouse that exhibited signs of neurologic dysfunction. (D) PrP amyloid plaques stained with α-PrP antiserum (RO73) in the caudate nucleus of a Tg(MoPrP-P101L) mouse that exhibited signs of neurologic dysfunction. Bar in B also applies to A = 50 μm. Bar in D also applies to C = 25 μm.

able or low levels of PrP^{Sc} in the brains of these Tg(MoPrP-P101L)H mice are consistent with the results of these transmission experiments, which suggest low titers of infectious prions. Although no PrP^{Sc} was detected in the brains of inoculated Tg196 mice exhibiting neurologic dysfunction by immunoassays after limited proteolysis, PrP amyloid plaques as well as spongiform degeneration were frequently found. The neurodegeneration found in inoculated Tg196 mice seems likely to result from a modification of mutant PrP^C that is initiated by mutant PrP^{Sc} present in the brain extracts prepared from ill Tg(MoPrP-P101L)H mice. In support of this explanation are the findings in some of the inherited human prion diseases as described above where neither protease-resistant PrP (63, 64) nor transmission to experimental rodents could be demonstrated (22). Furthermore, transmission of disease from Tg(MoPrP-P101L)H mice to Tg196 mice but not to

Swiss mice is consistent with earlier findings that demonstrate that homotypic interactions between PrP^C and PrP^{Sc} feature in the formation of PrP^{Sc}.

In other studies, modifying the expression of mutant and wtPrP genes in Tg mice permitted experimental manipulation of the pathogenesis of both inherited and infectious prion diseases. Although overexpression of the wtPrP-A transgene of greater than eightfold was not deleterious to the mice, it did shorten scrapie incubation times from ~145 to ~45 days after inoculation with Mo scrapie prions (65). In contrast, overexpression at the same level of a PrP-A transgene mutated at codon 101 produced spontaneous, fatal neurodegeneration between 150 and 300 days of age in two new lines of Tg(MoPrP-P101L) mice designated 2866 and 2247. Genetic crosses of Tg(MoPrP-P101L)2866 mice with gene-targeted mice lacking both PrP alleles (Prn-$p^{0/0}$) produced animals with a highly synchronous onset of illness between 150 and 160 days of age. The Tg(MoPrPP101L)2866/Prnp$^{0/0}$ mice had numerous PrP plaques and widespread spongiform degeneration in contrast to the Tg2866 and 2247 mice that exhibited spongiform degeneration but only a few PrP amyloid plaques. Another line of mice designated Tg2862 overexpresses the mutant transgene ~32-fold and develops fatal neurodegeneration between 200 and 400 days of age. Tg2862 mice exhibited the most severe spongiform degeneration and had numerous, large PrP amyloid plaques. While mutant PrP^C(P101L) clearly produces neurodegeneration, wtPrP^C profoundly modifies both the age of onset of illness and the neuropathology for a given level of transgene expression. These findings and those from other studies (38) suggest that mutant and wtPrP interact, perhaps through a chaperone-like protein, to modify the pathogenesis of the dominantly inherited prion diseases.

Proteins Modifying Prion Diseases That Are Not Encoded by PrP Genes

While the results of the foregoing investigations indicate that PrP transgenes modulate virtually all aspects of scrapie, including prion propagation, incubation time length, synthesis of PrP^{Sc}, the species barrier, and neuropathologic lesions, evidence for the role of proteins not encoded by PrP genes is beginning to emerge. In particular, studies with Tg(MHu2M) and Tg(HuPrP) mice argue for the existence of a species-specific factor as noted above that has provisionally been designated protein X (38). Investigations of prion strains in congenic mice suggest that a gene linked to but separate from PrP profoundly modifies the neuropathology of scrapie (66). We have provisionally designated this gene product as protein Y.

Protein X and the Transmission of Prions

Attempts to abrogate the prion species barrier between humans and mice by using an approach similar to that described for the abrogation of the species barrier between Syrian hamsters and mice were unsuccessful. Mice expressing HuPrP transgenes did not develop signs of CNS dysfunction more rapidly or frequently than non-Tg controls (38).

The successful breaking of the species barrier between humans and mice has its origins in a set of studies with Tg mice expressing chimeric PrP genes derived from SHa and Mo PrP genes (67). One SHa/MoPrP gene, designated MH2M PrP, contains five amino acid substitutions encoded by SHaPrP, while another construct designated MHM2 PrP has two substitutions. Tg(MH2M PrP) mice were susceptible to both SHa or Mo prions, whereas three lines expressing MHM2 PrP were resistant to SHa prions (40). The brains of Tg(MH2M PrP) mice dying of scrapie contained chimeric PrP^{Sc} and prions with an artificial host range favoring propagation in mice that express the corresponding chimeric PrP, and were also transmissible, at reduced efficiency, to non-Tg mice and hamsters. These findings provided additional genetic evidence for homophilic interactions between PrP^{Sc} in the inoculum and PrP^{C} synthesized by the host.

With the recognition that Tg(HuPrP) mice were not suitable recipients for the transmission of Hu prions, we constructed Tg(MHu2M) mice analogous to the Tg(MH2M) mice described above. Human PrP differs from mouse PrP at 28 of 254 positions (68), while chimeric MHu2MPrP differs at nine residues. The mice expressing the MHu2M transgene are susceptible to human prions and exhibit abbreviated incubation times of ~200 days (38). In these initial studies the chimeric MHu2M transgene encoded a Met at codon 129 and all three of the patients were homozygous for Met at this residue. Two of the cases were sporadic CJD and the third was an iatrogenic case that occurred after treatment with pituitary-derived human growth hormone (HGH). Whether it will be necessary to match the PrP genotype of the Tg(MHu2M) mouse with that of the CJD patient from whom the inoculum was derived, or some variations in sequence can be tolerated, remains to be established.

From Tg(SHaPrP) mouse studies, prion propagation is thought to involve the formation of a complex between PrP^{Sc} and the homotypic substrate PrP^{C} (39). Attempts to mix PrP^{Sc} with PrP^{C} have failed to produce nascent PrP^{Sc} (69), raising the possibility that proteins such as chaperones might be involved in catalyzing the conformational changes that feature in the formation of PrP^{Sc} (48). One explanation for the difference in susceptibility of Tg(MHu2M) and Tg(HuPrP) mice to Hu prions in mice may be that mouse chaperones catalyzing the refolding of PrP^{C} into PrP^{Sc} can readily interact with the MHu2MPrP^{C}/HuPrP^{CJD} complex but not with HuPrP^{C}/HuPrP^{CJD}. The identification of protein X is an important avenue of research since isolation of this protein or complex of proteins would presumably facilitate studies of PrP^Sc formation. To date, attempts to isolate specific proteins that bind to PrP have been disappointing (70). Whether or not identification of protein X will require isolation of a ternary complex composed of PrP^C, PrP^Sc, and protein X remains to be determined.

The sensitivity of Tg(MHu2M) mice to Hu prions suggests that a similar approach to the construction of Tg mice susceptible to bovine spongiform encephalopathy (BSE) and scrapie sheep prions may prove fruitful. The BSE epidemic has led to considerable concern about the safety for humans consuming beef and dairy products. Although epidemiologic studies over the past two decades argue that humans do not contract CJD from scrapie-infected sheep products (24–26), it is unknown whether any of the seven amino acid substitutions that distinguish bovine from sheep PrP render bovine prions permissive in humans (56). Whether Tg(MHu2M) mice are susceptible to bovine or sheep prions is unknown.

Protein Y and the Neuropathology of Prion Disease

Four lines of congenic mice were produced by crossing the PrP gene of the ILn/J mouse onto C57BL. The four lines of congenic mice are designated B6.I-4 for B6.I-*B2m^a*, B6.I-1 for B6.I-*Prnp^b*, B6.I-2 for B6.I-*Il-1a^d Prnp^b*, and B6.I-3 for B6.I-*B2m^a Prnp^b* (71). Neuropathologic examination of B6.I-1, B6.I-2, I/LnJ, and VM/Dk mice inoculated with 87V prions showed numerous PrP amyloid plaques in accord with an earlier report on VM/Dk mice (72). In B6.I-1 mice intense spongiform degeneration, gliosis, and PrP immunostaining were found in the ventral posterior lateral (VPL) nucleus of the thalamus, the habenula, and the raphe nuclei of the brain stem (66). These same regions showed intense immunoreactivity for PrP^Sc on histoblots. Unexpectedly, B6.I-2 and ILn/J mice exhibited only mild vacuolation of the thalamus and brain stem. These findings suggest that a locus near *Prnp* influences the deposition of PrP^Sc, and thus vacuolation, in the thalamus, the habenula, and raphe nuclei. We have provisionally designated the product of this gene protein Y.

Identification of the gene that encodes protein Y that is distinct from but near *Prnp* will be important. The gene Y product appears to control, at least in part, neuronal vacuolation and presumably PrP^Sc deposition in mice inoculated with scrapie prions. Isolation of protein Y should be helpful in dissecting the molecular events that feature in the pathogenesis of the prion diseases.

Prion Diversity

Prion Strains and Variations in Patterns of Disease

For many years, studies of experimental scrapie were performed exclusively with sheep and goats. The disease was first transmitted by intraocular inoculation (73) and later by intracerebral, oral, subcutaneous, intramuscular, and intravenous injections of brain extracts from sheep developing scrapie. Incubation periods of 1 to 3 years were common and often many of the inoculated animals failed to develop disease (74–76). Different breeds of sheep exhibited markedly different susceptibilities to scrapie prions inoculated subcutaneously, suggesting that the genetic background might influence host permissiveness (77).

The diversity of scrapie prions was first appreciated in goats inoculated with "hyper" and "drowsy" isolates (78). Subsequently, studies in mice demonstrated the existence of many scrapie "strains" (79–82), which continues to pose a fascinating conundrum. What is the macromolecule that carries the information required for each strain to manifest a unique set of biologic properties if it is not a nucleic acid?

There is good evidence for multiple "strains" or distinct isolates of prions as defined by specific incubation times, distribution of vacuolar lesions and patterns of PrP^{Sc} accumulation (83–86). The lengths of the incubation times have been used to distinguish prion strains inoculated into sheep, goats, mice, and hamsters. Dickinson and his colleagues (83, 87, 88) developed a system for "strain typing" by which mice with genetically determined short and long incubation times were used in combination with the F1 cross. For example, C57BL mice exhibited short incubation times of ~150 days when inoculated with either the Me7 or Chandler isolates; VM mice inoculated with these same isolates had prolonged incubation times of ~300 days. The mouse gene controlling incubation times was labeled *Sinc* and long incubation times were said to be a dominant trait because of prolonged incubation times in F1 mice. Prion strains were categorized into two groups based upon their incubation times: (1) those causing disease more rapidly in "short" incubation time C57BL mice and (2) those causing disease more rapidly in "long" incubation time VM mice.

PrP Gene Dosage Controls the Length of the Scrapie Incubation Time

More than a decade of study was required to unravel the mechanism responsible for the "dominance" of long incubation times; not unexpectedly, long incubation times were found not to be dominant traits. Instead, the apparent dominance of long incubation times is due to a gene dosage effect (66).

Our own studies began with the identification of a widely available mouse strain with long incubation times. ILn/J mice inoculated with RML prions were found to have incubation times exceeding 200 days (89), a finding that was confirmed by others (90). Once molecular clones of the PrP gene were available, we asked whether or not the PrP genes of short and long mice segregate with incubation times. A restriction fragment length polymorphism (RFLP) of the PrP gene was used to follow the segregation of MoPrP genes (*Prnp*) from short NZW or C57BL mice with long ILn/J mice in F1 and F2 crosses. This approach permitted the demonstration of genetic linkage between a *Prnp* and a gene modulating incubation times (*Prn-i*) (91). Other investigators have confirmed the genetic linkage, and one group has shown that the incubation time gene *Sinc* is also linked to PrP (92, 93). It now seems likely that the genes for PrP, *Prn-i*, and *Sinc* are all congruent; the term *Sinc* is no longer used (94). The PrP sequences of NZW with short and long scrapie incubation times, respectively, differ at codons 108 (L → F) and 189 (T→V) (95).

Although the amino acid substitutions in PrP that distinguish *Prnp^a* from *Prnp^b* mice argued for the congruency of *Prnp* and *Prn-i*, experiments with *Prnp^a* mice expressing *Prnp^b* transgenes demonstrated a "paradoxical" shortening of incubation times (96). We had predicted that these Tg mice would exhibit a prolongation of the incubation time after inoculation with RML prions based on (*Prnp^a* × *Prnp^b*) F1 mice, which do exhibit long incubation times. We described those findings as "paradoxical shortening"

TABLE 16.3. MoPrP-A expression is a major determinant of incubation times in mice inoculated with the RML scrapie prions.

Mice	Prn-p genotype (copies)	Prn-p transgenes (copies)	Alleles a	Alleles b	Incubation time[a] (days ± SEM)	n
Prnp$^{0/0}$	0/0		0	0	>600	4
Prnp$^{+/0}$	a/0		1	0	426 ± 18	9[a]
B6.I-1	b/b		0	2	360 ± 16	7[b]
B6.I-2	b/b		0	2	379 ± 8	10[b]
B6.I-3	b/b		0	2	404 ± 10	20
(B6 × B6.I-1)F1	a/b		1	1	268 ± 4	7
B6.I-1 × Tg(MoPrP-B$^{0/0}$)15	a/b		1	1	255 ± 7	11[c]
B6.I-1 × Tg(MoPrP-B$^{0/0}$)15	a/b		1	1	274 ± 3	9[d]
B6.I-1 × Tg(MoPrP-B$^{+/0}$)15	a/b	bbb/0	1	4	166 ± 2	11[c]
B6.I-1 × Tg(MoPrP-B$^{+/0}$)15	a/b	bbb/0	1	4	162 ± 3	8[d]
C57BL/6J (B6)	a/a		2	0	143 ± 4	8
B6.I-4	a/a		2	0	144 ± 5	8
non-Tg(MoPrP-B$^{0/0}$)15	a/a		2	0	130 ± 3	10
Tg(MoPrP-B$^{+/0}$)15	a/a	bbb/0	2	3	115 ± 2	18
Tg(MoPrP-B$^{+/+}$)15	a/a	bbb/bbb	2	6	111 ± 5	5
Tg(MoPrP-B$^{+/0}$)94	a/a	>30b	2	>30	75 ± 2	15[e]
Tg(MoPrP-A$^{+/0}$)B4053	a/a	>30a	>30	0	50 ±	16

[a] Data from ref. 9.

[b] Data from ref. 7.

[c] The homozygous Tg(MoPrP-B$^{+/+}$)15 mice were maintained as a distinct subline selected for transgene homozygosity two generations removed from the (B6 × LT/Sv)F2 founder. Hemizygous Tg(MoPrP-B$^{+/0}$)15 mice were produced by crossing the Tg(MoPrP-B$^{+/+}$)15 line with B6 mice.

[d] Tg(MoPrP-B$^{+/0}$)15 mice were maintained by repeated backcrossing to B6 mice.

[e] Data from ref. 96.

because we and others had believed for many years that long incubation times are dominant traits (83, 91). From studies of congenic and transgenic mice expressing different numbers of the *a* and *b* alleles of *Prnp* (Table 16.3), we now realize that these findings were not paradoxical; indeed, they result from increased PrP gene dosage (66). When the RML isolate was inoculated into congenic and transgenic mice, increasing the number of copies of the *a* allele was found to be the major determinant in reducing the incubation time; however, increasing the number of copies of the *b* allele also reduced the incubation time, but not to the same extent as that seen with the *a* allele (Table 16.3).

The discovery that incubation times are controlled by the relative dosage of *Prnp^a* and *Prnp^b* alleles was foreshadowed by studies of Tg(SHaPrP) mice in which the length of the incubation time after inoculation with SHa prions was inversely proportional to the transgene product, SHaPrP^C (39). Not only does the PrP gene dose determine the length of the incubation time, but also the passage history of the inoculum, particularly in *Prnp^b* mice (Table 16.4). The PrP^{Sc} allotype in the inoculum produced the shortest incubation times when it was the same as that of PrP^C in the host (96). The term *allotype* is used to describe allelic variants of PrP. To address the issue of whether gene products other than PrP might be responsible for these findings, we inoculated B6 and B6.I-4 mice carrying *Prnp$^{a/a}$* as well as I/Ln, and B6.I-2 mice (66, 71), with RML prions passaged in mice homozygous for either the *a* or *b* allele of *Prnp* (Table 16.4). CD-1 and NZW/LacJ mice produce prions containing PrP^{Sc}-A encoded by *Prnp^a* while I/LnJ mice produce PrP^{Sc}-B prions. The incubation times in the congenic mice reflected the PrP allotype, rather than other factors acquired during prion passage. The effect of the allotype barrier was small when measured in *Prnp$^{a/a}$* mice but was clearly demonstrable in *Prnp$^{b/b}$* mice. B6.I-2 congenic mice inoculated with prions from I/Ln mice had an incubation time of 237 ± 8 days

TABLE 16.4. Mismatching of PrP allotypes between PrP^{Sc} in the inoculum and PrP^C in the inoculated host extends prion incubation times in congenic mice.

Mice	Host Genotype	Host Donor	Inoculum genotype	Donor Incubation time	n
C57BL/6J (B6)	a/a	CD-1	a/a	143 ± 4	8
B6.I-4	a/a	NZW	a/a	144 ± 5	8
B6.I-4	a/a	I/Ln	b/b	150 ± 6	6
B6.I-2	b/b	CD-1	a/a	360 ± 16	8
B6.I-2	b/b	NZW	a/a	404 ± 4	20
B6.I-2	b/b	I/Ln	b/b	237 ± 8	17
I/LnJ[a]	b/b	CD-1	a/a	314 ± 13	11
I/LnJ	b/b	NZW	a/a	283 ± 21	8
I/LnJ	b/b	I/Ln	b/b	193 ± 6	16

[a] I/LnJ results previously reported (66).

compared with times of 360 ± 16 days and 404 ± 4 days for mice inoculated with prions passaged in CD-1 and NZW mice, respectively. Thus, previous passage of prions in $Prnp^b$ mice shortened the incubation time by ~40% when assayed in $Prnp^b$ mice, compared with those inoculated with prions passaged in $Prnp^a$ mice (97).

Overdominance

The phenomenon of *overdominance* in which incubation times in F1 hybrids are longer than those of either parent (98) contributed to the confusion surrounding control of scrapie incubation times. When the 22A scrapie isolate was inoculated into B6, B6.I-1, and (B6 × B6.I-1)F1, overdominance was observed: the scrapie incubation time in B6 mice was 405 ± 2 days, in B6.I mice 194 ± 10 days, and in (B6 × B6.I-1)F1 mice 508 ± 14 days (Table 16.5). Shorter incubation times were observed in Tg(MoPrP-B)15 mice that were either homozygous or hemizygous for the $Prnp^b$ transgene. Hemizygous Tg(MoPrP-B$^{+/0}$)15 mice exhibited a scrapie incubation time of 395 ± 12 days while the homozygous mice had an incubation time of 286 ± 15 days.

As with the results with the RML isolate (Table 16.4), the findings with the 22A isolate can be explained on the basis of gene dosage; however, the relative effects of the *a* and *b* alleles differ in two respects. First, the *b* allele is the major determinant of the scrapie incubation time with the 22A isolate, not the *a* allele. Second, increasing the number of copies of the *a* allele does not diminish the incubation but prolongs it: the *a* allele is inhibitory with the 22A isolate (Table 16.5). With the 87V prion isolate the inhibitory effect of the $Prnp^a$ allele is even more pronounced since only a

TABLE 16.5. MoPrP^C-A inhibits the synthesis of 22A scrapie prions.

Mice	Prnp genotype	Prnp transgenes (copies)	Alleles (copies) a	Alleles (copies) b	Incubation time (days ± SEM)	n
B6.I-1	b/b		0	2	194 ± 10	7
(B6 × B6.I-1)F1	a/b		1	1	508 ± 14	7
C57BL/6J (B6)	a/a		2	0	405 ± 2	8
non-Tg(MoPrP-B$^{0/0}$)15	a/a		2	0	378 ± 8	3[a]
Tg(MoPrP-B$^{+/0}$)15	a/a	bbb/0	2	3	318 ± 14	15[a]
Tg(MoPrP-B$^{+/0}$)15	a/a	bbb/0	2	3	395 ± 12	6[b]
Tg(MoPrP-B$^{+/+}$)15	a/a	bbb/bbb	2	6	266 ± 1	6[a]
Tg(MoPrP-B$^{+/+}$)15	a/a	bbb/bbb	2	6	286 ± 15	5[b]

[a] The homozygous Tg(MoPrP-B$^{+/+}$)15 mice were maintained as a distinct subline selected for transgene homozygosity two generations removed from the (B6 × LT/Sv)F2 founder. Hemizygous Tg(MoPrP-B$^{+/0}$)15 mice were produced by crossing the Tg(MoPrP-B$^{+/+}$)15 line with B6 mice.

[b] Tg(MoPrP-B$^{+/0}$)15 mice were maintained by repeated backcrossing to B6 mice.

few *Prnp^a* and (*Prnp^a* × *Prnp^b*)F1 mice develop scrapie after >600 days postinoculation (66).

The most interesting feature of the incubation time profile for 22A is the overdominance of the *a* allele of *Prnp* in prolonging incubation period. On the basis of overdominance, Dickinson and Outram (99) put forth the replication site hypothesis postulating that dimers of the *Sinc* gene product feature in the replication of the scrapie agent. The results in Table 16.5 are compatible with the interpretation that the target for PrP^{Sc} may be a PrP^C dimer or multimer. The assumptions under this model are that PrP^C-B dimers are more readily converted to PrP^{Sc} than are PrP^C-A dimers and that PrP^C-A:PrP^C-B heterodimers are even more resistant to conversion to PrP^{Sc} than PrP^C-A dimers. Increasing the ratio of PrP-B to PrP-A would lead to shorter incubation times by favoring the formation of PrP^C-B homodimers (Table 16.5). A similar mechanism may account for the relative paucity of individuals heterozygous for the Met/Val polymorphism at codon 129 of the human PrP gene in spontaneous and iatrogenic CJD (100). Alternatively, PrP^C-PrP^{Sc} interaction can be broken down into two distinct aspects: binding affinity and efficacy of conversion to PrP^{Sc}. If PrP-A has a higher affinity for 22A PrP^{Sc} than does PrP^C-B, but is inefficiently converted to PrP^{Sc}, the exceptionally long incubation time of *Prnp^{a/b}* heterozygotes might reflect reduction in the supply of 22A prions available for interaction with the PrP^C-B product of the single *Prnp^b* allele. Additionally, PrP^C-A may inhibit the interaction of 22A PrP^{Sc} with PrP^C-B leading to prolongation of the incubation time. This interpretation is supported by prolonged incubation times in Tg(SHaPrP) mice inoculated with mouse prions in which SHaPrP^C is thought to inhibit the binding of MoPrP^{Sc} to the substrate MoPrP^C (39).

Patterns of PrP^{Sc} Deposition

Besides measurements of the length of the incubation time, profiles of spongiform degeneration have also been used to characterize different prion strains (84, 101). With the development of a new procedure for in situ detection of PrP^{Sc}, designated histoblotting (102), it became possible to localize and quantify PrP^{Sc} as well as to determine whether or not "strains" produce different, reproducible patterns of PrP^{Sc} accumulation (86, 103).

Histoblotting overcame two obstacles that plagued PrP^{Sc} detection in brain by standard immunohistochemical techniques: the presence of PrP^C and weak antigenicity of PrP^{Sc} (104). The histoblot is made by pressing 10-μm-thick cryostat sections of fresh frozen brain tissue to nitrocellulose paper. To localize protease-resistant PrP^{Sc} in brain, the histoblot is digested with proteinase K to eliminate PrP^C, followed by denaturation of the undigested PrP^{Sc} to enhance binding of PrP antibodies. Immunohistochemical staining yields a far more intense, specific, and reproducible PrP signal than can be achieved by immunohistochemistry on standard tissue sections. The intensity of immunostaining correlates well with neurochemical estimates of PrP^{Sc} concentration in homogenates of dissected brain regions. PrP^C can be localized in histoblots of normal brains by eliminating the proteinase K digestion step.

Comparisons of PrP^{Sc} accumulation on histoblots with histologic sections showed that PrP^{Sc} deposition preceded vacuolation and only those regions with PrP^{Sc} underwent degeneration. Microdissection of individual brain regions confirmed the conclusions of the histoblot studies: those regions with high levels of PrP 27–30 had intense vacuolation (105). Thus, we concluded that the deposition of PrP^{Sc} is responsible for the neuropathologic changes found in the prion diseases.

While studies with both mice and Syrian hamsters established that each isolate has a specific signature as defined by a specific pattern of PrP^{Sc} accumulation in the brain (66, 86, 103), comparisons must be done on an isogenic background (37, 40). Variations in the patterns of PrP^{Sc} accumulation were found to be equally as great as those seen between two strains when a single strain is inoculated in mice expressing different PrP genes. Based on the initial studies that were performed in animals of a single genotype, we suggested that PrP^{Sc} synthesis occurs in specific populations of cells for a given distinct prion isolate.

Are Prion Strains Different PrP^{Sc} Conformers?

Explaining the problem of multiple distinct prion isolates might be accommodated by multiple PrP^{Sc} conformers that act as templates for the folding of de novo synthesized PrP^{Sc} molecules during prion "replication" (Fig. 16.4). Although it is clear that passage history can be responsible for the prolongation of incubation time when prions are passed between mice expressing different PrP allotypes (97) or between species (39), many scrapie strains show distinct incubation times in the same inbred host (106).

In recent studies we inoculated three strains of prions into congenic and Tg mice harboring various numbers of the *a* and *b* alleles of *Prnp* (66). The number of *Prnp*^{a} genes was the major determinant of incubation times in mice inoculated with the RML prion isolate and was inversely related to the length of the incubation time (Table 16.3). In contrast, the *Prnp*^{a} allele prevented scrapie in mice inoculated with 87V prions. *Prnp*^{b} genes were permissive for 87V prions and shortened incubation times in most mice inoculated with 22A prions (Table 16.5).

Experiments with the 87V isolate suggest that a genetic locus encoding protein Y, distinct from *Prnp*, controls the deposition of PrP^{Sc} and the attendant neuropathology. While each prion isolate produced distinguishable patterns of PrP^{Sc} accumulation in brain, a comparison of these patterns showed that those patterns found with RML and 22A prions in congenic *Prnp*^{b} mice were more similar than those with RML prions in *Prnp*^{a} and *Prnp*^{b} congenic mice. Thus, both the PrP genotype and prion isolate modify

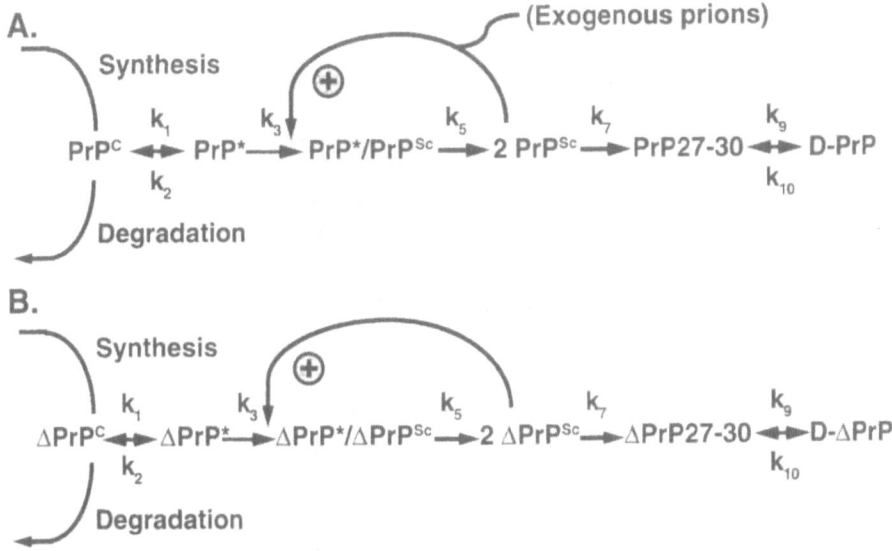

FIGURE 16.4. Models for the replication of prions. (a) Proposed scheme for the replication of prions in sporadic and infectious prion diseases. wtPrP^C is synthesized and degraded as part of the normal metabolism of many cells. Stochastic fluctuations in the structure of PrP^C can create (k_1) a rare, partially unfolded, monomeric structure, PrP*, that is an intermediate in the formation of PrP^{Sc}, but can revert (k_2) to PrP^C or be degraded prior to its conversion (k_3) into PrP^{Sc}. Normally, the concentration of PrP* is small and PrP^{Sc} formation is insignificant. In infectious prion diseases, exogenous prions enter the cell and stimulate conversion of PrP* into PrP^{Sc}. In the absence of exogenous prions, the concentration of PrP^{Sc} may eventually reach a threshold level in sporadic prion diseases after which a positive feedback loop would stimulate the formation of PrP^{Sc}. Limited proteolysis of the N-terminus of PrP^{Sc} produces (k_7) PrP 27-30 which can also be generated in scrapie-infected cells from a recombinant vector encoding PrP truncated at the N-terminus (124). Denaturation (k_9) of PrP^{Sc} or PrP 27-30 renders these molecules protease sensitive and abolishes scrapie infectivity; attempts to renature (k_{10}) these PrP^{Sc} or PrP 27-30 have been unsuccessful to date (7, 10). (b) Scheme for the replication of prions in genetic prion diseases. Mutant (Δ) PrP^C is synthesized and degraded as part of the normal metabolism of many cells. Stochastic fluctuations in the structure of ΔPrP^C are increased compared with wtPrP^C, which creates (k_1) a partially unfolded, monomeric structure, ΔPrP*, that is an intermediate in the formation of ΔPrP^{Sc}, but can revert (k_2) to ΔPrP^C or be degraded prior to its conversion (k_3) into ΔPrP^{Sc}. Limited proteolysis of the N-terminus of ΔPrP^{Sc} produces (k_7) ΔPrP 27-30, which in some cases may be less protease resistant than wtPrP 27-30 (125, 126). Adapted from (127).

the distribution of PrP^{Sc} and the length of the incubation time. These findings suggest that prion strain specified properties result from different affinities of PrP^{Sc} in the inocula for PrP^C-A and PrP^C-B allotypes encoded by the host.

Although the proposal for multiple PrP^{Sc} conformers is rather unorthodox, we already know that PrP can assume at least two profoundly different conformations: PrP^C and PrP^{Sc} (48). Of note, two different isolates from mink dying of transmissible mink encephalopathy exhibit different sensitivities of PrP^{Sc} to proteolytic digestion, supporting the suggestion that isolate-specific information might be carried by PrP^{Sc} (107–109). How many conformations PrP^{Sc} can assume is unknown. The molecular weight of a PrP^{Sc} homodimer is consistent with the ionizing radiation target size of $55,000 \pm 9,000$ daltons as determined for infectious prion particles independent of their polymeric form (110). If prions are oligomers of PrP^{Sc}, which seems likely, then this offers another level of complexity that in turn generates additional diversity.

Conclusion

Prions Are Not Viruses

The study of prions has taken several unexpected directions over the past few years. The discovery that prion diseases in humans are uniquely both genetic and infectious has greatly strengthened and extended the prion concept. To date, 18 different mutations in the human PrP gene all resulting in nonconservative substitutions have been found to either be linked genetically to or segregate with the inherited prion diseases. Yet, the transmissible prion particle is composed largely, if not entirely, of an abnormal isoform of the prion protein designated PrP^{Sc} (1). These findings argue that prion diseases should be considered pseudoinfections, since the particles transmitting disease appear to be devoid of a foreign nucleic acid and thus differ from all known microorganisms as well as viruses and viroids. Because much information, especially about scrapie of rodents, has been derived using experimental protocols adapted from virology, we continue to use terms such as *infection, incubation period, transmissibility,* and *end-point titration* in studies of prion diseases.

Do Prions Exist in Lower Organisms?

In *S. cervisiae*, ure2 and [URE3] mutants were described that can grow on ureidosuccinate under conditions of nitrogen repression such as glutamic acid and ammonia (111). Mutants of ure2 exhibit mendelian inheritance, whereas [URE3] is cytoplasmically inherited (112). The [URE3] phenotype can be induced by UV irradiation and by overexpression of ure2p, the gene product of ure2; deletion of ure2 abolishes [URE3]. The function of ure2p is unknown but it has substantial homology with glutathione-S-transferase; attempts to demonstrate this enzymic activity with purified ure2p have not been successful (113). Whether the [URE3] protein is a posttranslationally modified form of ure2p that acts on unmodified ure2p to produce more of itself remains to be established.

Another possible yeast prion is the [PSI] phenotype (112). [PSI] is a nonmendelian inherited trait that can be induced by expression of the PNM2 gene (114). Both [PSI] and [URE3] can be cured by exposure of the yeast to 3 mM guanidine hydrochloride (Gdn HCl). The mechanism responsible for abolishing [PSI] and [URE3] with a low concentration of Gdn HCl is unknown. In the filamentous fungus *Podospora anserina*, the het-s locus controls the vegetative incompatibility; conversion from the S^s to the s state seems to be a posttranslational, autocatalytic process (115).

If any of the above cited examples can be shown to function in a manner similar to prions in animals, then many new, more rapid and economical approaches to prion diseases should be forthcoming.

Common Neurodegenerative Diseases

The knowledge accrued from the study of prion diseases may provide an effective strategy for defining the etiologies and dissecting the molecular pathogenesis of the more common neurodegenerative disorders such as Alzheimer's disease, Parkinson's disease, and amyotrophic lateral sclerosis (ALS). Advances in the molecular genetics of Alzheimer's disease and ALS suggest that, like the prion diseases, an important subset is caused by mutations that result in nonconservative amino acid substitutions in proteins expressed in the CNS (116–123). Since people at risk for inherited prion diseases can now be identified decades before neurologic dysfunction is evident, the development of an effective therapy is imperative.

Future Studies

Tg mice expressing foreign or mutant PrP genes now permit virtually all facets of prion diseases to be studied and have created a framework for future investigations. Furthermore, the structure and organization of the PrP gene suggested that PrP^{Sc} is derived from PrP^C or a precursor by a posttranslational process. Studies with scrapie-infected cultured cells have provided much evidence that the conversion of PrP^C to PrP^{Sc} is a posttranslational process that probably occurs within a subcellular compartment bounded by cholesterol-rich membranes. The molecular mechanism of PrP^{Sc} formation remains to be elucidated, but chemical and physical studies have shown that the conformations of PrP^C and PrP^{Sc} are profoundly different.

224   G.C. Telling et al.

The study of prion biology and diseases seems to be a new and emerging area of biomedical investigation. While prion biology has its roots in virology, neurology, and neuropathology, its relationships to the disciplines of molecular and cell biology as well as protein chemistry have become evident only recently. Certainly, the possibility that learning how prions multiply and cause disease will open up new vistas in biochemistry and genetics seems likely.

Ackowledgments. We thank M. Baldwin, D. Borchelt, G. Carlson, F. Cohen, C. Cooper, S. DeArmond, R. Fletterick, R. Gabizon, M. Gasset, D. Groth, L. Hood, K. Hsiao, Z. Huang, R. Koehler, V. Lingappa, K.-M. Pan, D. Riesner, M. Scott, A. Serban, N. Stahl, A. Taraboulos, M. Torchia, and D. Westaway for their help in these studies. This work is supported by grants from the National Institutes of Health (NS14069, AG08967, AG02132, NS22786, and AG10770) and the American Health Assistance Foundation, as well as by gifts from the Sherman Fairchild Foundation, Bernard Osher Foundation and National Medical Enterprises.

References

1. Prusiner SB. Molecular biology of prion diseases. Science 1991;252:1515–22.
2. Prusiner SB. Novel proteinaceous infectious particles cause scrapie. Science 1982;216:136–44.
3. Alper T, Haig DA, Clarke MC. The exceptionally small size of the scrapie agent. Biochem Biophys Res Commun 1966;22:278–84.
4. Alper T, Cramp WA, Haig DA, Clarke MC. Does the agent of scrapie replicate without nucleic acid? Nature 1967;214:764–6.
5. Hunter GD. Scrapie: a prototype slow infection. J Infect Dis 1972;125:427–40.
6. Riesner D, Kellings K, Meyer N, Mirenda C, Prusiner SB. Nucleic acids and scrapie prions. In: Prusiner SB, Collinge J, Powell J, et al., eds. Prion diseases of humans and animals. London: Ellis Horwood, 1992:341–58.
7. Prusiner SB, McKinley MP, Bowman KA, et al. Scrapie prions aggregate to form amyloid-like birefringent rods. Cell 1983;35:349–58.
8. Büeler H, Aguzzi A, Sailer A, et al. Mice devoid of PrP are resistant to scrapie. Cell 1993;73:1339–47.
9. Prusiner SB, Groth D, Serban A, et al. Ablation of the prion protein (PrP) gene in mice prevents scrapie and facilitates production of anti-PrP antibodies. Proc Natl Acad Sci USA 1993;90:10608–12.
10. Prusiner SB, Groth D, Serban A, Stahl N, Gabizon R. Attempts to restore scrapie prion infectivity after exposure to protein denaturants. Proc Natl Acad Sci USA 1993;90:2793–7.
11. Sigurdsson B. Rida, a chronic encephalitis of sheep with general remarks on infections which develop slowly and some of their special characteristics. Br Vet J 1954;110:341–54.
12. Gajdusek DC. Unconventional viruses and the origin and disappearance of kuru. Science 1977;197:943–60.

13. Gajdusek DC. Subacute spongiform virus encephalopathies caused by unconventional viruses. In: Maramorosch K, McKelvey JJ Jr, eds. Subviral pathogens of plants and animals: viroids and prions. Orlando: Academic Press, 1985:483–544.

14. Gajdusek DC, Gibbs CJ Jr, Alpers M. Experimental transmission of a kuru-like syndrome to chimpanzees. Nature 1966;209:794–6.

15. Kirschbaum WR. Zwei eigenartige Erkrankungen des Zentralnervensystems nach Art der spastischen Pseudosklerose (Jakob). Z Ges Neurol Psychiatr 1924;92:175–220.

16. Meggendorfer F. Klinische und genealogische Beobachtungen bei einem Fall von spastischer Pseudosklerose Jakobs. Z Ges Neurol Psychiatr 1930;128:337–41.

17. Sparkes RS, Simon M, Cohn VH, et al. Assignment of the human and mouse prion protein genes to homologous chromosomes. Proc Natl Acad Sci USA 1986;83:7358–62.

18. Hsiao K, Baker HF, Crow TJ, et al. Linkage of a prion protein missense variant to Gerstmann-Sträussler syndrome. Nature 1989;338:342–5.

19. Prusiner SB. Inherited prion diseases. Proc Natl Acad Sci USA 1994;91:4611–4.

20. Masters CL, Gajdusek DC, Gibbs CJ Jr. Creutzfeldt-Jakob disease virus isolations from the Gerstmann-Sträussler syndrome. Brain 1981;104:559–88.

21. Prusiner SB. Scrapie prions. Annu Rev Microbiol 1989;43:345–74.

22. Tateishi J, Doh-ura K, Kitamoto T, et al. Prion protein gene analysis and transmission studies of Creutzfeldt-Jakob disease. In: Prusiner SB, Collinge J, Powell J, et al., eds. Prion diseases of humans and animals. London: Ellis Horwood, 1992:129–34.

23. Malmgren R, Kurland L, Mokri B, Kurtzke J. The epidemiology of Creutzfeldt-Jakob disease. In: Prusiner SB, Hadlow WJ, eds. Slow transmissible diseases of the nervous system, vol 1. New York: Academic Press, 1979:93–112.

24. Brown P, Cathala F, Raubertas RF, Gajdusek DC, Castaigne P. The epidemiology of Creutzfeldt-Jakob disease: conclusion of 15-year investigation in France and review of the world literature. Neurology 1987;37:895–904.

25. Harries-Jones R, Knight R, Will RG, Cousens S, Smith PG, Matthews WB. Creutzfeldt-Jakob disease in England and Wales, 1980–1984: a case-control study of potential risk factors. J Neurol Neurosurg Psychiatry 1988;51:1113–9.

26. Cousens SN, Harries-Jones R, Knight R, Will RG, Smith PG, Matthews WB. Geographical distribution of cases of Creutzfeldt-Jakob disease in England and Wales 1970–84. J Neurol Neurosurg Psychiatry 1990;53:459–65.

27. Borchelt DR, Scott M, Taraboulos A, Stahl N, Prusiner SB. Scrapie and cellular prion proteins differ in their kinetics of synthesis and topology in cultured cells. J Cell Biol 1990;110:743–52.

28. Westaway D, DeArmond SJ, Cayetano-Canlas J, et al. Degeneration of skeletal muscle, peripheral nerves, and the central nervous system in transgenic mice overexpressing wild-type prion proteins. Cell 1994;76:117–29.

29. Masters CL, Harris JO, Gajdusek DC, Gibbs CJ Jr, Bernouilli C, Asher DM. Creutzfeldt-Jakob disease: patterns of worldwide occurrence and the significance of familial and sporadic clustering. Ann Neurol 1978;5:177–88.

30. Gajdusek DC, Zigas V. Degenerative disease of the central nervous system in New Guinea—the endemic occurrence of "kuru" in the native population. N Engl J Med 1957;257:974–8.

31. Gajdusek DC, Zigas V. Clinical, pathological and epidemiological study of an acute progressive degenerative disease of the central nervous system among natives of the eastern highlands of New Guinea. Am J Med 1959;26:442–69.

32. Alpers M. Epidemiology and clinical aspects of kuru. In: Prusiner SB, McKinley MP, eds. Prions—novel infectious pathogens causing scrapie and Creutzfeldt-Jakob disease. Orlando: Academic Press, 1987:451–65.

33. Hsiao K, Prusiner SB. Inherited human prion diseases. Neurology 1990; 40:1820–7.

34. Brown P. The phenotypic expression of different mutations in transmissible human spongiform encephalopathy. Rev Neurol 1992;148:317–27.

35. Medori R, Tritschler H-J, LeBlanc A, et al. Fatal familial insomnia, a prion disease with a mutation at codon 178 of the prion protein gene. N Engl J Med 1992;326:444–9.

36. Hsiao KK, Scott M, Foster D, Groth DF, DeArmond SJ, Prusiner SB. Spontaneous neurodegeneration in transgenic mice with mutant prion protein. Science 1990;250:1587–90.

37. Hsiao KK, Groth D, Scott M, et al. Serial transmission in rodents of neurodegeneration from transgenic mice expressing mutant prion protein. Proc Natl Acad Sci USA 1994;91:9126–30.

38. Telling GC, Scott M, Hsiao KK, et al. Transmission of Creutzfeldt-Jakob disease from humans to transgenic mice expressing chimeric human-mouse prion protein. Proc Natl Acad Sci USA 1994;91:9936–40.

39. Prusiner SB, Scott M, Foster D, et al. Transgenetic studies implicate interactions between homologous PrP isoforms in scrapie prion replication. Cell 1990;63:673–86.

40. Scott M, Groth D, Foster D, et al. Propagation of prions with artificial properties in transgenic mice expressing chimeric PrP genes. Cell 1993;73:979–88.

41. Stahl N, Borchelt DR, Hsiao K, Prusiner SB. Scrapie prion protein contains a phosphatidylinositol glycolipid. Cell 1987;51:229–40.

42. Gordon WS. Advances in veterinary research. Vet Res 1946;58:516–20.

43. Parry HB. In: Oppenheimer DR, ed. Scrapie disease in sheep. New York: Academic Press, 1983.

44. Parry HB. Scrapie: a transmissible and hereditary disease of sheep. Heredity 1962;17:75–105.

45. Dickinson AG, Young GB, Stamp JT, Renwick CC. An analysis of natural scrapie in Suffolk sheep. Heredity 1965;20:485–503.

46. Wilesmith JW, Ryan JBM, Hueston WD, Hoinville LJ. Bovine spongiform encephalopathy: epidemiological features 1985 to 1990. Vet Rec 1992;130:90–4.

47. Stahl N, Baldwin MA, Teplow DB, et al. Structural analysis of the scrapie prion protein using mass spectrometry and amino acid sequencing. Biochemistry 1993;32:1991–2002.

48. Pan K-M, Baldwin M, Nguyen J, et al. Conversion of α-helices into β-sheets features in the formation of the scrapie prion proteins. Proc Natl Acad Sci USA 1993;90:10962–6.

49. Pattison IH. Experiments with scrapie with special reference to the nature of the agent and the pathology of the disease. In: Gajdusek DC, Gibbs CJ Jr,

Alpers MP, eds. Slow, latent and temperate virus infections, NINDB Monograph 2. Washington, DC: U.S. Government Printing, 1965:249–57.

50. Pattison IH. The relative susceptibility of sheep, goats and mice to two types of the goat scrapie agent. Res Vet Sci 1966;7:207–12.

51. Pattison IH, Jones KM. The possible nature of the transmissible agent of scrapie. Vet Rec 1967;80:1–8.

52. Bockman JM, Prusiner SB, Tateishi J, Kingsbury DT. Immunoblotting of Creutzfeldt-Jakob disease prion proteins: host species-specific epitopes. Ann Neurol 1987;21:589–95.

53. Wilesmith JW, Hoinville LJ, Ryan JBM, Sayers AR. Bovine spongiform encephalopathy: aspects of the clinical picture and analyses of possible changes 1986–1990. Vet Rec 1992;130:197–201.

54. Goldmann W, Hunter N, Martin T, Dawson M, Hope J. Different forms of the bovine PrP gene have five or six copies of a short, G-C-rich element within the protein-coding exon. J Gen Virol 1991;72:201–4.

55. Hope J, Reekie LJD, Hunter N, et al. Fibrils from brains of cows with new cattle disease contain scrapie-associated protein. Nature 1988;336:390–2.

56. Prusiner SB, Fuzi M, Scott M, et al. Immunologic and molecular biological studies of prion proteins in bovine spongiform encephalopathy. J Infect Dis 1993;167:602–13.

57. Scott M, Foster D, Mirenda C, et al. Transgenic mice expressing hamster prion protein produce species-specific scrapie infectivity and amyloid plaques. Cell 1989;59:847–57.

58. Westaway D, Cooper C, Turner S, Da Costa M, Carlson GA, Prusiner SB. Structure and polymorphism of the mouse prion protein gene. Proc Natl Acad Sci USA 1994;91:6418–22.

59. Büeler H, Fischer M, Lang Y, et al. Normal development and behaviour of mice lacking the neuronal cell-surface PrP protein. Nature 1992;356:577–82.

60. Barry RA, Prusiner SB. Monoclonal antibodies to the cellular and scrapie prion proteins. J Infect Dis 1986;154:518–21.

61. Kascsak RJ, Rubenstein R, Merz PA, et al. Mouse polyclonal and monoclonal antibody to scrapie-associated fibril proteins. J Virol 1987;61:3688–93.

62. Rogers M, Serban D, Gyuris T, Scott M, Torchia T, Prusiner SB. Epitope mapping of the Syrian hamster prion protein utilizing chimeric and mutant genes in a vaccinia virus expression system. J Immunol 1991;147:3568–74.

63. Brown P, Goldfarb LG, Kovanen J, et al. Phenotypic characteristics of familial Creutzfeldt-Jakob disease associated with the codon 178[Asn] *PRNP* mutation. Ann Neurol 1992;31:282–5.

64. Medori R, Montagna P, Tritschler HJ, et al. Fatal familial insomnia: a second kindred with mutation of prion protein gene at codon 178. Neurology 1992;42:669–70.

65. Telling GT, Haga T, Torchia M, Tremblay P, DeArmond SJ, Prusiner SB. Interactions between wild-type and mutant prior proteins modulate neurodegeneration in transgenic mice. Genes Dev 1996 (in press).

66. Carlson GA, Ebeling C, Yang S-L, et al. Prion isolate specified allotypic interactions between the cellular and scrapie prion proteins in congenic and transgenic mice. Proc Natl Acad Sci USA 1994;91:5690–4.

67. Scott MR, Köhler R, Foster D, Prusiner SB. Chimeric prion protein expression in cultured cells and transgenic mice. Protein Sci 1992;1:986–97.

68. Kretzschmar HA, Stowring LE, Westaway D, Stubblebine WH, Prusiner SB, DeArmond SJ. Molecular cloning of a human prion protein cDNA. DNA 1986;5:315–24.

69. Raeber AJ, Borchelt DR, Scott M, Prusiner SB. Attempts to convert the cellular prion protein into the scrapie isoform in cell-free systems. J Virol 1992;66:6155–63.

70. Oesch B, Teplow DB, Stahl N, Serban D, Hood LE, Prusiner SB. Identification of cellular proteins binding to the scrapie prion protein. Biochemistry 1990;29:5848–55.

71. Carlson GA, Ebeling C, Torchia M, Westaway D, Prusiner SB. Delimiting the location of the scrapie prion incubation time gene on chromosome 2 of the mouse. Genetics 1993;133:979–88.

72. Bruce ME, Dickinson AG, Fraser H. Cerebral amyloidosis in scrapie in the mouse: effect of agent strain and mouse genotype. Neuropathol Appl Neurobiol 1976;2:471–8.

73. Cuillé J, Chelle PL. Experimental transmission of trembling to the goat. CR Seances Acad Sci 1939;208:1058–60.

74. Dickinson AG, Stamp JT. Experimental scrapie in Cheviot and Suffolk sheep. J Comp Pathol 1969;79:23–6.

75. Hadlow WJ, Kennedy RC, Race RE. Natural infection of Suffolk sheep with scrapie virus. J Infect Dis 1982;146:657–64.

76. Hadlow WJ, Kennedy RC, Race RE, Eklund CM. Virologic and neurohistologic findings in dairy goats affected with natural scrapie. Vet Pathol 1980;17:187–99.

77. Gordon WS. Variation in susceptibility of sheep to scrapie and genetic implications. In: Report of scrapie seminar, ARS 91-53. Washington, DC: U.S. Department of Agriculture, 1966:53–67.

78. Pattison IH, Millson GC. Scrapie produced experimentally in goats with special reference to the clinical syndrome. J Comp Pathol 1961;71:101–8.

79. Dickinson AG, Fraser H. An assessment of the genetics of scrapie in sheep and mice. In: Prusiner SB, Hadlow WJ, eds. Slow transmissible diseases of the nervous system, vol 1. New York: Academic Press, 1979:367–86.

80. Bruce ME, Dickinson AG. Biological evidence that the scrapie agent has an independent genome. J Gen Virol 1987;68:79–89.

81. Kimberlin RH, Cole S, Walker CA. Temporary and permanent modifications to a single strain of mouse scrapie on transmission to rats and hamsters. J Gen Virol 1987;68:1875–81.

82. Dickinson AG, Outram GW. Genetic aspects of unconventional virus infections: the basis of the virino hypothesis. In: Bock G, Marsh J, eds. Novel infectious agents and the central nervous system. Ciba Foundation Symposium 135. Chichester, UK: John Wiley and Sons, 1988:63–83.

83. Dickinson AG, Meikle VMH, Fraser H. Identification of a gene which controls the incubation period of some strains of scrapie agent in mice. J Comp Pathol 1968;78:293–9.

84. Fraser H, Dickinson AG. Scrapie in mice. Agent-strain differences in the distribution and intensity of grey matter vacuolation. J Comp Pathol 1973;83:29–40.

85. Bruce ME, McBride PA, Farquhar CF. Precise targeting of the pathology of the sialoglycoprotein, PrP, and vacuolar degeneration in mouse scrapie. Neurosci Lett 1989;102:1–6.

86. Hecker R, Taraboulos A, Scott M, et al. Replication of distinct prion isolates is region specific in brains of transgenic mice and hamsters. Genes Dev 1992;6:1213–28.

87. Dickinson AG, Meikle VMH. Host-genotype and agent effects in scrapie incubation: change in allelic interaction with different strains of agent. Mol Gen Genet 1971;112:73–9.

88. Dickinson AG, Bruce ME, Outram GW, Kimberlin RH. Scrapie strain differences: the implications of stability and mutation. In: Tateishi J, ed. Proceedings of workshop on slow transmissible diseases. Tokyo: Japanese Ministry of Health and Welfare, 1984:105–18.

89. Kingsbury DT, Kasper KC, Stites DP, Watson JD, Hogan RN, Prusiner SB. Genetic control of scrapie and Creutzfeldt-Jakob disease in mice. J Immunol 1983;131:491–6.

90. Carp RI, Moretz RC, Natelli M, Dickinson AG. Genetic control of scrapie: incubation period and plaque formation in mice. J Gen Virol 1987;68:401–7.

91. Carlson GA, Kingsbury DT, Goodman PA, et al. Linkage of prion protein and scrapie incubation time genes. Cell 1986;46:503–11.

92. Hunter N, Hope J, McConnell I, Dickinson AG. Linkage of the scrapie-associated fibril protein (PrP) gene and Sinc using congenic mice and restriction fragment length polymorphism analysis. J Gen Virol 1987;68:2711–6.

93. Race RE, Graham K, Ernst D, Caughey B, Chesebro B. Analysis of linkage between scrapie incubation period and the prion protein gene in mice. J Gen Virol 1990;71:493–7.

94. Ziegler DR. In: O'Brien SJ, ed. Genetic maps—locus maps of complex genomes. 6th ed. Cold Spring Harbor, NY: Cold Spring Harbor Laboratory Press, 1993;4.42–5.

95. Westaway D, Goodman PA, Mirenda CA, McKinley MP, Carlson GA, Prusiner SB. Distinct prion proteins in short and long scrapie incubation period mice. Cell 1987;51:651–62.

96. Westaway D, Mirenda CA, Foster D, et al. Paradoxical shortening of scrapie incubation times by expression of prion protein transgenes derived from long incubation period mice. Neuron 1991;7:59–68.

97. Carlson GA, Westaway D, DeArmond SJ, Peterson-Torchia M, Prusiner SB. Primary structure of prion protein may modify scrapie isolate properties. Proc Natl Acad Sci USA 1989;86:7475–9.

98. Dickinson AG, Meikle VM. A comparison of some biological characteristics of the mouse-passaged scrapie agents, 22A and ME7. Genet Res 1969;13:213–25.

99. Dickinson AG, Outram GW. The scrapie replication-site hypothesis and its implications for pathogenesis. In: Prusiner SB, Hadlow WJ, eds. Slow transmissible diseases of the nervous system, vol 2. New York: Academic Press, 1979:13–31.

100. Palmer MS, Dryden AJ, Hughes JT, Collinge J. Homozygous prion protein genotype predisposes to sporadic Creutzfeldt-Jakob disease. Nature 1991; 352:340–2.

101. Fraser H. Neuropathology of scrapie: the precision of the lesions and their diversity. In: Prusiner SB, Hadlow WJ, eds. Slow transmissible diseases of the nervous system, vol 1. New York: Academic Press, 1979:387–406.

102. Taraboulos A, Jendroska K, Serban D, Yang S-L, DeArmond SJ, Prusiner SB. Regional mapping of prion proteins in brains. Proc Natl Acad Sci USA 1992;89:7620–4.

103. DeArmond SJ, Yang S-L, Lee A, et al. Three scrapie prion isolates exhibit different accumulation patterns of the prion protein scrapie isoform. Proc Natl Acad Sci USA 1993;90:6449–53.

104. DeArmond SJ, Mobley WC, DeMott DL, Barry RA, Beckstead JH, Prusiner SB. Changes in the localization of brain prion proteins during scrapie infection. Neurology 1987;37:1271–80.

105. Casaccia-Bonnefil P, Kascsak RJ, Fersko R, Callahan S, Carp RI. Brain regional distribution of prion protein PrP27–30 in mice stereotaxically microinjected with different strains of scrapie. J Infect Dis 1993;167:7–12.

106. Bruce ME, McConnell I, Fraser H, Dickinson AG. The disease characteristics of different strains of scrapie in *Sinc* congenic mouse lines: implications for the nature of the agent and host control of pathogenesis. J Gen Virol 1991;72: 595–603.

107. Marsh RF, Bessen RA, Lehmann S, Hartsough GR. Epidemiological and experimental studies on a new incident of transmissible mink encephalopathy. J Gen Virol 1991;72:589–94.

108. Bessen RA, Marsh RF. Biochemical and physical properties of the prion protein from two strains of the transmissible mink encephalopathy agent. J Virol 1992;66:2096–101.

109. Bessen RA, Marsh RF. Identification of two biologically distinct strains of transmissible mink encephalopathy in hamsters. J Gen Virol 1992;73:329–34.

110. Bellinger-Kawahara CG, Kempner E, Groth DF, Gabizon R, Prusiner SB. Scrapie prion liposomes and rods exhibit target sizes of 55,000 Da. Virology 1988;164:537–41.

111. Lacroute F. Non-mendelian mutation allowing ureidosuccinic acid uptake in yeast. J Bacteriol 1971;106:519–22.

112. Wickner RB. [URE3] as an altered URE2 protein: evidence for a prion analog in *Saccharomyces cerevisiae*. Science 1994;264:566–9.

113. Coschigano PW, Magasanik B. The *URE2* gene product of *Saccharomyces cerevisiae* plays an important role in the cellular response to the nitrogen source and has homology to glutathione *S*-transferases. Mol Cell Biol 1991;11:822–32.

114. Cox BS, Tuite MF, McLaughlin CS. The psi factor of yeast: a problem in inheritance. Yeast 1988;4:159–78.

115. Deleu C, Clavé C, Bégueret J. A single amino acid difference is sufficient to elicit vegetative incompatibility in the fungus *Podospora anserina*. Genetics 1993;135:45–52.

116. Schellenberg GD, Bird TD, Wijsman EM, et al. Genetic linkage evidence for a familial Alzheimer's disease locus on chromosome 14. Science 1992;258: 668–71.

117. Van Broeckhoven C, Backhovens H, Cruts M, et al. Mapping of a gene predisposing to early-onset Alzheimer's disease to chromosome 14q24.3. Nat Genet 1992;2:335–9.

118. Van Broeckhoven C, Haan J, Bakker E, et al. Amyloid β protein precursor gene and hereditary cerebral hemorrhage with amyloidosis (Dutch). Science 1990;248:1120–2.

119. Rosen DR, Siddique T, Patterson D, et al. Mutations in Cu/Zn superoxide dismutase gene are associated with familial amyotrophic lateral sclerosis. Nature 1993;362:59–62.

120. Mullan M, Houlden H, Windelspecht M, et al. A locus for familial early-onset Alzheimer's disease on the long arm of chromosome 14, proximal to the α1-antichymotrypsin gene. Nat Genet 1992;2:340–2.

121. Goate A, Chartier-Harlin M-C, Mullan M, et al. Segregation of a missense mutation in the amyloid precursor protein gene with familial Alzheimer's disease. Nature 1991;349:704–6.

122. St. George-Hyslop P, Haines J, Rogaev E, et al. Genetic evidence for a novel familial Alzheimer's disease locus on chromosome 14. Nat Genet 1992;2:330–4.

123. Levy E, Carman MD, Fernandez-Madrid IJ, et al. Mutation of the Alzheimer's disease amyloid gene in hereditary cerebral hemorrhage, Dutch type. Science 1990;248:1124–6.

124. Rogers M, Yehiely F, Scott M, Prusiner SB. Conversion of truncated and elongated prion proteins into the scrapie isoform in cultured cells. Proc Natl Acad Sci USA 1993;90:3182–6.

125. Monari L, Chen SG, Brown P, et al. Fatal familial insomnia and familial Creutzfeldt-Jakob disease: different prion proteins determined by a DNA polymorphism. Proc Natl Acad Sci USA 1994;91:2839–42.

126. Prusiner SB, Hsiao KK. Human prion diseases. Ann Neurol 1994;35:385–95.

127. Cohen FE, Pan K-M, Huang Z, Baldwin M, Fletterick RJ, Prusiner SB. Structural clues to prion replication. Science 1994;264:530–1.

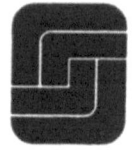

17

Posttranslational Modifications and Conformational Changes of PrP^{Sc} and Their Relationship to Infectivity

RICHARD RUBENSTEIN, RICHARD J. KASCSAK, CAROL L. SCALICI, REGINA FERSKO, ADRIENNE A. RUBENSTEIN, MICHAEL C. PAPINI, AND RICHARD I. CARP

PrP^{Sc} is a specific protein marker for slow infectious diseases known as the transmissible subacute spongiform encephalopathies (TSSE). Although PrP^{Sc} is closely associated with infectivity, it is not known if it is the infectious agent itself, a component of the agent, or merely adventitiously associated with infectivity. In the present study, proteinase K treatment or electrophoretic analysis of partially denatured PrP^{Sc} preparations reveals a dissociation between infectivity and demonstrable PrP^{Sc}. We demonstrate that the resistance of PrP^{Sc} to partial denaturation and of infectivity to inactivation differs markedly for two scrapie strains. The 139A mouse scrapie strain was more susceptible to inactivation compared with the 263K hamster scrapie strain. Our findings support other evidence that not all PrP^{Sc} is required for infectivity.

PrP^{Sc} was monitored for the posttranslational generation of advanced glycosylation end products (AGEs). Immunocytochemical studies identified PrP^{Sc}-associated AGEs within CNS amyloid deposits, suggesting a role for this modification in the pathogenesis and amyloidogenesis of TSSE.

Our studies combined with previous biologic analyses suggest that PrP^{Sc} cannot be the sole component associated with the infectious agent. In addition, the association of this protein with amyloidogenesis may involve posttranslational glycation.

The nature of the scrapie infectious agent remains one of the most intriguing mysteries in the field of TSSE. It is clear that whatever the nature of the agent, the PrP gene product plays a fundamental and critical role in the disease process. The incidence of genetically inherited forms of human prion disease appears to be related to a range of mutations in the PrP gene (1). This gene product also appears to be central in the species barrier phenomena and the regulation of incubation periods (1). Evidence indicates that the abnormal isoform of the PrP gene product, PrP^{Sc}, which is derived from the host-coded PrP^c, is pathogenic and an absolute requirement for development of clinical disease (1). However, the mechanism(s) responsible for the conversion of PrP^c to PrP^{Sc} and the relationship between the infectious agent and the PrP gene product remain unanswered.

Previous studies have sought to help resolve this issue by analyzing the physiochemical properties of the infectious agent. The strongest evidence supporting PrP^{Sc} as a component of the agent is the close association of PrP^{Sc} and infectivity in several different purification protocols (2–4). In addition, levels of PrP^{Sc} in specific organs have been correlated with levels of infectivity in those organs (5). While a considerable portion of abnormal PrP generated during the development of clinical disease does not appear to be directly associated with infectivity but instead with pathogenesis (6), infectivity has never been purified in the absence of PrP^{Sc}. The absolute requirement, however, of PrP^{Sc} for agent replication remains debatable. The failure of PrP knockout mice to exhibit clinical disease following scrapie agent inoculation (7) reinforces the role of abnormal PrP in the development of clinical pathology but does not clarify the role of PrP^{Sc} relative to the infectious agent or its replication (8).

The relationship of PrP^{Sc} and infectivity has also been investigated employing denaturation studies. Procedures that are aimed at denaturing PrP^{Sc} tend to correlate with decreased levels of infectivity (9). Recent studies by Brown et al. (10) have attempted to further characterize the nature of the scrapie agent using a similar approach. Scrapie infectivity associated with partially purified 263K PrP^{Sc} preparations was resistant to inactivation by boiling in the presence of sodium dodecyl sulfate (SDS) and 2-mercaptoethanol (2-ME). Infectivity correlated with gel separated PrP^{Sc} following polyacrylamide gel electrophoresis (PAGE). This is in contrast to earlier studies in which there was a substantial loss of infectivity following SDS and 2-ME treatment and a dissociation between PrP^{Sc} and infectivity (9, 11, 12). In the present study, the relationship between infectivity and PrP^{Sc} was further analyzed following a variety of treatment conditions. Differences among scrapie strains were demonstrated relative to the stability of both infectivity and the associated PrP^{Sc}.

Advanced glycosylation is the nonenzymatic addition of reducing sugars to the primary amino groups of proteins. This process describes the chemical pathway that leads from early glycation products, Schiff bases, and Amadori products to more advanced complex moieties termed *advance glycosylation end products* (AGEs). This posttranslational modification alters the functional, structural, and trafficking properties of the proteins to which they are bound. Such alterations play a role in protein turnover, tissue remodeling, and the pathophysiologic changes that occur in normal aging and neurodegenerative diseases.

Posttranslational modifications have been implicated in the conversion of the host-derived PrP^c into the abnormal pathogenic form, PrP^{Sc}. Since it has been reported that AGEs are associated with the aggregation and cross-linking of proteins, it is likely that they also play a role in amyloidogenesis. Therefore, the role of glycation in PrP^{Sc}-associated amyloidogenesis was investigated.

Materials and Methods

Scrapie Strains

The hamster-adapted scrapie strain, 263K, and the mouse-adapted scrapie strain, 139A, were kindly provided by Dr. Richard Kimberlin (SARDAS, Edinburgh, Scotland). The 139A scrapie strain was passaged and bioassayed in C57Bl/6J mice, obtained from Jackson Laboratories (Bar Harbor, Maine), or Compton White mice (obtained from our breeding colony). Female, weanling mice were used for all experiments. The 263K scrapie strain was passaged and bioassayed in female, weanling LVG/LAK hamsters obtained from Charles River Laboratories (Wilmington, Maine). Preparation of inoculum, injection, scoring, and sacrificing of animals were performed as previously described (13, 14). In the case of SDS-PAGE fragments, infectivity analysis was performed as described by Brown et al. (10).

PrP^{Sc} Purification

Partially purified fractions of PrP^{Sc} were isolated from frozen 263K-infected hamster brains and from frozen 139A-infected mouse brains using the procedure previously described by Hilmert and Diringer (3) that we have modified for use with the TL100 Beckman tabletop ultracentrifuge. The modifications include centrifugation of the proteinase K (PK)-treated pellet in the TL100.4 rotor at 250k xg for 2 hours. The pellet was resuspended in Tris-buffered saline (TBS) containing 0.1% Sarcosyl, layered over a 20% sucrose cushion and centrifuged at 250k xg for 2 hours. The resulting pellet was resuspended in 15 to 20 μL TBS containing 0.1% Sulfobetaine 3-14 (SB3-14) per gram of original brain material. Protein was measured using the Pierce Micro BCA protein assay reagent kit as per manufacturer's instructions.

PrP^{Sc} Treatments

In the initial experiments, approximately 0.4 and 4μg of the partially purified PrP^{Sc} preparations were treated with various combinations of SDS and 2-ME (final concentrations: 1% SDS–2.5% 2-ME, 2% SDS–5% 2-ME, 3%

SDS–7.5% 2-ME). For infectivity measurements by end-point titrations, samples were serially diluted in phosphate-buffered saline (PBS), whereas for SDS-PAGE analysis, all the samples were boiled for 5 min and loaded on 12% Laemmli gels (15). For all subsequent studies, final concentrations of SDS and 2-ME were 1% and 2.5%, respectively.

In addition to the SDS and 2-ME treatments described above, partially purified fractions of PrP^{Sc} were also subjected to the following treatments: (1) PK1—this treatment consisted of incubating a 4-µg aliquot of the partially purified PrP^{Sc} preparations with 100 µg/mL of PK for 1 hour at 37°C. Following the addition of phenylmethylsulfonyl fluoride (PMSF) to 10 mM, the sample was either bioassayed or electrophoresed after adding an equal volume of SDS-PAGE sample buffer (final concentrations: 1% SDS, 2.5% 2-ME, 6% glycerol, 0.001% bromophenol blue, in 62.5 mM Tris-HCl, pH 6.8) and boiling for 5 min; (2) PK2—a 4-µg aliquot of the PrP^{Sc} preparation was first boiled for 5 min in the presence of 1% SDS and 2.5% 2-ME. The sample was cooled to 4°C and then treated with 100 µg/mL PK for 1 hr at 37°C. After adding PMSF to 10 mM, the sample was either bioassayed or subjected to SDS-PAGE as described above.

Western Blot Analysis

Following SDS-PAGE, proteins were transferred to nitrocellulose and immunostained (16) with either the 3F4 monoclonal antibody (17) for 263K PrP^{Sc} or a rabbit anti-mouse PrP sera (17) for 139A PrP^{Sc}.

Western blotting for the analysis of AGEs was performed on brain homogenates and partially purified PrP^{Sc} fractions. Homogenates were prepared in 100 mM Tris-HCl pH 7.4 and aliquots treated with 2% SDS prior to electrophoresis.

The antibodies used for AGE characterization were obtained from Dr. Richard Bucala, The Picower Institute for Medical Research, Manhasset, New York. RU antibody was raised to AGE-modified RNase and recognized a common epitope on all AGE-modified proteins (18). This antibody is used at 1:100 for immunohistochemistry and 1:1,000 for enzyme-linked immunosorbent assays (ELISAs) and Western blots. The R1 antibody serves as a control and was generated to RNase, which was unmodified.

Immunohistochemistry

Animals to be perfused are given a 200-µL intraperitoneal injection of Nembutal containing 10 mg pentobarbital and perfused through the heart with PBS followed by periodate-lysine-paraformaldehyde fixative containing 0.5% paraformaldehyde (17). The brain is removed and postfixed for 48 hours in the same fixation. Tissue is cut into four sections by making coronal slices at levels 150, 300, and 450 (19). Tissue is postfixed for another 24 hours, washed in PBS, and processed for paraffin embedding. Various pretreatment procedures were used to enhance immunoreactivity:

1. Tissue sections are incubated with formic acid (96%–98%) for 15 min at room temperature followed by washing in PBS.
2. Tissue sections are treated with TUF (target unmasking formula) (Signet Laboratories, Inc.) at 90°C for 15 min followed by washing in PBS.
3. Tissue sections are treated with proteinase K (0.05%) (Boehringer Mannheim) in PBS for 30 min at 37°C followed by washing in PBS containing 100 mM PMSF.

For staining, sections are blocked in PBS containing 10% normal goat serum for 15 min at room temperature. If the primary antibody consists of a mouse antibody to be used on mouse tissue, an additional blocking step is included. Sections are blocked with unconjugated goat anti-mouse antibody from the same source as the antibody, which is used conjugated as part of the detection kit. Following this step, primary antibodies are added in 50 mM Tris pH 7.2 containing 1% normal goat serum, and the antibody remains on the sections at 4°C overnight. Staining is achieved using the DAKO LSAB kit, which employs biotinylated secondary antibody and streptavidin conjugated with alkaline phosphatase. Substrate is new fuchsin, which yields a red precipitate allowing counterstaining with hematoxylin.

Results

Association Between PrP^{Sc} and Infectivity

Purified PrP^{Sc} was prepared from hamster brains infected with the 263K hamster-adapted scrapie strain and mouse brains infected with the 139A mouse-adapted scrapie strain. Approximately 400 to 500 µg of PrP^{Sc} was obtained from 12 g of 263K-infected hamster brains and 300 to 400 µg of PrP^{Sc} from 12 g of 139A-infected mouse brains. Infectivity yields from these preparations varied between 8 to 9.5 $\log_{10} LD_{50}$ units per preparation. For most studies, a 4 µg (2 µL) aliquot was used that contained 6 to 7.5 $\log_{10} LD_{50}$ units of scrapie infectivity.

PrP^{Sc} was treated with various concentrations of SDS and 2-ME, boiled for 5 min, and examined by SDS-PAGE. The combinations of 1% SDS-2.5% 2-ME, 2% SDS-5% 2-ME and 3% SDS-7.5% 2-ME all had similar effects on PrP^{Sc} as examined by Western blotting (Fig. 17.1) and silver staining (not shown). At any given amount of PrP^{Sc}, increasing the SDS and 2-ME concentrations did not affect the intensity of Western blot staining. This indicated that all of the PrP^{Sc} migrated into the gel even at the lowest SDS- 2-ME concentrations. For subsequent studies, we used a 1% SDS and

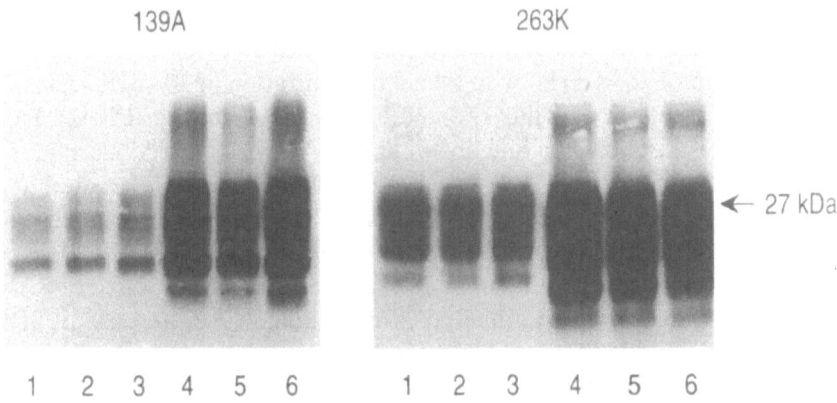

FIGURE 17.1. Western blot analysis of 139A and 263K PrP^{Sc} preparations following treatment with various SDS and 2-ME concentrations. Prior to electrophoresis, approximately 0.4 µg PrP^{Sc} (lanes 1–3) or 4.0 µg PrP^{Sc} (lanes 4–6) was boiled for 5 min in the presence of 1% SDS–2.5% 2-ME (lanes 1,4), 2% SDS–5% 2-ME (lanes 2,5), or 3% SDS–7.5% 2-ME (lanes 3,6).

2.5% 2-ME combination yielding a detergent to PrP^{Sc} ratio of 5 (w/w). These later concentrations are routinely used in sample preparation for Laemmli SDS-PAGE gels and were the concentrations used in the studies by Brown et al. (10).

The effects of proteolysis on PrP^{Sc} was examined for the 263K and 139A scrapie strains (Table 17.1). Although PK digestion is incorporated into the PrP^{Sc} purification protocol, an additional 60-min PK digestion of the partially purified PrP^{Sc} at 37°C (Table 17.1, PK1) reduced the infectivity of the 139A scrapie strain by 1 log. In contrast, a similar treatment of the 263K PrP^{Sc} preparation caused a slight increase in infectivity. The effect of PK treatment was also evident in Western blot analysis of the PrP^{Sc} prepara-

TABLE 17.1. Infectivity titers of 263K and 139A PrP^{Sc} preparations following various treatments.

Treatment	Infectivity ($\log_{10}$ LD$_{50}$/sample)[a]	
	263K	139A
None	7.2	6.7
PK1	7.9	5.7
1% SDS and 2.5% 2-ME	8.3	5.7
1% SDS and 2.5% 2-ME and boiled	8.4	4.2
PK2	≥ 5.9	≤ 2.2

[a] Infectivity titers were determined by end-point titration assays as described in Materials and Methods.

tions in which there was a decrease in 139A PrP^{Sc} immunostaining but no change in 263K PrP^{Sc} staining following PK treatment (Fig. 17.2, lane 2).

Treatment of 263K PrP^{Sc} with a final concentration of 1% SDS–2.5% 2-ME in the absence of boiling resulted in an increase in titer of over 1 log (Table 17.1), probably as a function of disaggregation of infectious units. In contrast, treatment of 139A PrP^{Sc} with 1% SDS–2.5% 2-ME resulted in a 1 log loss of infectivity (Table 17.1). Boiling the 139A PrP^{Sc} samples for 5 min after SDS–2-ME treatment resulted in an additional loss of 1.5 logs of infectivity compared with the unboiled samples. In contrast, infectivity for the 263K scrapie strain was unaffected by boiling in the presence of SDS and 2-ME (Table 17.1). Our results for 263K are similar to those obtained by Brown et al. (10) with this scrapie strain.

Additional studies were performed to examine differences in the stability of PrP^{Sc} and infectivity associated with the two scrapie strains. Following treatment of PrP^{Sc} by boiling in the presence of SDS and 2-ME as described above, the samples were treated with PK. Infectivity measurements (Table 17.1, PK2) were performed and PrP^{Sc} was analyzed by Western blotting and silver staining. Bioassays demonstrated that although equivalent infectivity titers were present in the samples prior to treatment, following the SDS–2-ME treatment and PK digestion, the two scrapie strains showed a differential reduction in titers. At least 5.9 logs of infectivity remained for 263K, while less than 2.2 logs of infectivity could be detected in 139A preparations. PrP^{Sc}, however, could not be detected by either silver staining or Western blotting (data not shown) in either (263K or 139A) preparation. We have previously determined that the limits of PrP^{Sc} detectability by Western blot analysis is approximately 20 to 50 ng. Based on our infectivity results and the limits of PrP^{Sc} detection, we can conclude that there are 10^6 PrP^{Sc} molecules/infectious unit. Our calculations are similar to those previously published (20). In addition, these results clearly point to the existence

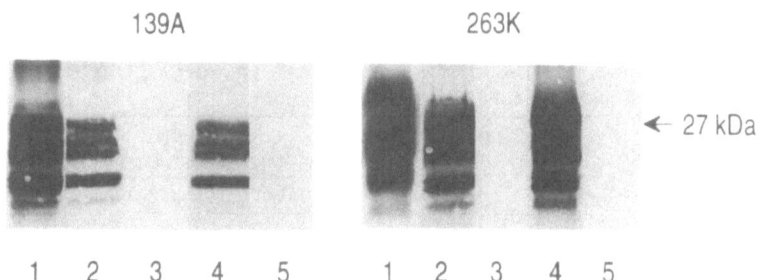

FIGURE 17.2. Western blot analysis of 139A and 263K PrP^{Sc} preparations following PK1 and PK2 treatments (see Materials and Methods). Prior to electrophoresis, the PrP^{Sc} samples were either untreated (lane 1), PK1 (lane 2) or PK2 (lane 3) treated. Additional samples were diluted, incubated for 24 hr at 37°C and concentrated following PK1 (lane 4) or PK2 treatment (lane 5).

of both infectious and noninfectious forms of PrP^{Sc}. If there was a direct correlation between PrP^{Sc} and infectivity, then a 2.5 log reduction in 263K infectivity titers (from 8.4 to 5.9 in the case of PK2) would be equivalent to a 1:500 dilution of the PrP^{Sc} preparation. Since we can dilute a 263K PrP^{Sc} preparation 3 logs and still maintain Western blot staining, the amount of PrP^{Sc} lost during PK2 was greater than can be accounted for by the 2.5-log reduction in infectivity.

We addressed the possibility that the infectivity that remained following PK2 was the result of PrP^{Sc} reassociation, either after PK inactivation or upon dilution of samples for bioassays. Therefore, following the PK2 treatment of PrP^{Sc} from 139A and 263K, the PK was inactivated with PMSF and the samples were diluted 1,000-fold to simulate the bioassay dilution conditions. These samples were incubated for 24 hours at 37°C and then concentrated to their original volume using a centricon-10 column (Amicon, Beverly, MA). Identical samples, which were not PK treated, were processed similarly. The concentrated samples were analyzed by Western blotting. As shown in Figure 17.2, both the untreated and PK1-treated PrP^{Sc} samples were recovered following concentration. In contrast, the PK2 treated samples were not detectable by Western blotting either before or following dilution and subsequent concentration. Since the results described in Figure 17.2 might have been hampered by the limits of PrP^{Sc} detectability, we repeated these studies starting with 10-fold more PrP^{Sc}. The results of these studies again showed no Western blot staining for PK2-treated samples following attempted reassociation (data not shown).

In an effort to more closely analyze the association of PrP^{Sc} and scrapie infectivity, we performed infectivity measurements of PrP^{Sc} following electrophoresis. SDS-PAGE was performed on 263K PrP^{Sc} after the addition of 1% SDS, 2.5% 2-ME, and boiling for 5 min. PrP^{Sc} containing 8.4 logs of scrapie infectivity was applied to each gel lane. Following SDS-PAGE, one lane of the gel was fixed and silver stained, another Western blotted, and a third quick frozen. After the PrP^{Sc} bands were identified, the frozen lane was cut into seven pieces as depicted in Figure 17.3. Each piece was assayed for scrapie infectivity by end-point titrations. The results, as shown in Figure 17.3, indicated that approximately 1% of the scrapie infectivity could be found in the gel and that infectivity was distributed in all gel fractions. None of the fractions revealed a peak of infectivity and the lowest levels of infectivity were seen in the area of PrP^{Sc} migration. The highest levels of infectivity were found both above (gel fragments 1–3) and below (gel fragment 7) the region of PrP^{Sc} staining. A similar study was performed with 139A PrP^{Sc} preparations. In these experiments, PrP^{Sc} containing 4.2 logs of infectivity was applied to the gel. End-point titrations following SDS-PAGE indicated that none of the five gel regions yielded a peak concentration of infectivity (Fig. 17.3). In addition, the majority of 139A scrapie infectivity was localized above (gel fragments 1–3) the region of PrP^{Sc} staining. It is interesting to note the difference in recovery of infectivity

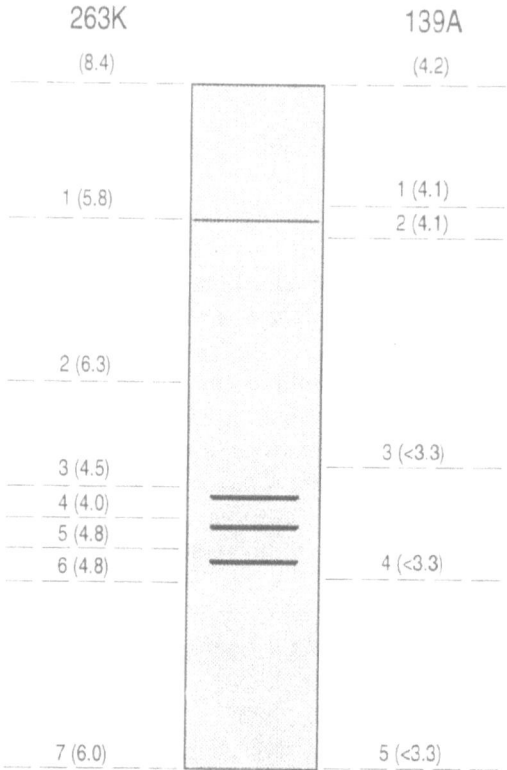

FIGURE 17.3. Diagram depicting infectivity distribution in a gel lane following SDS-PAGE of 263K or 139A PrP^{Sc}. Following electrophoresis, the gel was divided into several fractions and infectivity was measured by endpoint titrations. The numbers in parentheses represent the titers of each gel slice expressed as $\log_{10} LD_{50}$/sample. There are seven gel pieces for 263K and five for 139A. The stacking gel is represented by gel slice 1 (263K and 139A). Gel piece 2 (for 139A only) represents the interface between the stacking and resolving gel. Gel slices 4 to 6 (for 263K) and 4 (for 139A) represent the area of PrP^{Sc} migration as detected on companion Western blots.

between the two scrapie strains. In the case of 139A, between 1 and 2 logs of infectivity are lost during sample preparation and prior to gel loading. All of the infectivity applied to the gel was recovered in the gel fragments following SDS-PAGE. However, in the case of 263K, the loss of infectivity occurred following the SDS-PAGE procedure. The loss of infectivity may have been caused by the inability of infectivity to enter the gel efficiently. We have, therefore, examined this by embedding the PrP^{Sc} sample in agarose plugs and inserting these into the sample wells. The distribution of infectivity following SDS-PAGE utilizing the agarose plugs yielded results for both 139A and 263K that were similar to those described above for samples not embedded in agarose.

Advanced Glycosylation in Prion Diseases

The prion disease-specific protein, PrP^{Sc}, is not only closely associated with the infectious agent but also plays an important role in pathogenesis and amyloidogenesis. Posttranslational modifications have been implicated in the conversion of the host-derived precursor protein (PrP^C) into the abnormal pathogenic isoform. Immunohistochemistry was utilized to monitor the association of AGEs with PrP. Various scrapie strain-host strain combinations were examined. Animals were perfused with paraformaldehyde-lysine-periodate (PLP) fixative and tissue sections of brain were immunostained with various antibodies. Polyclonal rabbit anti-PrP and mouse monoclonal anti-PrP were used to identify PrP antigenic sites. Polyclonal RU antibody, obtained from Dr. R. Bucala, raised to AGEs bound to RNase served to identify brain associated AGEs. This antibody recognizes a conserved epitope associated with AGE modification. R1 antibody, raised to RNase without associated AGEs, and preimmune rabbit sera were used as control antibodies. In addition, tissue sections were stained with rabbit polyclonal antibody to glial fibrillary acidic protein (GFAP) as an additional positive control reagent to monitor astrocytosis.

These studies identified PrP^{Sc} associated AGEs within CNS amyloid deposits of MB mice infected with scrapie strain 87V and of Syrian hamsters infected with scrapie strain 263K (Fig. 17.4). In MB mice infected with scrapie strain 87V, plaque staining was scattered throughout the cerebral cortex (Fig. 17.4, panel A). Staining with antibody to AGEs (Fig. 17.4, panel B) revealed a similar pattern of staining, suggesting a correlation between PrP amyloid deposition and AGE modification. The control antibody for AGE staining (anti-RNase that has not been glycated) did not reveal any staining (Fig. 17.4, panel C). Syrian hamsters infected with scrapie strain 263K were also monitored for AGE accumulation. Sections stained with antibody to PrP revealed PrP amyloid plaque deposition located near the ventricles (Fig. 17.4, panel D). Staining with antibody to AGEs (Fig. 17.4, panel E) again revealed a similar pattern of staining. The antibody to PrP yielded diffuse neutrophil staining in addition to the plaque staining. Antibody to AGE reacted only with the plaques. No staining was observed in tissue sections from uninfected animals. AGEs appeared to be confined to the amyloid deposits and not to neurophil- or cell-associated PrP^{Sc}. Pretreatment of the tissue section prior to immunohistochemistry was required to obtain immunostaining. Pretreatment with 99% formic acid at room temperature for 10 min was the most effective means of enhancing staining. Such treatment has been shown previously to greatly enhance PrP

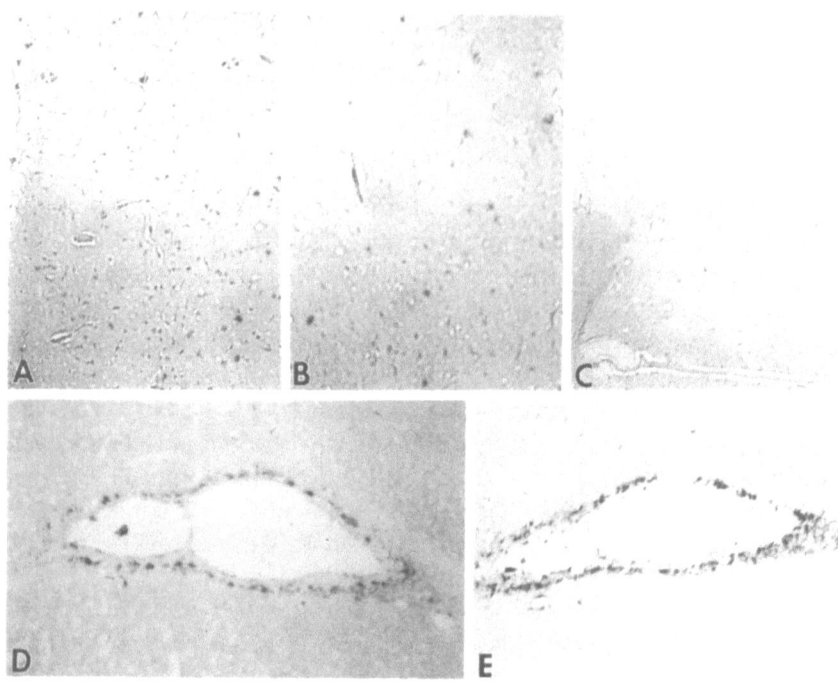

FIGURE 17.4. Immunohistochemical staining for AGEs in scrapie infected animals. MB mice infected with scrapie strain 87V (A–C) or Syrian hamsters infected with scrapie strain 263K (D, E) were perfused with PLP and tissue sections immunostained using rabbit antisera to PrP (A, D), rabbit antisera to AGE-RNase (B, E) or control antibody to RNase (C).

immunostaining. AGE immunostaining can be obtained following a variety of fixation procedures including 10% neutral formalin, 4% paraformaldehyde, 0.5% glutaraldehyde, and PLP containing 1% paraformaldehyde. However, all fixation procedures required pretreatment of the tissue with either formic acid or target unmasking fluid (TUF). The antibodies utilized for the visualization of PrP are particularly sensitive.

The reactivity of amyloid deposits in prion diseases with antibody to AGEs was confirmed in two additional systems (results not shown). Amyloid deposits in the cortex of MB mice infected with scrapie strain ME7 also display AGE immunoreactivity. Tissue sections obtained from a patient with sporadic Creutzfeldt-Jakob disease (CJD), obtained from Dr. P. Brown, Laboratory of Central Nervous Systems Studies, NIH, revealed cortical PrP amyloid deposits that immunoreact with antibody to AGEs. The modification of PrP amyloid with AGEs has now been identified in three different species: humans, mice, and hamsters.

Western blot analysis on brain homogenates (not shown) was used to further examine AGE modification of PrP^{Sc}. Homogenates were prepared from MB mice infected with 87V, Syrian hamsters infected with 263K, and a case of human sporadic CJD. Homogenates were blotted either untreated or following treatment with proteinase K (50 μg/mL 30 min at room temperature). Staining with rabbit antibody to PrP revealed the characteristic PrP banding pattern; three protein bands are present, ranging in molecular weight from 33–35 kd to 20–21 kd. These bands show a characteristic shift in molecular size following protease treatment.

Reactivity of these same samples with rabbit antibody to AGEs failed to reveal immunostaining. This result was in agreement with the immunohistochemical results that indicated that neither cell- nor neurophil-associated PrPs contained AGE modifications. Evidence suggests that only a fraction of PrP^{Sc} is modified with AGEs, that fraction closely associated with amyloid deposits. This result does not rule out the possibility that cell- and neurophil-associated PrP^{Sc} are glycated but have not matured to the extent that they are reactive with anti-AGE antibodies.

Many proteins, including PrP, appear to be susceptible to glycation in vitro. Solubilized (formic acid extracted) and aggregated (infectious) forms of PrP^{Sc} were incubated at 37°C with 500 mM D-glucose-6-phosphate or 500 mM D-ribose for 30 days. Proteins were then analyzed by ELISA assay for reactivity with a variety of antibodies (Table 17.2). Various monoclonal and polyclonal antibodies to PrP were inhibited from binding to glycated PrP^{Sc}. The epitope recognized by Mab 3F4 (amino acids 108–111) was most affected. Antibodies that recognized other epitopes at amino acid 169 (7G5) or amino acids 196–204 (94–5) were less affected but still inhibited from binding. Reactivity of the polyclonal rabbit antibody (78295) was also reduced. Antibody binding activity appeared to be more restricted on the aggregated versus the soluble forms of PrP^{Sc}. Both forms of PrP were immunoreactive with the anti-AGE antibody (RU) and failed to react with the control antibody (R1). The results indicate that glycation of PrP^{Sc} restricts anti-PrP antibody binding and may reflect protein aggregation and/or steric interference with antibody attachment. The interference with antibody binding may restrict the identification of glycated PrP^{Sc} with certain

TABLE 17.2. Advanced glycosylation of PrP^{Sc}.

Sample PrP incubation		Time (days)	Antibodies					
			7G5	3F4	94–5	78295	RU	R1
Control	A	30	1.24[a]	1.34	1.54	1.36	0.14	0.12
G-6-P	A	30	0.16	0.08	0.54	0.25	0.88	0.14
Ribose	A	30	0.24	0.06	0.51	0.28	0.96	0.17
Control	S	30	1.76	1.65	1.42	1.53	0.15	0.16
G-6-P	S	30	0.64	0.18	0.69	0.44	0.98	0.13
Ribose	S	30	0.54	0.19	0.75	0.35	1.56	0.17

[a] Optical density at 405 nm (average duplicate readings from two independent assays).
A, aggregated; S, soluble; G-6-P, D-glucose-6-phosphate.

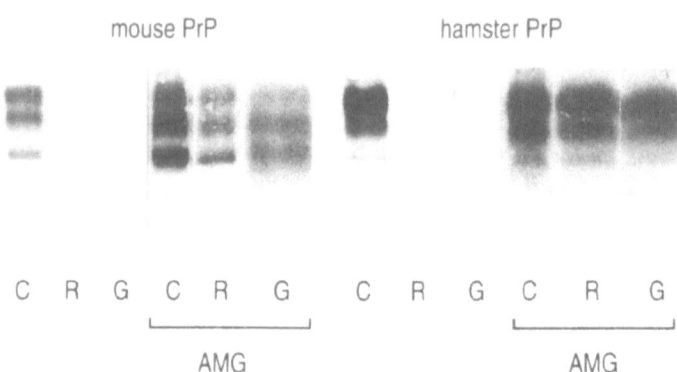

FIGURE 17.5. Western blot analysis of PrP^{Sc} glycation. Partially purified PrP^{Sc} preparations from mouse brains (infected with the 139A scrapie strain) or hamster brains (infected with the 263K scrapie strain) were incubated at 37°C without (C—control) or with sugars [500 mM D-ribose (R) or D-glucose-6-phosphate (G)] for 7 days in the absence or presence of 500 mM aminoguanidine (AMG).

antibodies. The N-terminal region of the protein appeared to be more affected than the C-terminal portion. It is not known if this reflects glycation processes which occur in vivo. Incubation of PrP^{Sc} protein with aminoguanidine (at 500 mM) during the glycation process (in the presence of 500 mM D-glucose-6-phosphate or 500 mM D-ribose) specifically inhibited AGE formation (results not shown). The prevention of glycation also prevented the inhibition of antibody binding to PrP^{Sc}, further indicating that this inhibition was specific to AGE modification. Aminoguanidine has previously been reported to interfere with Amadori product formation. Aminoguanidine, therefore, is an effective tool in both preventing AGE formation and providing evidence for the specificity of glycation-related phenomena.

Similar results to the ELISA was seen by Western blotting of partially purified PrP^{Sc} preparations from the brains of 139A-infected C57Bl/6J mice and 263K-infected Syrian hamsters (Fig. 17.5). Incubation of PrP^{Sc} with 500 mM D-ribose or 500 mM D-glucose-6-phosphate for 7 days showed loss of PrP^{Sc} immunoreactivity. The absence of staining is probably not due to the inability of PrP^{Sc} to migrate into the gel because (a) formic acid treatment of the glycated material prior to Western blotting did not increase PrP^{Sc} immunoreactivity; and (b) slot blot analysis of the glycated samples showed similar results to the Western blots. Furthermore, incubation of PrP^{Sc} with the sugars in the presence of equimolar concentrations of aminoquanidine prevented the loss of immunoreactivity (Fig. 17.5).

Scrapie infectivity of the glycated PrP^{Sc} was measured by incubation period bioassays. Both the 139A and 263K scrapie strains showed a 2-log loss of infectivity after a 7-day incubation with either D-ribose or D-glucose-6-phosphate (data not shown).

Discussion

Numerous studies have suggested an intimate association between PrP^{Sc} and infectivity. This study has utilized partial denaturation and inactivation to further investigate this correlation. Following boiling in the presence of SDS and 2-ME, scrapie strain 139A displayed a 4-log reduction in infectivity, whereas the titer of 263K was unaffected by the treatment. The greater instability of 139A was also reflected in the increased sensitivity of 139A PrP^{Sc} to protease digestion compared to 263K PrP^{Sc}. Our results are in agreement with those published by Brown et al. (10) in which infectivity associated with purified 263K PrP^{Sc} preparations was resistant to inactivation by boiling in the presence of SDS and 2-ME. This is in contrast to earlier studies (21) in which the same hamster scrapie strain (263K) was shown to exhibit a substantial loss of infectivity following such treatment. However, those studies were performed with scrapie samples in which infectivity was not very extensively purified and brain samples that probably contained less aggregated proteins. The infectivity of the 139A preparation was only reduced 1 log in the presence of SDS–2-ME but further reduced by approximately 1.5 logs upon subsequent boiling. Earlier studies (21) had suggested the relative instability of 139A infectivity even in low concentrations of SDS. The greater inherent stability to denaturation exhibited by scrapie strain 263K is probably a function of its greater resistance to proteolytic digestion (16, 22) and the more tightly packed conformation of its SAF (16) compared with 139A. The stability of 263K may be a consequence of species differences in PrP (hamster vs. mouse) or a scrapie strain specified effect. Strain-specific influences are likely to play a role in that similar differences in stability have been associated with scrapie strains propagated in the same strain of mice (22) and with transmissible mink encephalopathy strains propagated in hamsters (23).

The high level of stability of 263K infectivity was evidenced by the high recovery of infectivity following boiling in SDS–2-ME and subsequent PK digestion. While such treatment resulted in the recovery of substantial levels of infectivity in the absence of detectable PrP^{Sc} by Western blot, a complete dissociation between PrP^{Sc} and infectivity has not been demonstrated due to the insensitivity (lower limit is 20–50 ng) of the PrP^{Sc} detection system. Comparison of infectivity and maximum PrP^{Sc} concentrations in such a preparation indicated approximately 10^6 PrP^{Sc} molecules per unit of infectivity. This falls within the range of molecules reported previously (20). The 10^6 value per unit of infectivity is about 10-fold lower than that seen in the original partially purified preparations prior to boiling and protease digestion. This suggests that a considerable amount of PrP^{Sc} can be removed from these preparations without loss of infectivity. Other studies have indicated that a large portion of PrP^{Sc} is not required for infectivity. Treatment with high pH has allowed infectivity to be separated from much of the PrP^{Sc} associated with partially purified PrP^{Sc} preparations from transmitted CJD (24). Treatment of 263K-infected hamsters with amphotericin B dissociated the peak of agent replication (infectivity) from the peak of PrP^{Sc} production (12). Some authors have suggested that as few as two or three PrP^{Sc} molecules may be all that is required for infectivity (2). It is certainly evident from the present studies that protease resistance is in and of itself not sufficient to render PrP^{Sc} infectious.

The stability of 263K infectivity in the presence of SDS and 2-ME afforded the ability to investigate the electrophoretic properties of infectivity using PAGE. Brown et al. (10) reported that scrapie infectivity was found almost exclusively in the region of the gel containing PrP^{Sc}. Based on these findings they concluded that the protein alone (prion theory) was required for infectivity and that additional components were not necessary. Our studies, however, do not agree and demonstrate that scrapie infectivity is largely distributed in regions of the gel not associated specifically with the migration of PrP^{Sc}. It is unclear why a large portion of 263K infectivity was isolated at the bottom of the gel, well below the size range of PrP^{Sc}. This finding may indicate an inherent technical difficulty in this type of analysis, which may also have contributed to the findings of Brown et al (10). Such a distribution of infectivity may reflect conformational alterations as well as varying degrees of aggregation and denaturation within the preparation. Such experiments are complicated by the propensity of both PrP^{Sc} and infectivity to self-aggregate and to aggregate with other components (such as other proteins and nucleic acids) and our lack of understanding concerning the physical properties of the infectious agent.

The failure to identify chemical differences between the normal and abnormal isoforms of PrP suggests that the differences between these proteins must be defined on the basis of conformation. Within the protein only model, this conformational change would confer both infectious and pathogenic properties. Complete denaturation as would be required for accurate migration in an SDS-PAGE system would most likely destroy this conformational alteration and render the protein noninfectious. Likewise, complete denaturation conditions would separate the protein from any noncovalently associated additional component, e.g., the nucleic acid of the virus or virino theories, also rendering the agent noninfectious. It is therefore not surprising, regardless of one's bias relative to the nature of the agent, that mobility in SDS-PAGE is not informative in this regard.

This study compared the denaturation and inactivation of scrapie strains 263K and 139A under identical conditions and demonstrated that such strains vary relative to the resistance of infectivity and the stability and protease sensitivity of associated PrP^{Sc}. These findings are added to the list of well-documented differences among scrapie strains: incubation times, severity and distribution of pathologic changes, clinical manifestations, transmission to new species, and thermal inactivation properties (1). It is within the context of these biologic and physical differences among scrapie strains that the nature of the agent must be defined. The nature of this agent must be reconciled not only with the denaturation studies reported here, but also with the complexity of biologic differences among scrapie strains and the number and diversity of such strains. The evidence argues that the infectious agent must contain an informational component beyond the capacity of PrP^Sc alone.

The term *amyloidoses* refers to a variety of different disease conditions in which abnormally processed proteins are deposited in the form of β-pleated sheet structures (25). Such deposits are remarkably stable, resistant to solubilization or proteolytic digestion. They accumulate over long periods of time in a variety of different tissues depending on the disease and type of amyloidogenic protein involved. These deposits have been suggested to play a role in pathogenesis in certain conditions resulting from the interference with normal physiologic processes of affected vital organs. Amyloidoses represent a wide spectrum of clinical disorders and involve a diversity of amyloidogenic proteins (25). CNS amyloid deposits occur in certain neurodegenerative conditions including Alzheimer's disease, Down syndrome, and prion diseases of man and animals. The amyloidogenic proteins in these disorders are beta A4 (26) and PrP (27), respectively.

Glycation or advanced glycosylation describes the chemical pathway that leads from early glycation products, such as Schiff bases and Amadori products, to more advanced protein-bound moieties that possess distinct characteristics (28). Early glycation products result from the nonenzymatic addition of reducing sugars to the primary amino groups of proteins to form Schiff base adducts. These then slowly undergo rearrangements to produce Amadori products. Over weeks to months, Amadori products then undergo additional rearrangements to produce a more advanced and complex class of protein bound moieties called glycosylation end products (AGEs) (29). Proteins modified with AGEs possess distinct cross-linking and immunofluorescence properties. This cross-linking can lead to cell membrane deformability and tissue rigidity (28). AGE modification may also lead to alterations in the normal trafficking of such proteins.

Recently it has become evident that the aggregation and cross-linking properties of AGEs may play a role in the seeding of amyloidogenic proteins (30). AGE modifications have been associated with cross-link formation, decreased protein solubility, and increased protease resistance, properties often associated with amyloidogenic proteins. Much is known about the normal and abnormal processing of these proteins but the mechanisms by which the insoluble protein deposits accumulate within the CNS are unknown. Posttranslational modifications and conformational changes are believed to play a key role. We have now shown that AGE modifications are present in PrP^Sc. These highly aggregated proteins with slow turnover rates appear to be prime candidates for AGE modification within the CNS. One of the key questions is the role such modifications play in the disease process.

The propensity for AGEs to participate in protein aggregation and protein cross-linking also points to a role in amyloidogenesis. Amyloid formation bears a mechanistic resemblance to processes involving ordered protein aggregation-nucleation dependent polymerization (31). Polymerization of PrP peptides into fibril-like structures can also be assisted by the presence of seed molecules (32). The conversion of normal precursor PrP^{C} into pathogenic PrP^{Sc} has now been shown to take place through a nucleation-dependent seeding mechanism (33). The ability of AGE-modified proteins to act as "seeds" and facilitate aggregation and polymerization events may well play an important role in the aging process and in pathologic events associated with neurodegenerative diseases.

Acknowledgments. The authors wish to thank J. DeKolf for evaluating and preparing the manuscript. We are grateful to Ms. P. Merz for her comments and critical review of this manuscript. We also thank R. Weed for photographic assistance. These studies were supported in part by funds from the New York State Office of Mental Retardation and Developmental Disabilties, NIH grant NS-21349, and Alzheimer's Association grant IIRG-104.

References

1. Carp RI, Kascsak RJ, Rubenstein R. Pathogenesis of unconventional slow virus infections. In: Liberski PP, ed. Light and electron microscopic neuropathology of slow virus disorders. Boca Raton, FL: CRC Press, 1993:33–61.
2. Gabizon R, McKinley MP, Prusiner SB. Purified prion proteins and scrapie infectivity copartition into lysosomes. Proc Natl Acad Sci USA 1987;84: 4017–21.
3. Hilmert H, Diringer H. A rapid and efficient method to enrich SAF-protein from scrapie brains of hamsters. Biosci Rep 1984;4:165–70.
4. Prusiner SB, Bolton DC, Groth DF, Bowman KA, Cochran SP, McKinley MP. Further purification and characterization of scrapie prions. Biochemistry 1982; 21:6942–50.
5. Hope J, Manson J. The scrapie fibril protein and its cellular isoform. In: Chesebro BW, ed. Current topics in microbiology and immunology—transmissible spongiform encephalopathies. Berlin: Springer-Verlag, 1991;172:57–74.
6. Carp RI, Kascsak RJ, Rubenstein R, Merz PA. The puzzle of PrP^{Sc} and infectivity—do the pieces fit? TINS 1994;17:148–9.
7. Bueler H, Aguzzi A, Saller A, Greiner RA, Autenried P, Aguet M, Weissman C. Mice devoid of PrP are resistant to scrapie. Cell 1993;73:1339–47.
8. Chesebro B, Caughey B. Scrapie agent replication without the prion protein? Curr Biol 1993;3:696–8.
9. McKinley MP, Bolton DC, Prusiner SB. A protease-resistant protein is a structural component of the scrapie prion. Cell 1983;35:57–62.

10. Brown P, Liberski PP, Wolff A, Gajdusek DC. Conservation of infectivity in purified fibrillary extracts of scrapie-infected hamster brain after sequential enzymatic digestion or polyacrylamide gel electrophoresis. Proc Natl Acad Sci USA 1990;87:7240–44.

11. Braig HR, Diringer H. Scrapie: concept of a virus-induced amyloidosis in brain. EMBO J 1985;4:2309–12.

12. Xi YG, Ingrosso L, Ladogana A, Masullo C, Pocchiari M. Amphotericin B treatment dissociates in vivo replication of the scrapie agent from PrP accumulation. Nature 1992;356:598–601.

13. Carp RI, Callahan SM. In vitro interaction of scrapie agent and mouse peritoneal macrophages. Intervirology 1981;16:8–13.

14. Carp RI, Kim YS, Callahan SM. Pancreatic lesions and hypoglycemia-hyperinsulinemia in scrapie-infected hamsters. J Infect Dis 1990;161:462–6.

15. Laemmli UK. Cleavage of structural proteins during the assembly of the head of bacteriophage T4. Nature 1970;227:680–5.

16. Kascsak RJ, Rubenstein R, Merz PA, et al. Immunological comparison of scrapie associated fibrils isolated from animals infected with four different scrapie strains. J Virol 1986;59:676–83.

17. Kascsak RJ, Tonna-DeMasi M, Fersko R, Rubenstein R, Carp RI, Powers JM. The role of antibodies to PrP in the diagnosis of transmissible spongiform encephalopathies. Dev Biol Stand, Basel: Karger, 1993;80:141–51.

18. Nakamura Y, Horii Y, Nishino T, et al. Immunohistochemical localization of advanced glycosylation end products (AGEs) in coronary atheroma and cardiac tissue in diabetes mellitus. Am J Pathol 1994;143:1649–56.

19. Sidman RL, Angevine JB Jr, Pierce E. Atlas of the mouse brain and spinal cord. Cambridge, MA: Harvard University Press, 1971.

20. Prusiner SB. Molecular biology of prion diseases. Science 1991;252:1515–22.

21. Prusiner SB, Groth DF, Cochran P, Masraiz FR, McKinley MP, Martinez HM. Molecular properties, partial purification and bioassay by incubation period measurements of the hamster agent. Biochemistry 1980;19:4883–91.

22. Kascsak RJ, Rubenstein R, Carp RI. Evidence for biological and structural diversity among scrapie strains. In: Chesebro BW, ed. Current topics in microbiology and immunology—transmissible spongiform encephalopathies. Berlin: Springer-Verlag, 1991;172:139–52.

23. Bessen RA, Marsh RF. Biochemical and physical properties of the prion protein from two strains of the transmissible mink encephalopathy agent. J Virol 1992;66:2096–101.

24. Manuelidis L, Sklaviadis T, Manuelidis EE. Evidence suggesting that PrP is not the infectious agent in Creutzfeldt-Jakob disease. EMBO J 1987;6:341–47.

25. Glenner GG. Amyloid deposits and amyloidosis. N Engl J Med 1990;302:1283–343.

26. Glenner GG, Murphy MA. Amyloidosis of the nervous system. J Neurol Sci 1989;94:1–28.

27. Kitamoto T, Tateishi J, Tashima T. Amyloid plaques in Creutzfeldt-Jakob disease stain with prion protein antibodies. Ann Neurol 1986;20:204–8.

28. Bucala R, Vlassara H, Cerami A. Measurement of advanced glycosylation end products. In: Harding J, James J, Crabbe C, eds. Post translational modifications of proteins. Boca Raton: CRC Press, 1992:53–79.

29. Bucala R, Cerami A. Advanced glycosylation: chemistry, biology and implications for rabbits and aging. Adv Pharmacol 1992;23:1–34.
30. Vitek MP, Bhattacharya K, Glendening JM, et al. Advanced glycation end products contribute to amyloidosis in Alzheimer disease. Proc Natl Acad Sci USA 1994;91:4766–79.
31. Jarrett JT, Lansbury PT Jr. Seeding "one dimensional crystallization" of amyloid: a pathogenic mechanism in Alzheimer disease and scrapie. Cell 1993;73:1055–58.
32. Come JH, Fraser PE, Lansbury PT Jr. A kinetic model for amyloid formation in the prion diseases: importance of seeding. Proc Natl Acad Sci USA 1993;90:5959–63.
33. Kocisko DA, Come JH, Priola SA, et al. Cell-free formation of protease-resistant prion protein. Nature 1994;370:471–3.

18

Amyloidosis: The Key to the Epidemiology and Pathogenesis of Transmissible Spongiform Encephalopathies

HEINO DIRINGER, MICHAEL BEEKES, ELIZABETH BALDAUF, SVEN CASSENS, AND MUHSIN ÖZEL

The Concept of Transmissible Spongiform Encephalopathies (TSEs) as Virus-Induced Amyloidoses

Numerous results have made it very clear that TSEs are pathogenetically closely related to the many nontransmissible amyloidoses (1). These results include the discovery of a disease-specific amyloid fibril (SAF) (2–4), its protein constituent (5–11), and of the gene coding for the amyloid protein precursor (12–18). Further findings provide evidence for the association of familial forms of human TSE (see Chapter 1) and of the susceptibility of sheep breeds to scrapie (19) with specific mutations in the amyloid-coding gene. This gene has been shown to be identical with the *sinc* gene in mice (20–23) or the *sip* gene in sheep (24, 25) [some groups prefer to call it the prion gene (20, 21, 23)]. In the absence of this gene, neither amyloidosis, infectivity, nor clinical symptoms develop (27, 28). As the amyloidogenic protein is preferentially located at the surface of neurons (29–31), it becomes clear why these cells are particularly vulnerable to infection and why TSEs present as degenerative diseases of the brain.

The evidence for the critical involvement of an amyloidogenic protein anchored to the surface of neurons is overwhelming. It is also known that a single host gene has a strong influence on the pathogenesis (32, 33) of these transmissible diseases. Obviously, given the existence of a variety of agent strains (34) of viral size (35–38), the concept of a virus- [or virino- (39)] induced amyloidosis of the central nervous system (40) offers a principal model for the pathogenesis of TSE. This concept presents a good starting point for the elucidation of the pathogenesis of TSE and for the search for an epidemiologic link between TSE of man and animals (40). According to this theory (Fig. 18.1), TSEs are induced by an as yet undetected virus (or by the nucleic acid of a virino), which converts its receptor protein into an amyloid resulting in clinical disease. In other words; no amyloidosis without a virus, and no clinical disease without amyloidosis.

With this in mind, we will first analyze the development of research during the last 6 years to see whether, in the light of our new knowledge, the likelihood of BSE posing a threat to human health has increased or decreased. Second, we will reconsider whether or not it is appropriate to ignore the possibility of viruses rather than infectious proteins transmitting the diseases, despite the widespread acceptance of the prion theory.

Risk Assessment for TSE: From the Southwood Report (1989) Until 1995

Cattle—a Dead-End Host?

In 1988 the members of the Southwood committee summarized their assessment of the risk of a transfer of bovine spongiform encephalopathy to man as follows: "From present evidence, it is likely that cattle will prove to be a dead-end host for the disease agent and most unlikely that BSE will have any implications for human health. Nevertheless, if our assessments of these likelihoods are incorrect, the implications would be extremely serious." This was based on what was known about scrapie in sheep, transmissible mink encephalopathy (TME), chronic wasting disease in mule deer and elk, and human TSE. Single cases of new forms of TSE in Britain, in a nyala and a gemsbok, were known in 1988.

From epidemiologic studies (see Chapter 4), it was concluded that improper rendering had caused the oral transmission of scrapie from sheep to cattle. In analogy to TME in the carnivore mink, it was assumed that the ruminant cow was also likely to present a dead-end host. A peak number of 17,000 to 20,000 cases of BSE was estimated provided there was no recycling of infectious cattle material in the rendering process. Today it must be recognized that, due to recycling of infectious BSE material, the peak incidence has reached more than 140,000 cases so far, rather than the estimated 20,000. The present count still does not include all those infected animals that could not develop the disease because they were killed and channeled into the human food chain before they reached the mean age of onset of clinical symptoms (4 to 6 years). In 1992, when the endemic was at its peak, about 1% of the cattle population in the United Kingdom came down with clinical disease. Several measures had been taken in the United Kingdom to protect animals and humans against the transmission of BSE (41). The ruminant feed ban introduced in July 1988 resulted in a decline of clinical BSE cases as from 1993 (see Chapter 4).

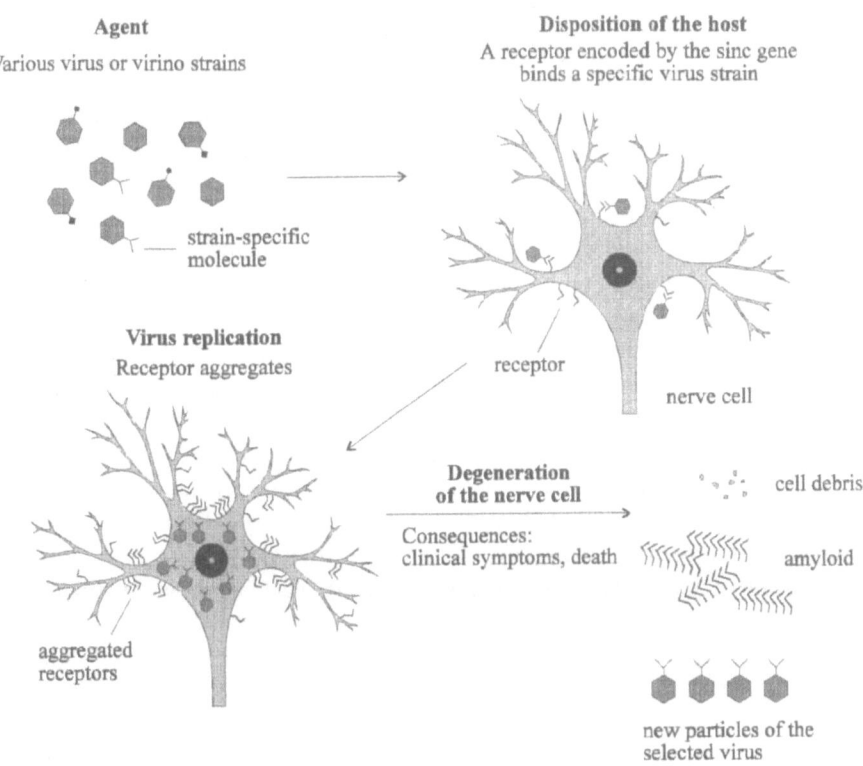

FIGURE 18.1. Concept of a virus-induced amyloidosis for scrapie, BSE, CJD and other transmissible spongiform encephalopathies. Adapted from Spektrum der Diringer and Özel (90).

However, horizontal (42) and maternal (43, 44) transmission are prevalent in the ruminant sheep. The disease is sustained in closed herds of the ruminant goat (45), and, as observed more recently, of the ruminant Kudu (46). Further, in January 1995 there were nine cases of BSE in cattle born in 1991, $2\frac{1}{2}$ to $3\frac{1}{3}$ years after the feed ban. By March 1995 the number had risen to 27 cases (personal communication). These observations strongly suggest that the ruminant cow, in contrast to the carnivore mink, is unlikely to be a dead-end host.

Indirect Transmission of Scrapie as a Virus Disease into Carnivores via Other Ruminants?

Scrapie in sheep can be transmitted horizontally via highly infectious placenta deposited on the pasture at lambing (42, 47). Natural oral transmission of this kind can also occur between sheep and goats (45, 48). In 1972 Pattison and colleagues published their observations on the spread of scrapie to sheep and goats by oral administration of fetal membranes from scrapie-affected sheep. They drew attention not only to the importance of the oral route for the horizontal intra- or interspecies transmissin of this prototype of TSE, but also to the key role of the female for transmission of the agent by stating that the oral route of infection with placenta "offers a possible explanation for our experience at Compton that contact spread of scrapie has never been observed during 18 years of experimentation with scrapie in sheep and goats, because only males or non-pregnant females have been used as scrapie-affected donor animals." Today, oral transmission from sheep to other ruminant species must be considered a possible, although rare, event, particularly as transmission of natural sheep scrapie to mice has shown that transmission between animal species by a peripheral route may be more efficient than by intracerebral (i.c.) inoculation (49). Cattle are especially good candidates for this type of natural interspecies transmission, as cows (see Chapters 4 and 8) have been shown to be susceptible to parenteral transmission of sheep scrapie. Rare cases of BSE in the past may have resulted from oral transmission by this route. One case of scrapie in cattle was described in 1884 (50). Recently, it was shown that spongiform encephalopathy resulting from primary transmission of sheep scrapie to cattle may not be identified by conventional histologic methods (51), and may therefore escape conventional surveillance systems.

Oral transmission of scrapie to cattle has resulted in the selection of a multipotent BSE agent, "multipotent" because the BSE agent retains its biologic properties in other species, whether it is transmitted accidentally by feed to various ruminants, to felines in zoos and to house cats, experimentally by parenteral routes to pigs, or experimentally by the oral route to sheep and goats. On the other hand, the BSE agent can be distinguished very clearly from the agent of natural sheep scrapie (52). The multipotent BSE agent has also been transmitted orally to mink (see Chapter 8) and by parenteral inoculation to primates (see Chapter 6). From a virologist's point of view all these examples of interspecies transmission make it impossible to ignore the potential existence of an indirect link between sporadic CJD cases and scrapie.

Linkage of Creutzfeldt-Jakob Disease (CJD) to Scrapie?

Direct and Indirect Transmission of TSE

Genetic studies of the human familial forms of TSE (see Chapter 1) suggest that at least the most common mutation in familial CJD at codon 200 may predispose an individual to contracting scrapie directly from sheep by consuming scrapie-contaminated food (40). This mutation is not fully penetrant (53, 54). Surprisingly, cases with this mutation were associated with scrapie in the past although the investigators were unaware of the genetic defect (40).

With this situation in mind, and with the knowledge we have gained about the indirect transmission of scrapie via cattle into other carnivores, it seems more than likely that one form of familial CJD and the sporadic cases of CJD always originate from direct or indirect transmissions from animals to man.

With the direct transmission of scrapie into families with the codon 200 mutation and the indirect route via ruminants feeding on pastures contaminated with scrapie-infected placenta of sheep, more than 95% of the CJD cases can be explained as zoonotic diseases with extremely rare occurrence (Fig. 18.2). In the case of the codon 200 cases, the rare mutation explains the rare occurrence of transmission. In the case of sporadic CJD, the rare transmission of scrapie into another ruminant explains the rare frequency of CJD in man. At present it is more difficult to associate the fully penetrant familial forms of CJD with scrapie. However, in those cases of familial TSE that are transmissible, an adapted strain of the agent may have been selected at a previous time by the specific genetic disposition of these families.

Neither of these two possible links between scrapie in sheep and CJD in man would have been detected by any of the epidemiology studies performed in the past, although a connection was already strongly suspected 25 years ago. Compelling evidence existed for the spread of kuru by mourning rites in New Guinea, which included cannibalism. It was also known that CJD, kuru, and the two animal disorders—scrapie and TME—were caused by filtrable agents with unconventional biologic, physical, and chemical properties. Furthermore, the scrapie agent causes a spongiform encephalopathy that is indistinguishable from CJD in several species of experimentally infected nonhuman primates. Logically, these findings led to the hypothesis that CJD might result from exposure of humans to the scrapie virus of sheep (55).

Twenty years ago, epidemiologic studies might have failed because the link is likely to be found either in a specific genetic disposition or in an indirect transmission via cattle. Now, with a more detailed understanding of TSE, it is untenable to constantly reiterate that epidemiologic studies have not shown an association between scrapie and CJD.

The possibility of a link between human disease and scrapie becomes more obvious when sporadic CJD and the 200 codon mutation of the familial forms are compared with other familial forms of human CJD, with iatrogenically transmitted cadaveric growth hormone cases and with kuru (Table 18.1). In view of a viral concept, the first two forms have to be described as orally transmitted infections across a species barrier. The latter three forms must be described as either vertical (fully penetrant familial forms), parenteral (growth hormone), or oral (kuru) infections with an adapted strain of the virus within a species.

Brown (56) conducted a study comparing the distribution of the age of clinical onset for the two supposedly "animal-derived" forms of CJD with the fully penetrant forms of CJD. Brown showed that the sporadic forms

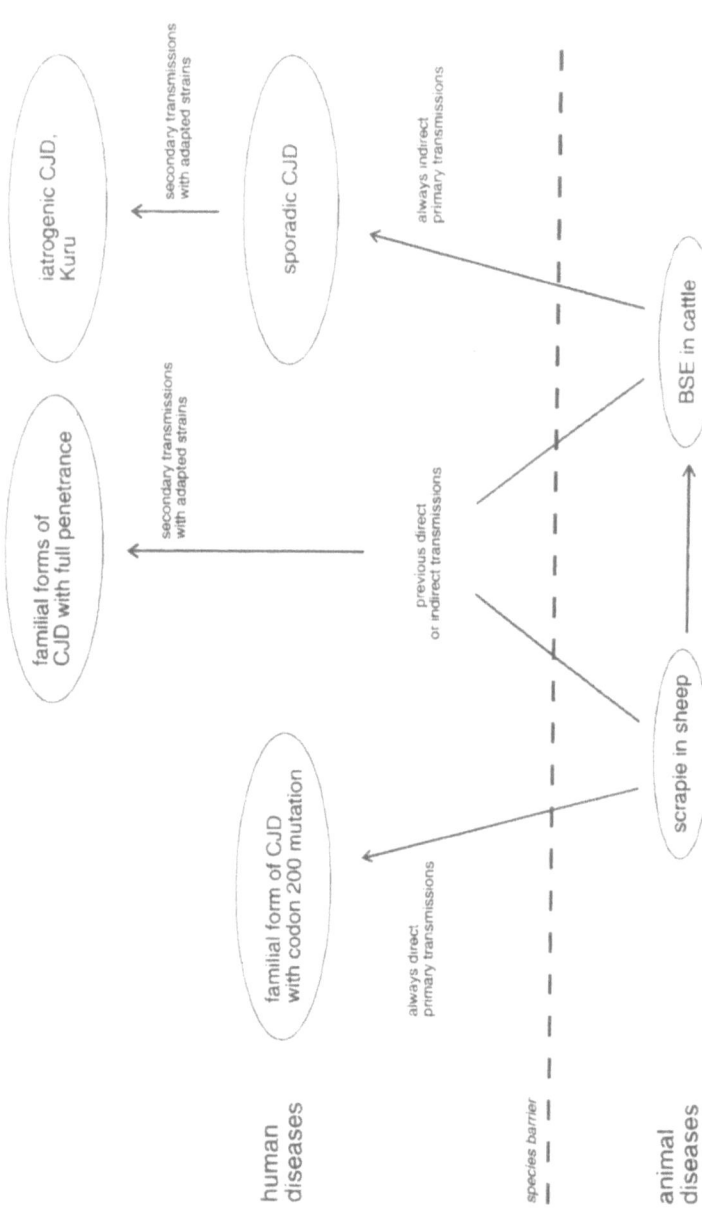

FIGURE 18.2. A model for the transmission pathways of TSE showing CJD as a possible zoonotic disease. Primary transmissions occur across the species barrier with unadapted virus. Secondary transmissions in man occur with adapted strains of virus.

TABLE 18.1. Forms of TSE grouped according to assumed primary and secondary transmissions with unadapted and adapted strains of virus.

Human TSE	Form of transmission	Mean age of onset	Characteristics	Transmissibility to nonhuman primates
Familial 200-codon CJD (oral)	Primary direct transmission from sheep	56	Dementia, short duration, virtually indistinguishable	Very good
Sporadic CJD (oral)	Primary indirect transmission via ruminants	62		
Iatrogenic growth hormone–derived CJD (parenteral)	Secondary transmissions from sporadic CJD (derived from primary transmissions from animals)	Age of onset often <30 years[a]	Cerebellar syndrome	Very good, highest rate
Kuru (oral)				
Familial forms Codon 178	Secondary transmissions with previously adapted agent strains	46	Absence of periodic electroencephalographic activity	More difficult
Codon 102		48	predominance of cerebellar symptoms, long duration	
Codon 117		38	long duration	
Codon 198		—	similar to 102-mutation	
Inserts		34	longest duration of illness	

Adapted from Brown et al. (57, 59).

[a] Depends on age at infection. Like other iatrogenic transmissions by transplants and invasive diagnostic or surgical methods, the secondary transmissions by growth hormone treatment often have very short incubation periods.

and the 200-mutation form of CJD almost coincide in age distribution and occur at a later age than the remaining familial forms. In one interpretation, Brown's observations were taken as evidence against the possibility that sporadic cases of CJD might occur by oral infection with a ruminant-derived agent (57). In contrast, his findings could also be explained by an oral route of infection and/or a change in host species as compared with the fully penetrant forms of familial CJD and the growth hormone cases (56). Both the "animal-derived" forms show an identical development of clinical symptoms, which is quite distinct from that observed in other familial forms of CJD, in the iatrogenic cases, and in kuru. Compared with other familial forms, the animal-derived forms of CJD also share a high rate of experimental transmissibility (58). They differ from any other familial form of CJD, iatrogenically induced CJD, and kuru, which, according to the virus theory, result from transmissions with species-adapted strains. The common characteristics of the animal-derived forms suggest a common route of infection, and could be explained by oral transmission of the sheep (200-mutation) or cattle agent (sporadic forms) to man.

Our conclusion is supported further by the fact that secondary infections through recycling of the cattle adapted agent can be identified as spongiform encephalopathies (see Chapter 3), other than primary transmissions of sheep scrapie to cattle (see Chapter 8), or, in other words, primary transmission of scrapie from sheep to cattle across the species barrier can be distinguished from secondary transmission, i.e. recycling of the BSE agent among cattle, by histopathology. Like our conclusions shown in Figure 18.2 and Table 18.1, this is to be expected and in full agreement with Pattison's observations on the scrapie agent breaching the species barrier, which he summerized as follows:

It has been noted that when scrapie is transmitted from one animal species to another (e.g., sheep/goat, goat/mouse, mouse/rat, etc.), it is likely that there will be:
a) A long incubation period.
b) Atypical clinical signs in the recipient animal.
c) Atypical histopathological lesions in the recipient animal.

Further, it has been noted that in the first passage between animals of the same species it is likely that:
a) The incubation period will shorten and become fixed.
b) The clinical signs and histopathological lesions characteristic for that species will be fixed.

Discrepancies Between the Distribution of Scrapie and CJD

Another steadily reiterated statement used to question the link between sporadic CJD and scrapie is that CJD has a worldwide distribution of about 1 case per million annually, even in countries that are considered to be free of scrapie, whereas scrapie is distributed unevenly. Let us analyze this general statement in more detail, always keeping in mind that the incubation period for orally transmitted low dose infections may exceed 25 to 30 years (see Chapter 1).

The Uneven Distribution of Scrapie

When a country is considered free of scrapie, this is not always based on intensive scientific surveys and may not take eradication measures into account. In countries with a large sheep-breeding industry, it is impossible to exclude the existence of scrapie with reasonable certainty without an intensive survey of the genetic resistance of the sheep breeds. Australia and New Zealand have been considered scrapie-free for about 40 years. Cases of scrapie have been described in impacted animals in both countries in the 1950s (59–61). Considering the long incubation period of CJD, the incidences of the human disease and scrapie in these countries at any one time cannot be compared directly and used as an argument against a possible link.

The Worldwide CJD Incidence of 1 Case per Million Annually

CJD has been known to be transmissible for about 25 years, but incubation periods may be as long as that or even longer. When transmissibility of CJD was first reported in 1968, Kirschbaum (62) reviewed 150 cases of CJD found worldwide since the first description in the 1920s. In 1979 a survey including 1,435 patients (55) counted 265 CJD cases in the United States alone between 1973 and 1977, more than the number of cases Kirschbaum collected over a period of more than 40 years. Meanwhile about 4,000 cases of CJD have been reported to the National Institutes of Health (63). Intensive surveillance started only 20 years ago. Since then the low incidence of 1 case per million annually has actually varied geographically between 0.3 to 2 cases. With this in mind, it seems unwarranted to use the different distributions of scrapie and CJD as an argument against a link between the two diseases. In addition, rare incidences of uncontrollable distribution of possibly unrecognized BSE-contaminated meat and cattle products might have occurred for decades. Further, the increase in travel and relocation may have contributed to an "equal" distribution of CJD, independent of the geographic origin of infection.

However, one may well expect some kind of clustering of CJD cases in countries with a high incidence of scrapie or in areas where a high number of consumers might have been exposed to contaminated food products. Clustering has indeed been described in neighboring villages in the United Kingdom (64), in neighborhoods 250 m apart (65), and for three cases of CJD in a family (genetic disposition not tested) and a close friend of the family (65). In two cases, a husband and wife died of CJD within a few years of each other (64, 66). The finding that the incidence of CJD in France was higher in Paris than in the countryside and, within the city, higher in some districts than in others (67, 68) may reflect the distribution of a food supply contaminated with the TSE agent. Similar clustering in large cities has also been observed in other countries (68).

The Situation in 1995 as Compared with 1989

There can be no doubt that after 6 years of research the statement made by the Southwood committee appears in a new light. The evidence available in 1988 allowed the statement that BSE was most unlikely to have any implications for human health. In 1995, however, this statement has become very questionable.

Can a Virus Be Excluded as the Cause of TSE?

Infectivity and amount of amyloid are the two measurable disease-specific parameters in TSE. In the hamster model of scrapie, the amount of amyloid can be detected qualitatively and quantitatively with a high sensitivity and good accuracy (Beekes et al., (69)). Recently, we have made use of this possibility to study the distribution of amyloid in the central nervous system of different species (70, 71). The method has also been applied to study the temporal development of amyloid and infectivity in spinal cords and brains of hamsters after an oral challenge with scrapie (69), and more recently after intraperitoneal infection (Baldauf et al., manuscript in preparation).

A constant ratio of infectivity to amyloid was found throughout the disease process (Table 18.2). The numerous analyses performed during the study allowed a good determination of this ratio as 1.5 to 3×10^4 LD_{50} per nanogram of amyloid (69).

Assuming a mean molecular weight of 25 kd for the protease-treated amyloid protein (72, 73), about 1 LD_{50} of infectivity is associated with 10^6 molecules of amyloid. Similar ratios have also been reported by others (74).

This result is in agreement with the observation in many laboratories that infectivity and amyloid are closely associated with each other and copurify. In contrast, however, to earlier data reported from our laboratory (75, 76), the present study gives no evidence for a lag-phase between the increase of infectivity and the increase in amyloid concentration in hamster brain after oral infection. An unprocessed scrapie hamster brain contains 3×10^9 LD_{50}, 100 mg of protein, and 100 mg of amyloid. Therefore, the maximum purification factor for the amyloid with respect to the total protein cannot exceed 1,000. If infectivity copurifies with amyloid, the concentration factor for infectivity in pure amyloid fractions is also limited to 1,000. This is insufficient for the detection of a possible virus. Table 18.3 presents the relative concentration (%) of viral protein and nucleic acid for various forms of a putative virus that could be hidden in pure amyloid fractions.

TABLE 18.2. Infectivity and TSE-specific amyloid in the brain and cervical spinal cord of hamsters orally infected with scrapie.

Tissue donor			Cervical spinal cord			Brain		
#	Incubation time (day)s	Clinical symptoms	Infectivity [log (LD$_{50}$/g)]	TSE-specific amyloid [log(ng/g)]	Infectivity/amyloid [log(LD$_{50}$/ng)]	Infectivity [log(LD$_{50}$/g)]	TSE-specific amyloid [log(ng/g)]	Infectivity/amyloid [log(LD$_{50}$/ng)]
1	100	–	6.9	2.6	4.3	5.8	2.0	3.8
2	105	–	6.9	2.6	4.3	6.2	2.1	4.1
3	105	–	6.4	2.0	4.4	6.8	3.0	3.8
4	109	–	7.7	3.5	4.2	7.4	2.9	4.5
5	114	–	7.7	3.7	4.0	7.1	3.0	4.1
6	114	–	6.8	2.9	3.9	7.4	2.9	4.5
7	119	–	7.4	3.2	4.2	7.2	2.9	4.3
8	123	–	8.1	4.4	3.7	8.7	4.4	4.3
9	123	–	7.6	3.9	3.7	7.8	3.7	4.1
10	127	–	7.9	3.7	4.2	8.1	3.7	4.4
11	133	+	8.0	3.9	4.1	8.4	3.9	4.5
12	133	+	8.7	4.4	4.3	8.5	4.3	4.2
Means (± S.E.M.):					4.1 ± 0.1			4.2 ± 0.1

Mean error of the determination of the infectivity: ±0.4 logs.
Mean error of the determination of the TSE-specific amyloid: ±0.2 logs.

TABLE 18.3. Viral protein and nucleic acid content expected in pure amyloid fractions.

Viral protein					Viral nucleic acid		
		Virus protein with various triangulation numbers T				Viral nucleic acids with various sizes	
Particle/LD$_{50}$ ratio	Relative size of virus protein (kd)	T1 (% w/w)	T3 (% w/w)	T4 (% w/w)	Particle/LD$_{50}$ ratio	Relative size of viral nucleic acid (dalton)	Viral nucleic acid (% w/w)
1:1	8	0.001	0.004	0.005	1:1	10^5	0.0004
1:1	25	0.006	0.018	0.024	1:1	10^6	0.004
10:1	8	0.01	0.04	0.05	10:1	10^5	0.004
10:1	25	0.06	0.18	0.24	10:1	10^6	0.04
100:1	8	0.1	0.4	0.5	100:1	10^5	0.04
100:1	25	0.6	1.8	2.4	100:1	10^6	0.4

Small icosahedral viruses are often of the T1 configuration, in which 60 peptides are arranged on 20 surfaces to form the viral coat. Let us assume a molecular weight of 5 kd [smallest viral coat protein known (77)] or 25 kd (mean size of protease treated amyloid protein) for the viral coat protein, and a particle to LD_{50} ratio of only 1. If 1 LD_{50} is associated with 3×10^4 ng of amyloid, then only 0.001% to 0.006% of a pure amyloid fraction from scrapie hamster brain would represent viral protein. If we had a particle to LD_{50} ratio of 100, then 0.1% to 0.6% of a pure amyloid fraction would represent proteinaceous viral material. The situation would be somewhat better if the disease were caused by larger viruses with a T3 or T4 configuration. These calculations make it obvious that putative viral particles or proteins would be difficult to detect due to their low amount and the excess of amyloid.

Concerning the content of viral nucleic acids, the situation is no better. Again assuming particle-to-infectivity-ratios of 1 and 100, and a viral nucleic acid of conventional size (M = 10^6 kd), 0.004% to 0.4% of the material in a pure amyloid fraction would represent viral nucleic acid. Tenfold lower concentrations are to be expected if the viral nucleic acid were as small as 10^5 kd. So far, we have not been able to obtain infectious fractions with more than 10 to 20 mg of amyloid per brain containing more than 10^8 LD_{50} and less than 500 ng and 200 ng of contaminating host-derived protein and nucleic acid, respectively. Therefore, technical problems rather than sweeping arguments ["In the absence of any evidence for viruses, this hypothesis (virus theory) looks to be untenable" (78)] explain the lack of physical evidence for a virus at present.

Is There an Advantage for a Virus in Inducing an Amyloidosis?

Like many other viruses, the scrapie agent enters its host via the gastrointestinal tract (79). It also accumulates in lymphoid tissue of the intestine (80), where Peyer's patches play an important role in virus uptake (81). The agent is phagocytosed by dendritic cells (82), replicates in these cells (83), and induces amyloid formation (84, 85). The major route of the scrapie agent into the central nervous system seems to be along sympathetic fibers of the splanchnic nerve after intragastric (84) and oral (69) infection.

A virus infecting its host via the gastrointestinal tract has to be extremely resistant to digestive fluids and enzymes. The scrapie agent meets these requirements. To orally infect mice with scrapie about 10^5 i.c. LD_{50} is necessary (86). This seems to be a very high dose of infectivity, but we have to consider the highly hostile environment in which the virus has to survive.

When rats drank saline containing 10^{11} plaque-forming units (pfu) of T7 bacteriophage (diameter of virus head $\sim 30\,nm$), about 10^3 pfu were found in the thoracic lymph (87). Only 1 out of 10^8 particles successfully entered the lymphatic system of the host.

If a T1 category virus with its shell of 60 peptides induces 10^6 molecules of amyloid, even a small fraction of the amyloid would suffice as a protective wrap for the delicate viral shell during its journey through the gastrointestinal tract. Under these circumstances, a relatively low amount of 10^5 i.c. LD_{50} (as compared to 10^8 pfu of the unprotected T7 phage) may be sufficient to induce disease via the oral route.

Further, a protective amyloid shell could allow the virus to hide in a relatively large amount of food at low concentrations. Digestion of the tissue would then lead to concentration of the resistant virus to a higher effective dose.

In summary, the induction of amyloid might provide several advantages for a virus infecting its hosts via the gastrointestinal route.

Small Virus-Like Particles

Small particles structurally resembling viruses have been isolated from scrapie hamster brains (88) as well as from several brain samples from CJD patients (89). They were not found in normal controls. The pentagonal stuctures (Fig. 18.3) are barely detectable in concentrated filtered extracts of preparations of scrapie-associated fibrils (89). Further purification and characterization of these particles will be necessary to determine their significance for TSE in animals and humans (90).

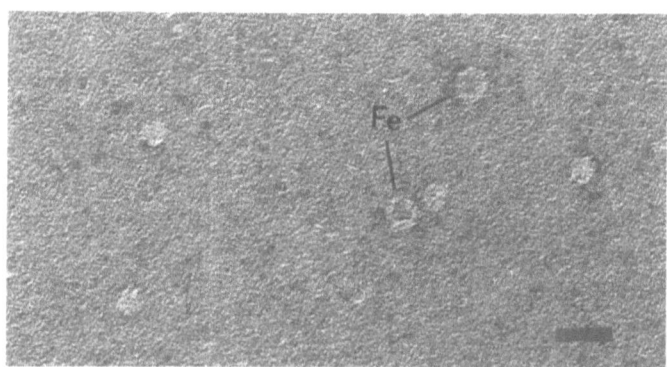

FIGURE 18.3. Negatively stained (1% uranyl acetate) sample of a fraction isolated from scrapie hamster brain showing two ferritin particles and four small virus-like particles. Magnification: 425,000-fold, bar = 20 nm.

Summary

While the concept of TSE as virus-induced amyloidoses offers a key to the understanding of these diseases, it also explains some of the barriers encountered in the search for the causative agent. Links between human and animal TSEs that have escaped conventional epidemiology may become more apparent in the light of the virus hypothesis.

For a virus that enters its hosts through the gastrointestinal tract, the induction of amyloid may constitute a protective mechanism. Unfortunately, an excess of amyloid could greatly impede the detection of virus-specific proteins or nucleic acids.

The explanation of TSE in terms of virus-induced amyloidoses presents a useful and viable approach to the elucidation of this group of diseases within the realm of classical biochemistry. It should not be discounted merely because the challenge of visualizing a cleverly camouflaged agent appears too imposing.

References

1. Diringer H. Hidden amyloidoses. Exp Clin Immunogenet 1992;9:212–29.
2. Merz PA, Somerville RA, Wisniewski HM, Iqbal K. Abnormal fibrils from scrapie-infected brain. Acta Neuropathol 1981;54:63–74.
3. Merz PA, Somerville RA, Wisniewski HM, Manuelidis L, Manuelidis EE. Scrapie-associated fibrils in Creutzfeldt-Jakob disease. Nature 1983;306:474–6.
4. Merz PA, Rohwer RG, Kascsak R, et al. Infection-specific particle from the unconventional slow virus diseases. Science 1984;225:437–40.
5. Bolton DC, McKinley MP, Prusiner SB. Identification of a protein that purifies with the scrapie prion. Science 1982;218:1309–11.
6. McKinley MP, Bolton DC, Prusiner SB. A protease-resistant protein is a structural component of the scrapie prion. Cell 1983;35:57–62.
7. Diringer H, Gelderblom H, Hilmert H, Ozel M, Edelbluth C, Kimberlin RH. Scrapie infectivity, fibrils and low molecular weight protein. Nature 1983; 306:476–8.
8. Barry RA, McKinley MP, Bendheim PE, Lewis GK, DeArmond SJ, Prusiner SB. Antibodies to the scrapie protein decorate prion rods. J Immunol 1985; 135:603–13.
9. DeArmond SJ, McKinley MP, Barry RA, Braunfeld MB, McColloch JR, Prusiner SB. Identification of prion amyloid filaments in scrapie-infected brain. Cell 1985;41:221–35.
10. Merz PA, Kascsak RJ, Rubenstein R, Carp RI, Wisniewski HM. Antisera to scrapie-associated fibril protein and prion protein decorate scrapie-associated fibrils. J Virol 1987;61:42–9.
11. Wiley CA, Burrola PG, Buchmeier MJ, et al. Immuno-gold localization of prion filaments in scrapie-infected hamster brains. Lab Invest 1987;57:646–56.
12. Chesebro B, Race R, Wehrly K, et al. Identification of scrapie prion protein-specific mRNA in scrapie-infected and uninfected brain. Nature 1985;315: 331–3.

13. Oesch B, Westaway D, Waelchli M, et al. A cellular gene encodes scrapie PrP 27–30 protein. Cell 1985;40:735–46.
14. Basler K, Oesch B, Scott M, et al. Scrapie and cellular PrP isoforms are encoded by the same chromosomal gene. Cell 1986;46:417–28.
15. Locht C, Chesebro B, Race R, Keith JM. Molecular cloning and complete sequence of prion protein cDNA from mouse brain infected with the scrapie agent. Proc Natl Acad Sci USA 1986;83:6372–6.
16. Robakis NK, Sawh PR, Wolfe GC, Rubenstein R, Carp RI, Innis MA. Isolation of a cDNA clone encoding the leader peptide of prion protein and expression of the homologous gene in various tissues. Proc Natl Acad Sci USA 1986;83: 6377–81.
17. Westaway D, Prusiner SB. Conservation of the cellular gene encoding the scrapie prion protein. Nucleic Acids Res 1986;14:2035–44.
18. Liao YC, Lebo RV, Clawson GA, Smuckler EA. Human prion protein cDNA: molecular cloning, chromosomal mapping, and biological implications. Science 1986;233:364–7.
19. Hunter N, Foster JD, Hope J. Natural scrapie in British sheep: breeds, ages and PrP gene polymorphisms. Vet Rec 1992;130:389–92.
20. Carlson GA, Kingsbury DT, Goodman PA, et al. Linkage of prion protein and scrapie incubation time genes. Cell 1986;46:503–11.
21. Westaway D, Goodman PA, Mirenda CA, McKinley MP, Carlson GA, Prusiner SB. Distinct prion proteins in short and long scrapie incubation period mice. Cell 1987;51:651–62.
22. Hunter N, Hope J, McConnell I, Dickinson AG. Linkage of the scrapie-associated fibril protein (PrP) gene and Sinc using congenic mice and restriction fragment length polymorphism analysis. J Gen Virol 1987;68:2711–6.
23. Scott M, Foster D, Mirenda C, et al. Transgenic mice expressing hamster prion protein produce species-specific scrapie infectivity and amyloid plaques. Cell 1989;59:847–57.
24. Hunter N, Foster JD, Dickinson AG, Hope J. Linkage of the gene for the scrapie-associated fibril protein (PrP) to the Sip gene in Cheviot sheep. Vet Rec 1989;124:364–6.
25. Foster JD, Hunter N. Partial dominance of the sA allele of the Sip gene for controlling experimental scrapie. Vet Rec 1991;128:548–9.
26. Goldmann W, Hunter N, Benson G, Foster JD, Hope J. Different scrapie-associated fibril proteins (PrP) are encoded by lines of sheep selected for different alleles of the Sip gene. J Gen Virol 1991;72:2411–7.
27. Bueeler H, Fischer M, Lang Y, et al. Normal development and behaviour of mice lacking the neuronal cell-surface PrP protein [see comments]. Nature 1992;356:577–82.
28. Bueeler H, Aguzzi A, Sailer A, et al. Mice devoid of PrP are resistant to scrapie. Cell 1993;73:1339–47.
29. Stahl N, Borchelt DR, Hsiao K, Prusiner SB. Scrapie prion protein contains a phosphatidylinositol glycolipid. Cell 1987;51:229–40.
30. Stahl N, Baldwin MA, Burlingame AL, Prusiner SB. Identification of glycoinositol phospholipid linked and truncated forms of the scrapie prion protein. Biochemistry 1990;29:8879–84.
31. Stahl N, Borchelt DR, Prusiner SB. Differential release of cellular and scrapie prion proteins from cellular membranes by phosphatidylinositol-specific phospholipase C. Biochemistry 1990;29:5405–12.

32. Masters CL, Gajdusek DC, Gibbs CJ Jr. The familial occurrence of Creutzfeldt-Jakob disease and Alzheimer's disease. Brain 1981;104:535–58.
33. Hunter N. Natural transmission and genetic control of susceptibility of sheep to scrapie. Curr Topics Microbilo Immunol 1991;172:165–80.
34. Bruce ME, McConnell I, Fraser H, Dickinson AG. The disease characteristics of different strains of scrapie in Sinc congenic mouse lines: implications for the nature of the agent and host control of pathogenesis. J Gen Virol 1991;72:595–603.
35. Eklund CM, Hadlow WJ, Kennedy RC. Some properties of the scrapie agent and its behavior in mice. Proc Soc Exp Biol Med 1963;112:974–9.
36. Gibbs CJ. Search for infectious etiology in chronic and subacute degenerative diseases of the central nervous system. Curr Topics Microbiol Immunol 1967;40:44–58.
37. Tateishi J, Koga M, Sato Y, Mori R. Properties of the transmissible agent derived from chronic spongiform encephalopathy. Ann Neurol 1980;7:390–1.
38. Diringer H, Kimberlin RH. Infectious scrapie agent is apparently not as small as recent claims suggest. Biosci Rep 1983;3:563–8.
39. Dickinson AG, Outram GW. The scrapie replication-site hypothesis and its implications on pathogenesis. In: Prusiner SB, Hadlow WJ, eds. Slow transmissible diseases of the nervous system, vol 2. New York: Academic Press, 1979:13–31.
40. Diringer H, Beekes M, Oberdieck U. The nature of the scrapie agent: the virus theory. Ann NY Acad Sci 1994;724:246–58.
41. British Ministry of Agriculture, Fisheries and Food. Bovine spongiform encephalopathy in Great Britain. Appendix 1: Chronology of events. In: A progress report (November 1994). British Ministry of Agriculture Fisheries and Food. Surbton, Surrey: MAFF 1994.
42. Pattison IH, Millson GC. Experimental transmission of scrapie to goats and sheep by the oral route. J Comp Pathol 1961;71:171–6.
43. Matthews JD. Ataxias and transmissible agents. Lancet 1967;1:851.
44. Dickinson AG, Young GB, Stamp JT, Renwick CC. An analysis of natural scrapie in Suffolk sheep. Heredity 1965;20:485–503.
45. Hourrigan J, Klingsporn A, Clark WW, deCamp M. Epidemiology of scrapie in the United States. In: Prusiner SB, Hadlow WJ, eds. Slow transmissible diseases of the nervous system, vol 1. New York: Academic Press, 1979:331–56.
46. Kirkwood JK, Cunningham AA, Wells GAH, Wilesmith JW, Barnett JEF. Spongiform encephalopathy in a herd of greater kudu (*Tragelaphus strepsiceros*): epidemiological observations. Vet Rec 1993;133:360–4.
47. Pattison IH, Hoare MN, Jebbett JN, Watson WA. Further observations on the production of scrapie in sheep by oral dosing with foetal membranes from scrapie-affected sheep. Br Vet J 1974;130:55–7.
48. Pattison IH, Hoare MN, Jebbett JN, Watson WA. Spread of scrapie to sheep and goats by oral dosing with foetal membranes from scrapie-affected sheep. Vet Rec 1972;90:465–8.
49. Kimberlin RH. Modified pathogenesis of scrapie at first intracerebral passage from sheep to mice (P 59-1). In: IXth International Congress of Virology. Abstracts. Norwich. Page Bros, 1993:318.
50. Sarradet M. Un cas de tremblante sur un boeuf. Rev Veterinaire 1883;7:310–2.

51. Cutlip RC, Miller JM, Race RE, et al. Intracerebral transmission of scrapie to cattle. J Infect Dis 1994;169:814–20.
52. Bruce ME, Chree A, McConnell J, Foster J, Pearson G, Fraser H. Agent strain variation in BSE and Scrapie. In: Bradley R, Marchant B, eds. Transmissible spongiform encephalopathies: Proceedings of a consultation on BSE with the scientific veterinary committee of the commission of the European Communities (Ref.: F.II.3—JC/0003); Brussels: European Commission, 1993: 189–204.
53. Hsiao K, Meiner Z, Kahana E, et al. Mutation of the prion protein in Libyan Jews with Creutzfeldt-Jakob disease. N Engl J Med 1991;324: 1091–7.
54. Goldfarb LG, Brown P, Mitrova E, et al. Creutzfeldt-Jacob disease associated with the PRNP codon 200Lys mutation: an analysis of 45 families. Eur J Epidemiol 1991;7:477–86.
55. Masters CL, Harris JO, Gajdusek DC, Gibbs CJ Jr, Bernoulli C, Asher DM. Creutzfeldt-Jakob disease: patterns of worldwide occurrence and the significance of familial and sporadic clustering. Ann Neurol 1979;5:177–88.
56. Brown P. The phenotypic expression of different mutations in transmissible human spongiform encephalopathy. Rev Neurol 1992;148:317–27.
57. Spongiform Encephalopathy Advisory Committee. Transmissible spongiform encephalopathies. A summary of present knowledge and research. London: HMSO Publication Centre, 1994:46.
58. Brown P, Gibbs CJ, Rodgers-Johnson P, et al. Human spongiform encephalopathy: the National Institutes of Health series of 300 cases of experimentally transmitted disease. Ann Neurol 1994;35:513–29.
59. Bull LB, Murnane D. An outbreak of scrapie in British sheep imported into Victoria. Aust Vet J 1958;34:213–5.
60. Brash AG. First outbreak of scrapie reported in New Zealand. NZ J Agriculture 1952;85:305–6.
61. Brash AG. Scrapie in imported sheep in New Zealand. NZ Vet J 1952;1: 27–30.
62. Kirschbaum WR. Creutzfeldt-Jakob disease. New York: Elsevier, 1968.
63. Gajdusek DC. Infectious and noninfectious amyloidoses of the brain: Systemic amyloidoses as predictive models in transmissible and nontransmissible amyloidotic neurodegeneration of Creutzfeldt-Jakob disease and aging brain and Alzheimer's disease. In: Calne DB, ed. Neurodegenerative diseases. Philadelphia: WB Saunders, 1994:301–17.
64. Matthews WB. Epidemiology of Creutzfeldt-Jakob disease in England and Wales. J Neurol Neurosurg Psychiatry 1975;38:210–3.
65. Will RG, Matthews WB. Evidence for case-to-case transmission of Creutzfeldt-Jakob disease. J Neurol Neurosurg Psychiatry 1982;45:235–8.
66. Garzuly F, Jellinger K, Pilz P. Subakute spongioese Encephalopthie (Jakob-Creutzfeldt-Syndrom). Klinisch-morphologische Analyse von 9 Faellen. (Subacute spongiform encephalopathy (Jakob-Cretzfeldt syndrome). Clinico-pathological evaluation of 9 cases). Arch Psychiatr Nervenkrank 1971;214:207–27.
67. Brown P, Cathala F. Creutzfeldt-Jakob disease in France: I. Retrospective study of the Paris area during the ten-year period 1968–1977. Ann Neurol 1979;5: 189–92.

68. Brown P, Cathala F, Gajdusek DC. Creutzfeldt-Jakob disease in France: III. Epidemiological study of 170 patients dying during the decade 1968–1977. Ann Neurol 1979;6:438–46.

69. Beekes M, Baldauf E, Diringer H. Sequential appearance and accumulation of pathognomonic markers in the central nervous system of hamsters orally infected with scrapie. J Gen Virol 1996.

70. Beekes M, Baldauf E, Cassens S, Diringer H, Keyes P, Scott AC, et al. Western blot mapping of disease-specific amyloid in various animal species and humans with transmissible spongiform encephalopathies using a high-yield purification method. J Gen Virol 1995;76:2567–76.

71. Beekes M, Baldauf E, Cassens S, et al. Western blot mapping of disease-specific amyloid in various animal species and humans with transmissible spongiform encephalopathies using a high-yield purification method. J Gen Virol 1995; 76:2567–76.

72. Kascsak RJ, Rubenstein R, Merz PA, et al. Immunological comparison of scrapie-associated fibrils isolated from animals infected with four different scrapie strains. J Virol 1986;59:676–83.

73. Rubenstein R, Kascsak RJ, Merz PA, et al. Detection of scrapie-associated fibril (SAF) proteins using anti-SAF antibody in non-purified tissue preparations. J Gen Virol 1986;67:671–81.

74. Rubenstein R, Carp RI, Ju W, et al. Concentration and distribution of infectivity and PrPSc following partial denaturation of a mouse-adapted and a hamster-adapted scrapie strain. Arch Virol 1994;139:301–11.

75. Czub M, Braig HR, Diringer H. Replication of the scrapie agent in hamsters infected intracerebrally confirms the pathogenesis of an amyloid-inducing virosis. J Gen Virol 1988;69:1753–6.

76. Czub M, Braig HR, Diringer H. Pathogenesis of scrapie: study of the temporal development of clinical symptoms, of infectivity titres and scrapie-associated fibrils in brains of hamsters infected intraperitoneally. J Gen Virol 1986; 67:2005–9.

77. Nakashima Y, Konigsberg W. Reinvestigation of a region of the fd bacteriophage coat protein sequence. J Mol Biol 1974;88:598–600.

78. Prusiner SB. The prion diseases. Sci Am 1995;272:48–51,54–7.

79. Hadlow WJ, Kennedy RC, Race RE. Natural infection of Suffolk sheep with scrapie virus. J Infect Dis 1982;146:657–64.

80. Hadlow WJ, Race RE, Kennedy RC, Eklund CM. Natural infection of sheep with scrapie virus. In: Prusiner SB, Hadlow WJ, eds. Slow transmissible diseases of the nervous system, vol 2. New York: Academic Press, 1979: 3–12.

81. Morrison LA, Fields BN. Parallel mechanisms in neuropathogenesis of enteric virus infections. J Virol 1991;65:2767–72.

82. Ehlers B, Rudolph R, Diringer H. The reticuloendothelial system in scrapie pathogenesis. J Gen Virol 1984;65:423–8.

83. Ehlers B, Diringer H. Dextran sulphate 500 delays and prevents mouse scrapie by impairment of agent replication in spleen. J Gen Virol 1984;65: 1325–30.

84. Kitamoto T, Muramoto T, Mohri S, Doh-Ura K, Tateishi J. Abnormal isoform of prion protein accumulates in follicular dendritic cells in mice with Creutzfeldt-Jakob disease. J Virol 1991;65:6292–5.

85. McBride PA, Eikelenboom P, Kraal G, Fraser H, Bruce ME. PrP protein is associated with follicular dendritic cells of spleens and lymph nodes in uninfected and scrapie-infected mice. J Pathol 1992;168:413–8.
86. Kimberlin RH, Walker CA. Pathogenesis of scrapie in mice after intragastric infection. Virus Res 1989;12:213–20.
87. Mims CA. Aspects of the pathogenesis of virus diseases. Bacteriol Rev 1964;28:30–71.
88. Özel M, Diringer H. Small virus-like structure in fractions from scrapie hamster brain. Lancet 1994;343:894–5.
89. Özel M, Xi Y-G, Baldauf E, Diringer H, Pocchiari M. Small virus-like structure in brains from cases of sporadic and familial Creutzfeldt-Jakob disease. Lancet 1994;344:923–4.
90. Diringer H, Özel M. Übertragbare spongiforme Enzephalopathien—wodurch werden sie verursacht. Spektr Wissensch 1995;3:52–4.

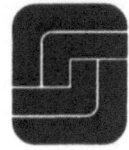

19

Effect of Amphotericin B on Different Experimental Strains of Spongiform Encephalopathy Agents

Maurizio Pocchiari, Loredana Ingrosso, and Anna Ladogana

Scrapie is a transmissible disease of the central nervous system (CNS) naturally occurring in sheep and goats, characterized by the formation of a modified, partly proteinase-resistant protein of the host, which tends to aggregate as amyloid fibrils and accumulate in the brain of infected individuals. We have previously reported that treatment with the polyene antibiotic amphotericin B (AmB) prolongs the incubation period of hamsters infected intracerebrally or intraperitoneally with strain 263K of scrapie. In this chapter, we show that AmB also prolongs the incubation periods of Armenian hamsters experimentally infected with the 263K strain of scrapie but not in Chinese hamsters. We also report the beneficial effect of AmB in mice injected with the KFu strain of CJD.

Spongiform encephalopathies are neurodegenerative diseases of humans [Creutzfeldt-Jakob disease (CJD), Gerstmann-Sträussler-Scheinker syndrome (GSS), and kuru] and animals [scrapie and bovine spongiform encephalopathy (BSE)] characterized by the formation of a partially proteinase-resistant protein (PrP-res) that derives from a conformational modification of an endogenous cellular protease-sensitive isoform (PrP-sen). In the brain of infected individuals, PrP-res aggregates and accumulates as amyloid fibrils (1).

In experimental animal models, sulfated polyanions (2–7), Congo red (8), and the polyene antibiotic AmB (9–15) have given encouraging results. These drugs prolong or sometimes even prevent the appearance of the disease by delaying the formation of PrP-res and/or inhibiting scrapie replication (5, 6, 13, 16–19). However, they are effective only when given either before or soon after the injection of the agent and are completely useless when administered at the appearance of clinical signs of the disease (10).

AmB differs from polyanions in its unique beneficial effect when the scrapie agent is given intracerebrally. Interestingly, however, the beneficial antiscrapie effect of AmB is limited to the 263K hamster-adapted strain

(10–12) and the murine strain C506 (14). AmB treatment is ineffective with other hamster or mouse-adapted strains of scrapie (13, 20), suggesting that it interferes with strain-specific (that is, 263K and C506) components of the scrapie agent. In this report we investigate the effect of AmB therapy in three different species of hamsters (Syrian, Armenian, and Chinese) injected with two Syrian hamster-adapted strains of scrapie (263K and 139H). Moreover, we examine the effect of AmB on a murine model of CJD.

Materials and Methods

Infection of Animals

Weanling outbred golden Syrian hamsters (*Mesocricetus auratus*), inbred Armenian (*Cricetus cricetus*) and Chinese hamsters (*Cricetulus barabensis griseus*), and mice (strain C57BL) were purchased from Charles River, Calco, Como, Italy. Hamsters were injected with the 263K strain (21) or the 139H strain (22) of scrapie; mice were injected with the KFu strain of Creutzfeldt-Jakob (23). The inocula were obtained from 10% (wt/vol) phosphate-buffered saline (PBS) suspensions of pooled brains from clinically affected animals. The brain suspensions were centrifuged at 500 g for 10 minutes; the supernatants were collected and stored at −70°C in 5-mL aliquots.

Before use, samples were thawed and vigorously shaken by vortexing; 0.05 mL (hamsters) or 0.03 mL (mice) were injected intracerebrally into the left hemisphere by using sterile syringes with 26-gauge needles. Animals were housed in stainless (Syrian hamsters) or in plastic cages (Chinese and Armenian hamsters, and mice) with water and food ad libitum, observed 5 days per week, and scored for the presence of clinical signs.

Amphotericin B Treatment

Amphotericin B (Sigma, catalogue no. A9528), was freshly prepared for each administration dissolving the powder (50 mg) in 10 mL of sterile distilled water. AmB was then diluted appropriately with 5% glucose, and 0.5 mL (1 mg of AmB, hamsters) or 0.25 mL (0.5 mg of AmB, mice) were injected intraperitoneally once a week alternatively in the lower right and left quadrants of the abdomen.

Treatment of 263K-infected animals was started the same day of inoculation and was maintained for 7 (Syrian hamsters), 15 (Armenian hamsters), and 30 (Chinese hamsters) weeks. Treatment of 139H-injected animals was started the same day of inoculation and was maintained for 16 (Syrian hamsters), 22 (Armenian hamsters), and 30 (Chinese hamsters) weeks. CJD-infected mice were treated for 12 weeks. Controls were nontreated scrapie (hamsters) or CJD (mice) inoculated animals. There is no difference in the mean incubation periods of scrapie-infected hamsters receiving injection of saline solution and nontreated hamsters (10).

Results

Properties of 263K and 139H Scrapie Strains in Different Species of Hamster

Intracerebral infection of the 263K and 139H Syrian hamster-adapted strains of scrapie in Armenian and Chinese hamsters induces the disease with different clinical (Table 19.1) and incubation periods (Fig. 19.1). The 263K strain has approximately a 50% shorter incubation period than the 139H strain in Syrian hamsters. However, the first passage of the 263K strain in Armenian and Chinese hamsters consistently produces a longer incubation period than 139H. This is particularly evident in Chinese hamsters where the incubation period of 263K is over 100 days longer than 139H infected animals. Clinical signs at onset and during the course of the disease differ consistently among the three species of hamsters infected with either the 263K or 139H strain of scrapie and, in the same animal species, between 263K and 139H injected animals. Thus, hypersensitivity to noise and aggressive behavior is primarily observed in 263K-infected Syrian hamsters but not in Armenian or Chinese hamsters infected with the same inocula, while obesity and hyperexcitability occur in 139H infected Syrian and Armenian hamsters, but not in Chinese hamsters. During the course of the disease, all species of hamsters present neurologic abnormalities (regardless of the strain of scrapie injected) but of different types and intensity. Examples of this are head tremor and wobbling gait observed only in 263K-infected Syrian hamsters and swaying gait in 263K-infected Chinese hamsters.

Effect of Amphotericin B on Different Scrapie-Host Strain Combinations

AmB treatment prolongs the incubation period of 263K-infected Syrian or Armenian hamsters but not of 263K-infected Chinese hamsters nor of all species of hamsters infected with the 139H strain (Fig. 19.2a,b). The mean increase in the incubation period in AmB-susceptible animals compared with nontreated controls was 31.9 days for 263K-infected Syrian hamsters (56.7% of control, Fig. 19.2c) and 43.1 days for 263K-infected Armenian hamsters (25.9% of control, Fig. 19.2c). Clinical signs and duration of the disease did not differ between nontreated and AmB-treated animals.

TABLE 19.1. Clinical characteristics of Syrian, Armenian, and Chinese hamsters infected with the 263K or 139H strain of scrapie.

Scrapie strain	Host	Incubation period (weeks)	Duration of the disease (weeks)	Clinical signs	
				At onset	During course
263K	Syrian	8–9	2–3	Hypersensitivity to noise; irritable and aggressive when handled; tremor of the head	Head tremor, wobbling gait, spontaneous and frequent back rolls
263K	Armenian	22–24	4–5	Stiffness of the tail, arched back; "trotting"-like gait to on tiptoe of the hind limbs; out of the cage, the animals run without control and fall off the bench	The animals tend to run and to execute circular movements, changing direction frequently from clockwise to counterclockwise, more evident when disturbed by handling or, less frequently, by noise; ruffled fur
263K	Chinese	48–50	8	Insidious onset, swaying pelvic gait	Hunched posture with rocking movements
139H	Syrian	16–18	6–7	Obesity, hyperexcitability; urinate when handled; animals roll up into a ball, jump on the hind limbs if disturbed	When disturbed by light noise, animals jump violently for about 1–2 minutes and then remain exhausted until the appearance of the next paroxysm; they can have up to 5 attacks in 10 minutes; diffuse tremors in the whole body
139H	Armenian	20–22	3–5	Obesity, hyperexcitability, increased speed of normal activities; run in a circle and frequently fall off the bench	Increased hyperexcitability and worsening of motor disturbances
139H	Chinese	34–36	4–5	Increased speed of normal activities, trotting gait, running in a circle	Worsening of motor disturbances, waddling gate

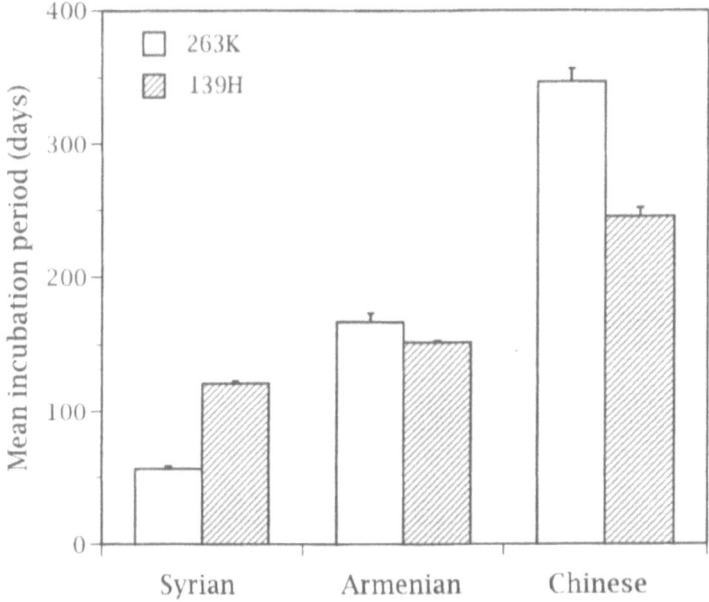

FIGURE 19.1. Mean incubation periods ± SD (bars) of Syrian, Armenian, and Chinese hamsters intracerebrally injected with the 263K (open box) or 139H (dashed box) strain of scrapie.

Effect of Amphotericin B Regarding Experimental Creutzfeldt-Jakob Disease

Mice infected with the KFu strain of CJD show typical plasticity of the tail, arched back, and abduction of the hind limbs when held by the tail. At later stages of the disease, they show bradykinesia and a decrease in body weight, and finally they die after about 2 weeks.

CJD-infected mice treated with AmB showed an 18.5-day increase in the mean incubation period compared with nontreated CJD injected controls (Fig. 19.3). The clinical signs and the duration of the disease in AmB-treated mice did not show any remarkable differences with respect to nontreated CJD-infected animals.

Discussion

The extended incubation periods seen at the first passage of the 263K or 139H strains of scrapie in Armenian and Chinese hamsters are similar to those previously reported (24, 25). This phenomenon is referred to as the *species barrier effect* (26) and depends on the interaction between the scrapie agent and PrP-sen. Differences in the amino acid sequence of PrP-sen in Syrian and Armenian or Chinese hamsters (24) are therefore responsible for the long incubation period in these animals.

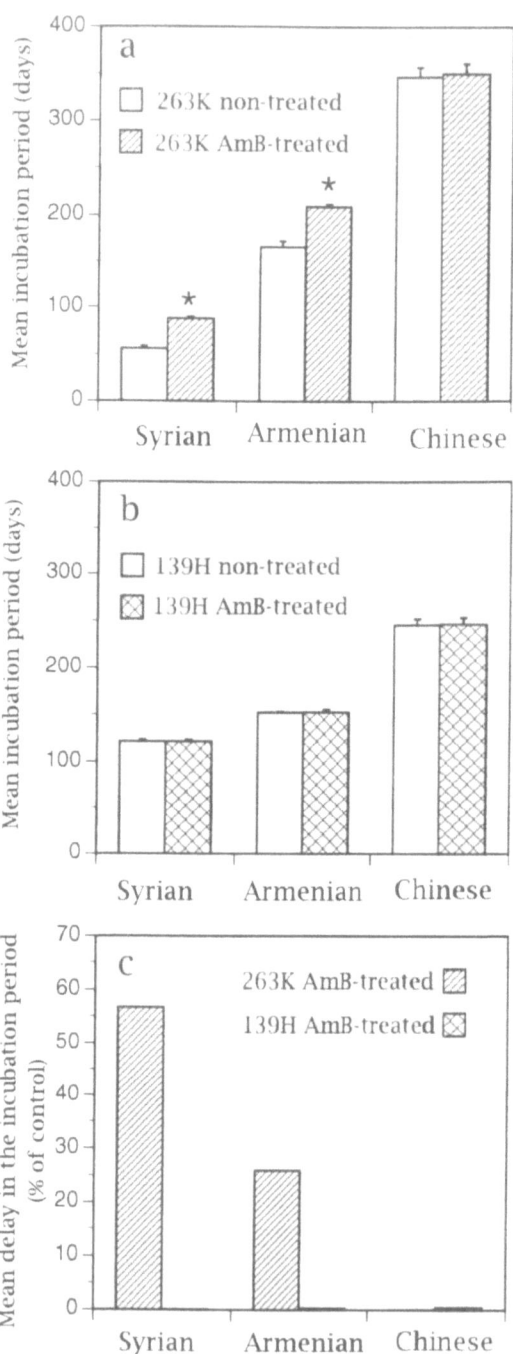

FIGURE 19.2. Mean incubation periods ± SD (bars) of amphotericin B (AmB)-treated and nontreated Syrian, Armenian, and Chinese hamsters intracerebrally injected with the (a) 263K or (b) 139H strain of scrapie. In (c) the results are expressed as the percentage of nontreated incubation periods. *Statistically different from control at $p < .001$ (two-tailed Student's t-test).

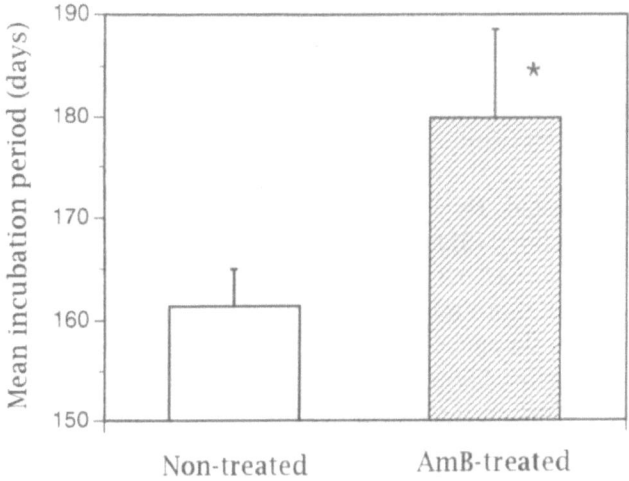

FIGURE 19.3. Mean incubation periods ± SD (bars) of amphotericin B (AmB)-treated (dashed box) and nontreated (open box) C57BL mice intracerebrally injected with the KFu strain of Creutzfeldt-Jakob disease (CJD) agent. *Statistically different from control at $p < .001$ (two-tailed Student's t-test).

The variability of clinical signs, incubation periods, and duration of the disease in these three different species of hamsters infected with the 263K or 139H strains of scrapie confirms the notion that both the host and the scrapie agent genomes contribute to the phenotypic characteristics of scrapie disease (27). It is noteworthy that each host/scrapie strain combination used in this experiment had a different pattern of disease (see Table 19.1 and Fig. 19.1). It is under investigation whether new mutant strains of scrapie (from 263K, 139H, or both) will emerge in the Armenian or Chinese hamsters after subsequent homologous passages.

The anti-scrapie effect of AmB is achieved through the delay of PrP-res accumulation in the brain without affecting scrapie replication (13). Thus, AmB may preclude binding of the infectious agent to a particular site of PrP-sen that is responsible for the conformational change from PrP-sen to PrP-res (the "conversion" site) without affecting the binding of the agent to a different moiety of the PrP-sen molecule, which is responsible for the entry and initiation of replication into the target cell (the "replication" site) (for a comprehensive review, see 1).

Recently, McKenzie and colleagues (15) failed to appreciate the dissociation between scrapie replication and PrP-res accumulation in the brain of AmB-treated hamsters. The authors affirm that, in contrast to our view, PrP-res accumulation and agent replication are linked, reinforcing the hypothesis that PrP-res is a major, if not the sole, component of the scrapie agent. In their report, however, the scrapie agent titer was measured in the brain homogenate while the PrP-res accumulation was measured in purified brain fractions. This experimental design, i.e., comparison of biologic measures in different fractions of the brain, has been repeatedly criticized (28–30). In our study (13), we avoided this by measuring scrapie infectivity and PrP-res accumulation in the same purified brain fraction. Moreover, the authors did not take into account that residual AmB in brain homogenate of treated animals may be responsible for the delay of the incubation period of recipient hamsters. Hence, a bias in the measurement of scrapie infectivity may have occurred (31). A similar delay of the incubation period caused by residual AmB has been observed in hamsters inoculated with brain homogenate and, more remarkably, spleen homogenate (11). AmB accumulates in the brain and, 100-fold more, in the spleen of treated hamsters (12). The discrepancy between the measurement of scrapie titer using tissue homogenate versus purified sample is well illustrated by spleen infectivity data (taken at 40 days after intracerebral scrapie inoculation) of nontreated and AmB-treated hamsters. The incubation periods of recipient hamsters inoculated with spleen homogenates were 118.6 ± 3.2 and >155 days, respectively (11). In contrast, using purified spleen (AmB-free) samples as inocula, the incubation periods were 86.2 ± 2.1 and 86.3 ± 2.3 days, respectively, for nontreated and AmB-treated hamsters (13). Thus, previously published data (13) on the dissociation between scrapie replication and PrP-res accumulation in the brain of AmB-treated hamsters are still valid.

It is interesting that AmB treatment had a beneficial anti-scrapie effect in 263K infected Syrian and Armenian hamsters but not in Chinese hamsters. These findings might indicate that in Chinese hamsters the disease is caused by a mutant strain that is resistant to AmB. Although this new mutant strain has a very low rate of replication, it may have a better affinity than the parental 263K strain. This, however, does not exclude the persistence of the original 263K in the brain of Chinese hamsters, which would retain its properties when back-passaged from Chinese into Syrian hamsters (25).

An alternative explanation is that the inhibition of PrP-res formation following AmB treatment depends on the binding of AmB to the 263K/PrP-sen complex rather than to the 263K alone. A difference in the structure of either the strain of scrapie or the PrP-sen would then nullify the beneficial effect of AmB. Further work is needed to unravel this point.

AmB treatment is ineffective with the 139H strain regardless of which species of hamster has been infected. This result confirms previous observations (13) and strengthens the notion that the 139H strain of scrapie is resistant per se to AmB treatment.

The beneficial effect observed in experimental CJD is of great interest and may give some hope for therapy in humans. Apart from ouabain treatment (32), which increased the mean interval between infection and death of intramuscularly injected CJD mice [44 days of delay with respect to 308 ± 48 (SD) days of untreated mice], we are not aware of any other drugs showing a beneficial effect in the experimental murine model of CJD (32, 33). The findings that AmB treatment prolonged the clinical phase of experimental CJD in an African monkey (9) but did not ameliorate the neurologic or cognitive signs in two CJD patients (34) may suggest that, as occurs with scrapie, some strains of CJD are AmB-susceptible, while others are not. Although clinical trials of potential anti-CJD drugs are difficult because of the extreme rarity of the disease, AmB therapy in a large cohort of CJD patients should be encouraged.

Acknowledgments. We thank Ms. Deborah Wool for editorial assistance. We also thank Franco Varano for his invaluable help in the organization of the animal facility, and Maurizio Bonanno for assistance in animal care. This work was partially supported by a grant from C.N.R. n. 94.00996.CT04.

References

1. Pocchiari M. Prions and related neurological diseases. Mol Aspects Med 1994;15:195–291.
2. Ehlers B, Diringer H. Dextran sulphate 500 delays and prevents mouse scrapie by impairment of agent replication in spleen. J Gen Virol 1984;65:1325–30.
3. Ehlers B, Rudolph R, Diringer H. The reticuloendothelial system in scrapie pathogenesis. J Gen Virol 1984;65:423–8.
4. Farquhar CF, Dickinson AG. Prolongation of scrapie incubation period by an injection of dextran sulphate 500 within the month before or after infection. J Gen Virol 1986;67:463–73.
5. Kimberlin RH, Walker CA. Suppression of scrapie infection in mice by heteropolyanion 23, dextran sulfate, and some other polyanions. Antimicrob Agents Chemother 1986;30:409–13.
6. Diringer H, Ehlers B. Chemoprophylaxis of scrapie in mice. J Gen Virol 1991;72:457–60.
7. Ladogana A, Casaccia P, Ingrosso L, Cibati M, Salvatore M, Xi YG, et al. Sulphate polyanions prolong the incubation period of scrapie-infected hamsters. J Gen Virol 1992;73:661–5.
8. Ingrosso L, Ladogana A, Pocchiari M. Congo red prolongs incubation period of scrapie-infected hamsters. J Virol 1995;69:506–8.
9. Amyx H, Salazar AM, Gajdusek DC, Gibbs CJ Jr. Chemotherapeutic trials in experimental slow virus disease. Neurology 1984;34(suppl 1):149.
10. Pocchiari M, Schmittinger S, Masullo C. Amphotericin B delays the incubation period of scrapie in intracerebrally inoculated hamsters. J Gen Virol 1987; 68:219–23.
11. Pocchiari M, Casaccia P, Ladogana A. Amphotericin B: a novel class of antiscrapie drug. J Infect Dis 1989;160:795–802.
12. Casaccia P, Ladogana A, Xi YG, Ingrosso L, Pocchiari M, Silvestrini MC, et al. Measurement of the concentration of amphotericin B in brain tissue of scrapie-infected hamsters with a simple and sensitive method. Antimicrob Agents Chemother 1991;35:1486–8.
13. Xi YG, Ingrosso L, Ladogana A, Masullo C, Pocchiari M. Amphotericin B treatment dissociates in vivo replication of the scrapie agent from PrP accumulation. Nature 1992;356:598–601.

14. Demaimay R, Adjou K, Lasmézas C, et al. Pharmacological studies of a new derivative of amphotericin B, MS-8209, in mouse and hamster scrapie. J Gen Virol 1994;75:2499–503.

15. McKenzie D, Kaczkowski J, Marsh R, Aiken J. Amphotericin B delays both scrapie agent replication and PrP-res accumulation early in infection. J Virol 1994;68:7534–6.

16. Caughey B, Race RE. Potent inhibition of scrapie associated PrP accumulation by Congo red. J Neurochem 1992;59:768–71.

17. Caughey B, Raymond GJ. Sulfated polyanion inhibition of scrapie-associated PrP accumulation in cultured cells. J Virol 1993;67:643–50.

18. Caughey B, Ernst D, Race RE. Congo red inhibition of scrapie agent replication. J Virol 1993;67:6270–2.

19. Gabizon R, Meiner Z, Halimi M, Ben-Sasson SA. Heparin-like molecules bind differentially to prion-proteins and change their intracellular metabolic fate. J Cell Physiol 1993;157:319–25.

20. Carp RI. Scrapie. Unconventional infectious agent. In: Specter S, ed. Neuropathogenic viruses and immunity. New York: Plenum Press, 1992:111–36.

21. Kimberlin RH, Walker C. Characteristics of a short incubation model of scrapie in the golden hamster. J Gen Virol 1977;34:295–304.

22. Kimberlin RH, Walker CA, Fraser H. The genomic identity of different strains of mouse scrapie is expressed in hamsters and preserved on reisolation in mice. J Gen Virol 1989;70:2017–25.

23. Tateishi J, Ohta M, Koga M, Sato Y, Kuroiwa Y. Transmission of chronic spongiform encephalopathy with Kuru plaques from humans to small rodents. Ann Neurol 1979;5:581–4.

24. Lowenstein DH, Butler DA, Westaway D, McKinley MP, DeArmond SJ, Prusiner SB. Three hamster species with different scrapie incubation times and neuropathological features encode distinct prion proteins. Mol Cell Biol 1990;10:1153–63.

25. Hecker R, Taraboulos A, Scott M, et al. Replication of distinct scrapie prion isolates is region specific in brains of transgenic mice and hamsters. Genes Dev 1992;6:1213–28.

26. Dickinson AG. Scrapie in sheep and goats. Front Biol 1976;44:209–41.

27. Bruce ME. Scrapie strain variation and mutation. Br Med Bull 1993;49:822–38.

28. Bolton DC, Rudelli RD, Currie JR, Bendheim PE. Copurification of Sp33-37 and scrapie agent from hamster brain prior to detectable histopathology and clinical disease. J Gen Virol 1991;72:2905–13.

29. Jendroska K, Heinzel FP, Torchia M, et al. Proteinase-resistant prion protein accumulation in Syrian hamster brain correlates with regional pathology and scrapie infectivity. Neurology 1991;41:1482–90.

30. Prusiner SB. Molecular biology and transgenetics of prion diseases. Crit Rev Biochem Mol Biol 1991;26:397–438.

31. Pocchiari M, Xi YG, Casaccia P, Ladogana A, Masullo C. Pharmacological manipulation of scrapie infectivity. In: Bradley R, Savey M, Marchant B, eds. Sub-acute spongiform encephalopathies. Dordrecht: Kluwer Academic, 1991:187–93.

32. Brown P. Chemotherapy of unconventional virus infections of the central nervous system. In: Field HJ, ed. Antiviral agents: the development and assessment of antiviral chemotherapy, vol 2. Boca Raton: CRC Press, 1988:13–28.

33. Tateishi J. Antibiotics and antivirals do not modify experimentally-induced Creutzfeldt-Jakob disease in mice. J Neurol Neurosurg Psychiatry 1981;44: 723–4.
34. Masullo C, Macchi G, Xi YG, Pocchiari M. Failure to ameliorate Creutzfeldt-Jakob disease with amphotericin B therapy. J Infect Dis 1992;165:784–5.

Part IV

Pathogenesis, Molecular Biology and Genetics, and Immunohistochemistry

20

Genetics of Human Spongiform Encephalopathies: Current Status

Lev G. Goldfarb

Creutzfeldt-Jakob disease (CJD) and other transmissible spongiform encephalopathies (TSEs) represent a group of disorders that are inheritable, sporadic, and infectious. Major efforts have been undertaken in the past several years to elucidate the nature of these disorders. Novel molecular biology techniques have become important tools for analysis of the genetic mechanisms controlling these disorders and in studies of the infectious agent. CJD is a common neurologic disorder seen in any population with a frequency of around 1 per million per year, or approximately 1:10,000 in a life time. A portion of these cases are familial, and Gerstmann-Sträussler-Scheinker disease (GSS) and fatal familial insomnia (FFI) are exclusively familial disorders. All have been experimentally transmitted to nonhuman primates and laboratory rodents. The disease-causing protein (PrP) accumulating in the brain of these patients is a protease-resistant insoluble form of a normal precursor protein molecule of unknown function that is encoded by the PRNP gene on chromosome 20. A number of mutations identified in this gene are responsible for various phenotypes of spongiform encephalopathy in its familial form, and a polymorphism at codon 129 determines susceptibility to the infectious and perhaps sporadic forms of CJD. Transmissible spongiform encephalopathies (TSEs) are transmissible amyloidoses, in which the host encoded protein has a propensity to acquiring a beta-sheet conformation and falling into amyloid; accumulation of amyloid eventually destroys the neurons and induces the deadly disease.

Genetic Control of Disease Expression in Familial CJD, GSS, and FFI

Familial spongiform encephalopathies are autosomal dominant disorders associated with mutations in the PRNP gene. Seventeen different alleles have been shown to be associated with familial disease. Many attempts were undertaken to ascribe a discrete phenotype to each of the known alleles, but the phenotypes are variable even within a single family and there always is a variety of phenotypes associated with a single allele. Research progress is bringing new knowledge about mechanisms determining the phenotype and there is a chance that in the future it will be possible to make better predictions about phenotypes based on the knowledge of genotype-phenotype interactions. Table 20.1 lists the known mutations and corresponding phenotypes in familial CJD and FFI.

The PRNP gene has an unstable region of five repeating sequences between codons 51 and 91, of which the first 27-bp sequence is followed by four 24-bp variant repeats (1). In a group of families with clinical, neuropathologic, and transmission features of CJD two to nine additional repeats of 24-bp sequences were inserted in this area (1–3). The disease in patients showing ins48-bp, ins120-bp, ins144-bp, ins168-bp, and ins216-bp alleles typically starts earlier than in patients carrying other alleles, and the duration of illness is much longer. It is characterized by progressive dementia, ataxia and myoclonus, and, in most cases, a widespread spongiform degeneration. Specific features of this group of patients include abnormal behavior, depression, and aggressiveness early in the disease and cortical atrophy in many cases of otherwise typical CJD. One of the members of a ins168-bp family had no spongiform change, whereas the mother of this patient had diffuse spongiform degeneration (4). In the ins144-bp family, three pathologically examined patients had characteristic spongiform degeneration, but in the fourth only cortical atrophy and no

TABLE 20.1. PRNP gene mutations and corresponding phenotypes in familial Creutzfeldt-Jakob disease and fatal familial insomnia.

Mutation	Age	Duration	Clinical presentation	Dominant pathology	Primary localization
ins48 bp ins120 bp ins144 bp ins168 bp ins216 bp	35	8 years	Dementia	Spongiosis	Cerebral cortex
D178N-V129	45	22 months	Memory loss	Spongiosis	Cerebral cortex
D178N-M129	49	15 months	Insomnia	Dystrophy	Thalamus
V180I	79	15 months	Rigidity, tremor	Spongiosis	Cerebral cortex
E200K-M129	55	8 months	Dementia	Spongiosis	Cerebral cortex
V210I	63	4 months	Rigidity, tremor	Spongiosis	Cerebral cortex
M232R	65	4 months	Dementia	Spongiosis	Cerebral cortex

spongiosis or any other characteristic features were found, except for moderate increase in the number of astrocytes (5). Methionine at position 129 was detected in the allele with extra repeats in most of these cases.

Polymorphism at codon 129 has attracted much attention in the last several years and became the subject of the most intensive research. First reported as a mutation in a number of patients with TSE who were homozygous for valine at position 129 (6), it was later found to occur as a neutral polymorphism with the frequency of alleles in the Caucasian population 0.62 for methionine and 0.38 for valine (7). Mounting evidence has now proven that the presence of either amino acid at position 129 does not cause disease, but this polymorphism is able to modify effects of other mutations and may determine predisposition to TSE in patients with no mutations. For example, in the large ins144-bp family, the age at death in patients who were homozygous for the methionine 129 allele was significantly lower than in heterozygotes, which suggests a protective role of valine at position 129 influencing the age of onset and duration of illness (8).

This same position 129 polymorphism plays an even more dramatic role in determining the phenotype in two distinct syndromes, one segregating with the D178N-V129 allele and expressed as dementing illness with severe and diffuse spongiform degeneration; the other, which is linked to the same D178N mutation located in the allele with methionine at position 129, does not show dementia or spongiform change in most of the patients and is instead expressing untreatable fatal insomnia, dysautonomia, and selective atrophy of thalamic nuclei (9).

Deletion of a single octapeptide of several different types was identified in neurologic patients and control individuals with a similar frequency (10). But a deletion of one specific type (R3–R4) was observed in a CJD patient without any other mutations in the coding region (11). A deletion of this same type was also detected in three other clinically and pathologically confirmed patients with familial and apparently sporadic CJD (12). In several families, the same R3–R4 deletion has been found in CJD patients carrying other pathogenic mutations (13) and in a patient with iatrogenic CJD (14).

A group of patients with the V180I allele exhibits CJD with a unique phenotype of very late onset (66 to 85) and rapidly progressive dementia, tremor, and rigidity, and pathologic pictures of prominent spongiform degeneration in the cerebral cortex and subcortical ganglia (15). It is of interest that the V180I-V129 allele produced multiple PrP-reactive amyloid plaques in addition to spongiform degeneration in the cerebral cortex and subcortical ganglia, but not in the cerebellum. Another late-onset syndrome is associated with the M232R mutation (16). These patients also suffered progressive dementia and, in addition, had myoclonus and periodic synchronous discharges on EEG. In a single patient, a double mutation of V180I and M232R was found; the age of disease onset in this patient was 85 (16).

The E200K-M129 allele is the most prevalent mutation in familial TSE. It has been identified so far in 68 families worldwide, with a total of more than 140 patients. This mutation has been implicated in the high-incidence clusters of CJD in Slovakia (17) and Chile (18), and in a population of Libyan Jews living in Israel (19, 20). In some of these areas, CJD is the fifth most common cause of death. Late onset at age 55 and an average duration illness of only 8 months are characteristic for the E200K phenotype, which evolves with dementia, ataxia, and myoclonus associated with diffuse and severe spongiform degeneration as the classic features of this form, as well as that of sporadic CJD (21). The unusually high prevalence of CJD in Libyan Jews had been previously attributed to consumption of sheep brain and raw eye balls that could be contaminated with the scrapie agent (22). Molecular studies helped recognize the genetic cause of this disorder and turn research efforts in a completely different direction.

The penetrance rate (relative number of mutation carriers developing the disease) has been estimated for several different alleles and the rate for the E200K-M129 allele was found to be only 56% (23). This is much lower than that for other studied alleles and explains the observation that both parents of affected individuals are quite often unaffected (15, 23). A somewhat reduced penetrance has also been shown for the V210I allele (24). In a series of 27 CJD families carrying various alleles the average penetrance rate was estimated as 84% (25).

The group of phenotypes designated as GSS disease presented in Table 20.2 is different from the CJD phenotypes in having a slower progression coupled with the presence of a cerebral/cerebellar amyloidosis with characteristic multicentric amyloid plaques (26). These patients have cerebellar dysfunction in some types of disease, dementing illness and severe spastic paraparesis in others. Neuropathologic findings consist of amyloid plaques and variable degree of spongiform degeneration, primarily in the cerebel-

TABLE 20.2. PRNP gene mutations and corresponding phenotypes in Gerstmann-Sträussler-Scheinker disease.

Allele	Age	Duration	Clinical presentation	Dominant pathology	Primary localization
P102L-M129	48	6 years	Ataxia	Plaques	Cerebellum
P105L-V129	44	9 years	Spasticity	Plaques	Cerebral cortex
A117V	38	3 years	Dementia	Plaques	Cerebral cortex
Y145X	38	12 years	Dementia, ataxia	Plaques	Cerebral cortex, cerebellum
F198S-V129	52	6 years	Dementia	Plaques Tangles	Cerebral cortex
Q217R-V129	52	6 years	Dementia	Plaques Tangles	Cerebral cortex
ins192bp-V129	44	12 months	Dementia	Plaques spongiosis	Cerebral cortex

lum. Multicentric amyloid plaque is the hallmark of GSS. Brain homogenates from GSS patients transmitted disease to nonhuman primates and rodents through intracerebral inoculation.

P102L was the first mutation to be identified in human TSE (27) and four other mutations have since been discovered to be associated with GSS. GSS mutations do not overlap with the mutations identified in CJD.

The P102L-M129 allele is the most prevalent cause of GSS. More than 20 affected families worldwide are known to carry this disease allele, including the original family described by Gerstmann, Sträussler, and Scheinker (28). Ataxia is the most prominent clinical sign in the corresponding phenotype, and some workers have made a point of differentiating it from primarily dementing forms of GSS (27). Ataxia is followed by progressive dementia later in the course of illness. This form of disease was reproduced experimentally in transgenic mice by introducing the corresponding human pathogenic mutation (29, 30).

The P105L-V129 allele was associated with GSS-type disease in several Japanese families (31). A feature of this disease is the presence of severe spasticity reminiscent of familial spastic paraparesis, but characteristic for GSS multicentric amyloid plaques are seen in the cerebral cortex. The cerebellum is well preserved. Unusual topography of these lesions may depend on the presence of valine at position 129. Coupling of the P105L mutation with V129 in all five affected families is especially remarkable because the V129 allele occurs in only 4% of the general Japanese population (32).

The phenotype corresponding to the A117V mutation is different from all previously discussed mutations in causing a dementia with an earlier age of onset and a faster progression of illness (33). Dementia is accompanied by dysarthria, rigidity, and tremor. Multicentric amyloid plaques are numerous in the cerebral cortex and subcortical ganglia, but not in the cerebellum.

The largest and probably best studied GSS family is the Indiana kindred with a recently identified F198S-V129 allele (34). This kindred and a Swedish family with the Q217R-V129 genotype (35) are similarly distinctive in that all affected members in these families showed widespread Alzheimer's disease-type neurofibrillary tangles in the cerebral cortex and subcortical nuclei (36). Uni- and multicentric amyloid plaques characteristic for GSS colocalized with plaques having the rich neuritic component often seen in Alzheimer's disease patients. Both variants of plaques were immunoreactive with anti-PrP but not with anti-beta-amyloid antibodies. The age of onset (mean at age 52) was shown to be dependent on codon 129 genotype: the average age of onset in a group of patients with V/V129 homozygosity was 44.7 years, whereas in M/V129 heterozygotes it was 60.3 years ($p < .05$) (32). Together with similar results from another large CJD family with the ins144-bp mutation (8), these observations suggest that homozygosity at codon 129 promotes an earlier disease onset in familial TSE. V129 might also be responsible for plaque formation in at least the phenotype corresponding to F198S-V129 because valine was detected in the plaques even when the patient was heterozygous (34).

Six analyzed patients in a family having a 192-bp insert show valine at position 129 of the elongated allele. The phenotype in this family was different from other families with extra repeat mutations in having abundant multicentric amyloid plaques in the cerebellum, a prominent feature of classic GSS (37).

The variability of phenotypes caused by different mutations probably depends on differential targeting of neuron populations: the rate and pattern of PrP accumulation determines the rate and pattern of lesion formation (38). Other phenotypic variants of TSE may be expected; for example, a variant involving degeneration of skeletal muscles has not been found in humans, but has been shown to exist in transgenic mice overexpressing wild-type PrP (39).

Transmission of familial disease to experimental animals has been achieved in families with ins120-bp, ins168-bp, ins192-bp, P102L-M129, E200K-M129, D178N-V129, and D178N-V129 in a total of 41 patients from 36 unrelated families (40).

Genetic Control of Susceptibility to Iatrogenic and Sporadic CJD

Homozygosity for either amino acid at position 129 was identified as a predisposing factor in a majority of sporadic and iatrogenic cases of CJD (32, 42–43). Homozygotes for the allele with methionine were overrepresented among patients with sporadic CJD (42) and iatrogenic CJD with CNS route of infection (43); the valine allele was overrepresented in iatrogenic CJD with a peripheral route of infection in many studies (6, 41, 43). The most plausible explanation for this phenomenon is that the structurally different molecules in a heterozygote do not polymerize as easily as identical molecules. The question as to why methionine homozygosity is overrepresented in sporadic forms and valine homozygosity in growth hormone-associated iatrogenic forms has not been addressed yet.

References

1. Owen F, Poulter M, Shah T, et al. An in-frame insertion in the prion protein gene in familial Creutzfeldt-Jakob disease. Mol Brain Res 1990;7:273–6.
2. Goldfarb LG, Brown P, McCombie WR, et al. Transmissible familial Creutzfeldt-Jakob disease associated with five, seven, and eight extra octapeptide coding repeats in the PRNP gene. Proc Natl Acad Sci USA 1991;88:10926–30.

3. Owen F, Poulter M, Collinge J, et al. A dementing illness associated with a novel insertion in the prion protein gene. Mol Brain Res 1992;13:155–7.
4. Brown P, Goldfarb LG, Brown WT, et al. Clinical and molecular genetic study of a large German kindred with Gerstmann-Sträussler-Scheinker syndrome. Neurology 1991;41:375–9.
5. Collinge J, Brown J, Hardy J, Mullan M, et al. Inherited prion disease with 144 base pair gene insertion. Brain 1992;115:687–710.
6. Goldfarb LG, Brown P, Goldgaber D, et al. Patients with Creutzfeldt-Jakob disease and kuru lack the mutation in the PRIP gene found in Gerstmann-Sträussler-Scheinker syndrome, but they show a different double-allele mutation in the same gene. Am J Hum Genet 1989;45(Supp 1):A189.
7. Owen F, Poulter M, Collinge J, Crow TJ. Codon 129 changes in the prion protein gene in Caucasians. Am J Hum Genet 1990;46:1215–6.
8. Baker HF, Poulter M, Crow TJ, et al. Amino acid polymorphism in human prion protein and age at death in inherited prion disease. Lancet 1991;337:1286.
9. Goldfarb LG, Petersen RB, Tabaton M, et al. Fatal familial insomnia and familial Creutzfeldt-Jakob disease: disease phenotype determined by a DNA polymorphism. Science 1992;258:806–8.
10. Palmer M, Mahal SP, Campbell TA, et al. Deletions on the prion protein gene are not associated with CJD. Hum Mol Genet 1993;2:541–4.
11. Diedrich JF, Knopman DS, List JF, et al. Deletion in the prion protein gene in a demented patient. Hum Mol Genet 1992;1:443–4.
12. Cervenakova L, Brown P, Piccardo P, et al. 24-Nucleotide deletion in the PRNP gene: analysis of associated phenotypes. In: Dormont D, Court L, eds. Transmissible subacute spongiform encephalopathies. Paris: Elsevier, 1996 (in press).
13. Bosque PJ, Vnencak-Jones CL, Johnson MD, et al. A PrP gene codon 178 base substitution and a 24-bp interstitial deletion in familial Creutzfeldt-Jakob disease. Neurology 1992;42:1864–70.
14. Goldfarb LG, Brown P, Gajdusek DC. The molecular genetics of human transmissible spongiform encephalopathy. In: Prusiner SB, Collinge J, Powell J, Anderton B, eds. Prion diseases of humans and animals. Chichester, UK: Ellis Horwood, 1992:139–53.
15. Kitamoto T, Ohta M, Doh-ura K, et al. Novel missense variants of prion protein in Creutzfeldt-Jakob disease or Gerstmann-Sträussler syndrome. Biochem Biophys Res Commun 1993;191:709–14.
16. Kitamoto T, Tateishi J. Human prion diseases with variant prion protein. Philos Trans R Soc Lond B 1994;343:391–8.
17. Goldfarb LG, Mitrova E, Brown P, Toh B-H, Gajdusek DC. Mutation in codon 200 of scrapie amyloid protein gene in two clusters of Creutzfeldt-Jakob disease in Slovakia. Lancet 1990;336:514.
18. Brown P, Galves S, Goldfarb LG, Gajdusek DC. Familial Creutzfeldt-Jakob disease in Chile is associated with the codon 200 mutation of the PRNP amyloid precursor gene on chromosome 20. J Neurol Sci 1992;112:65–7.
19. Goldfarb LG, Korczyn AD, Brown P, Chapman J, Gajdusek DC. Mutation in codon 200 of scrapie amyloid precursor gene linked to Creutzfeldt-Jacob disease in Sephardic Jews of Libyan and non Libyan origin. Lancet 1990;336:637.
20. Hsaio KH, Meiner Z, Kahana E, et al. Mutation of the prion protein in Libyan Jews with Creutzfeldt-Jakob disease. N Engl J Med 1991;324:1091–7.

21. Brown P, Goldfarb LG, Gibbs CJ Jr, Gajdusek DC. The phenotypic expression of different mutations in transmissible familial Creutzfeldt-Jakob disease. Eur J Epidemiol 1991;7:469-76.

22. Alter M. Creutzfeldt-Jakob disease: hypothesis for high incidence in Libyan Jews in Israel. Science 1974;186:848.

23. Goldfarb LG, Brown P, Mitrova E, et al. Creutzfeldt-Jakob disease associated with the PRNP codon 200Lys mutation: an analysis of 45 families. Eur J Epidemiol 1991;7:477-486.

24. Ripoll L, Laplanche J-L, Salzmann M, et al. A new point mutation in the prion protein gene at codon 210 in Creutzfeldt-Jakob disease. Neurology 1993; 43:1934-8.

25. Masters CL, Gajdusek DC, Gibbs CJ. The familial occurrence of Creutzfeldt-Jakob disease and Alzheimer's disease. Brain 1981;104:535-58.

26. Masters CL, Gajdusek DC, Gibbs C Jr. Creutzfeldt-Jakob disease virus isolations from the Gerstmann-Sträussler syndrome, with an analysis of the various forms of amyloid plaque deposition in the virus-induced spongiform encephalopathies. Brain 1981;104:559-88.

27. Hsiao K, Baker HF, Crow TJ, et al. Linkage of a prion protein missense variant to Gerstmann-Sträussler syndrome. Nature 1989;338:342-5.

28. Kretzschmar HA, Honold G, Seitelberger F, et al. Prion protein mutation in family first reported by Gerstmann, Sträussler, and Scheinker. Lancet 1991; 337:1160.

29. Hsiao KK, Scott M, Foster D, et al. Spontaneous neurodegeneration in transgenic mice with mutant prion protein. Science 1990;250:1587-90.

30. Hsiao KK, Groth D, Scott M, et al. Serial transmission in rodents of neurodegeneration from transgenic mice expressing mutant prion protein. Proc Natl Acad Sci USA 1994;91:9126-30.

31. Kitamoto T, Amano N, Terao Y, et al. A new inherited prion disease (PrP-P105L) mutation showing spastic paraparesis. Ann Neurol 1993;34:808-13.

32. Doh-ura K, Kitamoto T, Sakaki Y, Tateishi J. CJD discrepancy. Nature 1991;353:801-2.

33. Hsiao KK, Cass C, Schellenberg GD, et al. A prion protein variant in a family with the telencephalic form of Gerstmann-Sträussler-Scheinker syndrome. Neurology 1991;41:681-4.

34. Dlouhy SR, Hsiao K, Farlow MR, et al. Linkage of the Indiana kindred of Gerstmann-Sträussler-Scheinker disease to the prion protein gene. Nature Genetics 1992;1:64-7.

35. Hsiao K, Dlouhy SR, Farlow MR, et al. Mutant prion proteins in Gerstmann-Sträussler-Scheinker disease with neurofibrillary tangles. Nature Genetics 1992;1:68-71.

36. Ghetti B, Tagliavini F, Masters CL, et al. Gerstmann-Sträussler-Scheinker disease. Neurofibrillary tangles and plaques with PrP-amyloid coexist in an affected family. Neurology 1989;39:1453-61.

37. Goldfarb LG, Brown P, Vrbovska A, et al. An insert mutation in the chromosome 20 amyloid precursor gene in a Gerstmann-Sträussler-Scheinker family. J Neurol Sci 1992;111:189-94.

38. DeArmond SJ, Prusiner SB. The neurochemistry of prion diseases. J Neurochem 1993;61:1589-601.

39. Westaway D, DeArmond SJ, Cayetano-Canlas J, et al. Degeneration of skeletal muscle, peripheral nerves, and the central nervous system in transgenic mice overexpressing wild-type prion proteins. Cell 1994;76:117–29.
40. Brown P, Gibbs CJ Jr, Rodgers-Johnson P, et al. Human spongiform encephalopathy: the National Institutes of Health series of 300 cases of experimentally transmitted disease. Ann Neurol 1994;35:513–29.
41. Collinge J, Palmer MS, Dryden AJ, et al. Genetic predisposition to iatrogenic Creutzfeldt-Jakob disease. Lancet 1991;337:1441–2.
42. Palmer MS, Dryden AJ, Hughes JT, Collinge J. Homozygous prion protein genotype predisposes to sporadic Creutzfeldt-Jakob disease. Nature 1991; 352:340–2.
43. Brown P, Cervenakova L, Goldfarb LG, et al. Iatrogenic Creutzfeldt-Jakob disease: an example of the interplay between ancient genes and modern medicine. Neurology 1994;44:291–3.

21

PrP Allelic Variants Associated with Natural Scrapie

Peter B.G.M. Belt, Alex Bossers, Bram E.C. Schreuder, and Mari A. Smits

To determine whether an association between specific combinations of polymorphisms in the PrP gene of sheep and natural scrapie exists, we applied denaturing gradient gel electrophoresis (DGGE) to analyze the PrP genes of 69 scrapie-affected and 176 healthy sheep predominantly of the Texel breed. We found six PrP allelic variants representing different combinations of polymorphisms at codons 136, 154, 171, and 211. Based on the triplet sequences present at codons 136, 154, and 171, these alleles were designated PrP^{VRQ}, PrP^{ARR}, PrP^{ARQ}, PrP^{ARH}, PrP^{AHQ}, and $PrP^{ARQQ211}$. Among the 245 sheep, 15 different PrP genotypes or allele pairs were present. The distribution of the allelic variants revealed that the PrP^{VRQ} allele is associated with a high incidence of natural scrapie and the PrP^{ARR} allele with a low incidence. The PrP^{ARQ} and PrP^{ARH} alleles were not associated with disease incidence. Nevertheless, a number of scrapie-affected sheep were homozygous for the PrP^{ARQ} allele. Whether in these sheep factors other than the PrP genotype play a more decisive role in inducing the disease than in sheep with a PrP^{VRQ} allele remains to be established. The frequencies of the PrP^{AHQ} and the $PrP^{ARQQ211}$ alleles were too low to draw any conclusions.

Scrapie is a fatal neurodegenerative disease of the central nervous system (CNS) that occurs naturally in sheep. It is the archetype of a group of disorders known as transmissible spongiform encephalopathies (TSEs) or prion diseases. Creutzfeldt-Jakob disease (CJD), Gerstmann-Sträussler-Scheinker (GSS) syndrome, and fatal familial insomnia (FFI) of man and bovine spongiform encephalopathy (BSE) are also members of this group. The etiology of scrapie has been debated for many years but the precise etiology remains unknown. Some considered natural scrapie as a genetic disorder, caused by an autosomal recessive gene, that just happens to be experimentally transmissible (1). Others describe scrapie as a virus-like disease, the outcome of which is influenced by host genetic components (2–4). The current view is that scrapie is an infectious disease where host genetic factors play an important role. By experimental infection of sheep with scrapie, Dickinson and Outram (5) demonstrated that both disease susceptibility and incubation period are controlled by a single sheep gene.

The PrP gene of the host is the gene that is most closely associated with features of prion diseases such as susceptibility, incubation period, incidence, and pathology (6, 7). In man, at least 10 missense, 1 nonsense, and 7 insert mutations in the PrP gene segregate with prion diseases (7). A common polymorphism at codon 129 is associated with susceptibility to iatrogenic and sporadic CJD (8, 9). This polymorphism also determines the phenotype of the disease in familial CJD and FFI patients who have a PrP mutation at codon 178 (10). This suggests that specific combinations of polymorphisms within the PrP gene are associated with particular features of prion diseases.

In the PrP gene of sheep, several polymorphisms have been detected (11–13). A polymorphism at codon 136 is associated with scrapie susceptibility in both experimental (12, 14, 15) and natural scrapie (13, 16). A polymorphism at codon 171 is also associated with susceptibility of Cheviot sheep for experimental scrapie (14) and Suffolk sheep for natural scrapie (17). Other polymorphisms, including polymorphisms at codons 112 and 154, are rare and have not been associated with scrapie.

In this study we identified and characterized different combinations of polymorphisms (allelic variants) of the sheep PrP gene and we investigated their association with natural scrapie. To this end we determined the PrP allelic variants of 69 scrapie-affected and 176 healthy sheep by denaturing gradient gel electrophoresis (DGGE). This technique allows separation of DNA molecules differing by as little as a single base change.

Materials and Methods

Sheep

We used 210 sheep of the Texel breed; 34 were scrapie-affected, 91 were healthy and derived from scrapie-free flocks, and 85 were healthy and derived from scrapie-affected flocks. In addition, we used 35 scrapie-affected sheep of seven other breeds. The scrapie-affected sheep from 18 different flocks were collected throughout the Netherlands from August 1991 to October 1992, with ages between 2 and 5 years. All scrapie cases were confirmed by histology and immunocytochemistry as described elsewhere (18). The age-matched healthy control sheep derived from scrapie-free flocks (free from scrapie for the last 5 years) were collected from 15 different flocks in the Netherlands in the same period. The age-matched healthy control sheep derived from scrapie-affected flocks were collected from 13 different flocks in the Netherlands. The 35 scrapie-affected sheep of seven other breeds were collected in the Nertherlands in the same period as the Texel sheep and were of the same age (detailed data will be published elsewhere). About 4% to 8% of the Dutch flocks suffer from scrapie, with an incidence of 1 in 100 sheep/year (19). From all sheep, we collected blood samples in ethylenediaminetetraacetic acid (EDTA) vacutainers (Venoject, Terumo Europe N.V.) and they were frozen at $-20°C$ until further use.

DNA Amplification

High-molecular-weight DNA was isolated from blood as described by Sambrook et al. (20). The PrP open reading frame, which resides on a single exon (11), was amplified by polymerase chain reaction (PCR). Amplification reactions were done in 50 µL containing 100 ng genomic DNA, 25 mM Tris-HCl pH 8.7, 2 mM $MgCl_2$, 0.005% gelatine, 200 µM dNTPs, 1 µM of primers, and 1 unit of Taq DNA polymerase (Perkin-Elmer-Cetus). The sequences of the oligonucleotide primers and their positions in the PrP gene have been described (21). The amplification reactions were performed in a Perkin-Elmer-Cetus DNA Thermal Cycler for 35 cycles of 1 min at 92°C, 1.5 min at 58°C, and 1.5 min at 72°C with a hot start (80°C).

DGGE Analysis

Sheep genomic DNA was amplified as described above and 15 µL was digested with the restriction enzyme HinfI in a total volume of 20 µL. Subsequently, the amplified DNA was denatured for 10 min at 100°C and renatured at room temperature. The DNA was separated on DGGE gels containing 6% polyacrylamide (37.5:1, acrylamide:bisacrylamide) with a linear gradient from top to bottom of 45% to 65% denaturant (100% denaturant = 7 M urea/40% formamide (vol/vol), gel length = 17 cm) in 40 mM Tris-acetate, 1 mM EDTA, pH 7.4. Electrophoresis was done at 50 V for 24 hours at 60°C in 40 mM Tris-acetate, 1 mM EDTA, pH 7.4 in an electrophoresis tank as described by Myers et al. (22). Gels were stained with ethidium bromide and examined by UV transillumination.

Cloning and Sequencing

For cloning and sequencing purposes, the coding region of the PrP allelic variants was amplified by PCR using Vent DNA polymerase (New England Biolabs) as described previously (21). Amplified fragments were inserted into the EcoRI and XbaI site of pGEM7 (Promega). Clones, representing different allelic variants of the PrP gene, were distinguished from each other by DGGE analysis as described above. Sequencing was done using the chain termination reaction (23). At least two independent clones of each allelic variant were sequenced to exclude PCR artefacts.

Statistical Analysis

Results were analyzed using the χ^2 test of association by comparing geno-type frequencies in the scrapie-affected sheep and the healthy control sheep.

Results

Identification of PrP Allelic Variants and Novel Polymorphisms by DGGE Analysis

To identify and characterize allelic variants of the PrP gene of sheep and to identify hitherto unknown polymorphisms, we applied DGGE on PCR-amplified DNA from codon position 1 to 223 of the PrP-gene (11). A single DGGE run allowed us to detect polymorphisms in a region that covers almost 90% of the coding region for PrP 27-30 (21). On the basis of the DGGE patterns, we identified five different PrP allelic variants in 210 scrapie-affected and healthy sheep of the Texel breed. We cloned and sequenced these variants to identify the base changes responsible for their characteristic mobilities in the DGGE gel (Fig. 21.1, lanes 2–6). Based on the sequence differences found at codon positions 136 [alanine (A) or valine (V)], 154 [arginine (R) or A], and 171 [glutamine (Q), histidine (H), or R], the first variant was designated PrP^{ARR} (Fig. 21.1, lane 2). The second variant was designated PrP^{ARQ} and the third PrP^{VRQ} (Fig. 21.1, lanes 3 and

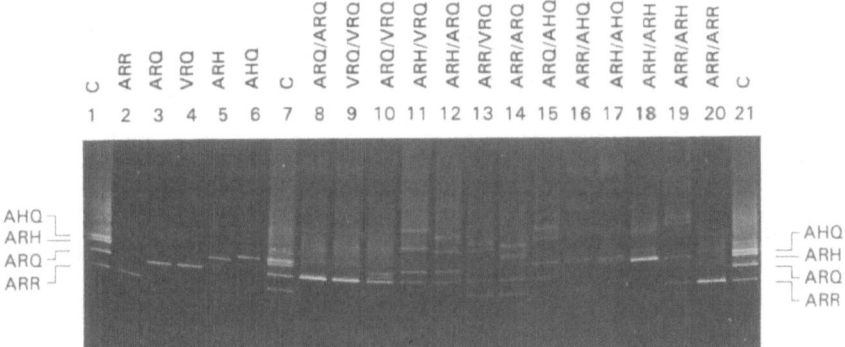

FIGURE 21.1. DGGE analysis of PCR amplified DNA. Lanes 2–6: DGGE patterns obtained from five cloned and sequenced allelic variants. Lanes 8–20: DGGE patterns representing 13 different genotypes present in sheep of the Texel breed. Lanes 1, 7, and 21 are control lanes that contain four combined allelic variants that were not denatured and renatured before loading. Allele variants are indicated on the right and on the left. Genotypes are indicated above the lanes. Reproduced with permission of the Journal of General Virology (21).

4). Although the homoduplex bands of the PrP^{ARQ} and PrP^{VRQ} variants could not be distinguished from each other, they could be distinguished when heteroduplex molecules were formed with these variants (compare, for example, lanes 11 and 12). The fourth variant was designated PrP^{ARH} (Fig. 21.1, lane 5). From the sequence analysis it also appeared that two silent mutations occurred in the PrP^{ARH} variant, one at codon 231 (AGG → CGG) and one at codon 237 (CTC → CTG). These mutations are, however, located outside the fragment analyzed by DGGE and therefore cannot contribute to the shift in electrophoretic mobility. The fifth allelic variant was designated PrP^{AHQ} (Fig. 21.1, lane 6).

We also analyzed the PrP genes of 35 scrapie-affected sheep of seven other breeds. In this group we found four DGGE patterns identical to those found in the sheep of the Texel breed, and one DGGE pattern that differed (Fig. 21.2). The homoduplex bands of this latter variant resembled the PrP^{ARH}/PrP^{ARQ} or the PrP^{ARH}/PrP^{VRQ} genotypes (Fig. 21.2, lane 6). The heteroduplex bands revealed however a hitherto unknown DGGE mobility pattern (compare Fig. 21.1, lanes 11 and 12, and Fig. 21.2, lanes 6 and 7). By cloning and sequencing we found that this sheep contained one PrP^{VRQ} allele and a PrP^{ARQ} allele with an additional G → A transition at codon 211 resulting in an R to Q substitution. The latter allelic variant was therefore designated PrP^{ARQQ211}.

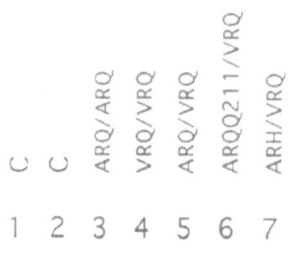

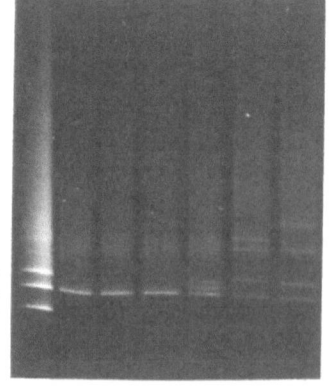

FIGURE 21.2. DGGE analysis of PCR amplified DNA. Lanes 3–7: DGGE patterns representing the five different genotypes present in the 35 scrapie-affected sheep from breeds other than the Texel breed. Lane 1 contains three combined allelic variants that were not denatured and renatured before loading. Lane 2 contains the PrP^{VRQ} variant. The haplotypes are indicated on the left. Genotypes are indicated above the lanes.

TABLE 21.1. Frequencies of PrP genotypes in scrapie-affected and healthy control sheep derived from scrapie-free flocks (h-sff) or scrapie-affected flocks (h-saf) of the Texel breed.

Genotype	Scrapie		h-sff		h-saf	
	n	%	n	%	n	%
PrP^{VRQ}/PrP^{VRQ}	3	9	0	0	0	0
PrP^{VRQ}/PrP^{ARQ}	12	35	4	4	4	5
PrP^{VRQ}/PrP^{ARH}	15	44	2	2	4	5
PrP^{VRQ}/PrP^{AHQ}	0	0	0	0	2	2
PrP^{VRQ}/PrP^{ARR}	1	3	4	4	11	13
PrP^{ARR}/PrP^{ARR}	0	0	15	17	8	9
PrP^{ARR}/PrP^{ARQ}	0	0	22	24	23	27
PrP^{ARR}/PrP^{ARH}	0	0	6	7	7	8
PrP^{ARR}/PrP^{AHQ}	0	0	2	2	3	4
PrP^{ARQ}/PrP^{ARQ}	3	9	10	11	3	4
PrP^{ARQ}/PrP^{ARH}	0	0	14	15	12	14
PrP^{ARQ}/PrP^{AHQ}	0	0	4	4	1	1
PrP^{ARH}/PrP^{ARH}	0	0	6	7	4	5
PrP^{ARH}/PrP^{AHQ}	0	0	2	2	2	2
PrP^{AHQ}/PrP^{AHQ}	0	0	0	0	1	1

Reproduced with permission of the Journal of General Virology (21).

PrP Allelic Variants and Their Association with Natural Scrapie

To investigate the association of the PrP allelic variants with natural scrapie, we identified the PrP alleles of 210 sheep of the Texel breed. Of these sheep, 34 were scrapie-affected, 91 were healthy and derived from scrapie-free flocks, and 85 were healthy and derived from scrapie-affected flocks. In addition, we determined the PrP genotypes of 35 scrapie-affected sheep of seven other breeds. In the Texel breed we found 14 different DGGE patterns (12 patterns are shown in Fig. 21.1, lanes 8–20), all of which could be reproduced by mixing two of the five cloned and sequenced alleles (data not shown). Since heteroduplex bands with the PrP^{ARQ} and PrP^{VRQ} variants could be distinguished from each other and homoduplex bands not (see above), all samples showing homoduplex bands were mixed with another variant and analyzed again by DGGE. After this analysis (data not shown) we ended up with 15 different sets of PrP alleles (genotypes) in the sheep of the Texel breed.

The frequencies of these genotypes in the three groups of sheep are summarized in Table 21.1. The PrP^{VRQ} variant was present in 91% of the scrapie-affected sheep, in 10% of the healthy sheep derived from scrapie-free flocks, and in 25% of the healthy sheep derived from scrapie-affected flocks. The PrP^{VRQ} homozygotes were present exclusively in the group of scrapie-affected sheep. We concluded that, at least in the Texel breed, the PrP^{VRQ} allele is associated with a high incidence of natural scrapie (Table 21.2, $\chi^2 = 73.18$, $p < .0001$ for scrapie vs. healthy sheep derived from scrapie-free flocks and $\chi^2 = 46.16$, $p < .0001$ for scrapie vs. healthy sheep derived from scrapie-affected flocks). There is also a slight but significant difference in the presence of the PrP^{VRQ} allele between the healthy sheep from scrapie-affected flocks and scrapie-free flocks (Table 21.2, $\chi^2 = 5.70$, $p < .0170$ for healthy sheep derived from scrapie-affected flocks vs. healthy sheep derived from scrapie-free flocks). In the Texel breed only 9% of the scrapie-affected sheep did not possess a PrP^{VRQ} allele; instead, these animals were homozygous for PrP^{ARQ}. However, this PrP^{ARQ} allele did not show any significant association with disease incidence (Table 21.2) since it is almost equally distributed among the three groups.

The PrP^{ARR} variant was present in only 3% (one sheep) of the scrapie-affected sheep, in 54% of the healthy sheep derived from scrapie-free flocks, and in 61% of the healthy sheep derived from scrapie-affected flocks. The PrP^{ARR} homozygotes were found exclusively in the groups of the healthy sheep. We concluded that, at least in the Texel breed, the PrP^{ARR} variant is associated with a low incidence of natural scrapie (Table 21.2, $\chi^2 = 26.77$, $p < .0001$ for scrapie vs. healthy sheep derived from scrapie-free flocks and $\chi^2 = 33.36$, $p < .0001$ for scrapie vs. healthy sheep derived from scrapie-affected flocks). As the PrP^{ARQ} variant, the PrP^{ARH} did not show any significant association with disease incidence (Table 21.2). Since the PrP^{ARR}/PrP^{ARQ} and PrP^{ARR}/PrP^{ARH} genotypes were only present in the healthy groups and the PrP^{VRQ}/PrP^{ARQ} and PrP^{VRQ}/PrP^{ARH} genotypes preferentially in the scrapie-affected group, the PrP^{VRQ} and PrP^{ARR} alleles seem to be dominant over the PrP^{ARQ} and PrP^{ARH} alleles. The occurrence of the PrP^{AHQ} variant in the three groups was too low to draw any conclusions.

In the 35 scrapie-affected sheep of seven other breeds we found five different genotypes. Of these sheep, 69% were homozygous or heterozy-

TABLE 21.2. χ^2 test of association by comparing PrP genotype frequencies in the scrapie-affected sheep, the healthy control sheep derived from scrapie-free flocks (h-sff), and the healthy control sheep derived from scrapie-affected flocks (h-saf) of the Texel breed, calculated by comparing homo- and heterozygotes in the scrapie-affected and healthy control groups.[a]

Variant	Scrapie ↔ h-sff		Scrapie ↔ h-saf		h-sff ↔ h-saf	
	χ^2	p	χ^2	p	χ^2	p
PrP^{VRQ}	73.18	<.0001	46.16	<.0001	5.70	.0170
PrP^{ARR}	26.77	<.0001	33.36	<.0001	4.29	.1169
PrP^{ARQ}	2.33	.3113	2.32	.3139	4.08	.1303
PrP^{ARH}	5.22	.0735	3.55	.1697	0.43	.8065
PrP^{AHQ}	1.89	.1687	3.89	.1427	1.10	.5755

[a] XX, XY and YY, where X = the given allelic variant and Y = all other possible variants.

TABLE 21.3. PrP genotypes of scrapie-affected sheep in other breeds than the Texel breed.

Genotype	n	%
PrP^{VRQ}/PrP^{VRQ}	8	23
PrP^{VRQ}/PrP^{ARQ}	9	26
PrP^{VRQ}/PrP^{ARH}	6	17
PrP^{VRQ}/PrP^{ARQQ211}	1	3
PrP^{ARQ}/PrP^{ARQ}	11	31

gous for PrP^{VRQ} and none possessed the PrP^{ARR} allele (Table 21.3). This strengthens our conclusion that the PrP^{VRQ} allele is also associated with a high incidence of scrapie and the PrP^{ARR} allele with a low incidence of scrapie in other breeds. Similar to the Texel breed, the scrapie-affected sheep of the other breeds that did not possess a PrP^{VRQ} allele were homozygous for the PrP^{ARQ} allele (Table 21.3). The occurrence of the PrP^{ARQQ211} variant was too low to draw any conclusions.

Discussion

In this chapter we present evidence that specific combinations of polymorphisms within the PrP gene of sheep, rather than single polymorphisms, are associated with the incidence of scrapie. We have used DGGE analysis to screen 245 sheep for the presence of various PrP alleles. We used 210 sheep of the Texel breed, 34 were scrapie-affected, 91 were healthy and derived from scrapie-free flocks, and 85 were healthy and derived from scrapie-affected flocks. In addition, we used 35 scrapie-affected sheep of seven other breeds. We found six different allelic variants with polymorphisms at codons 136, 154, 171, and/or 211. Based on the triplet sequences present at codons 136, 154, and 171, the allelic variants were designated PrP^{VRQ}, PrP^{ARR}, PrP^{ARQ}, PrP^{ARH}, PrP^{AHQ}, and PrP^{ARQQ211}. With these six allelic variants we found a total of 15 different PrP genotypes. The PrP^{VRQ} allele is associated with a high incidence of natural scrapie and the PrP^{ARR} allele with a low incidence. The PrP^{VRQ} and PrP^{ARR} alleles are antagonists in determining disease susceptibility and seem to be dominant over the PrP^{ARQ} and PrP^{ARH} alleles. Whether sheep with the PrP^{VRQ} allele investigated in this study are just highly susceptible to the scrapie agent, or whether they have the potential to develop the disease spontaneously similar to the inherited prion diseases (24) cannot be determined from these experiments.

All scrapie-affected sheep that did not possess a PrP^{VRQ} allele were homozygous for PrP^{ARQ}. Texel sheep homozygous for PrP^{ARQ}, however, were also present in the healthy control groups. In addition, we found no association between the PrP^{ARQ} allele and disease incidence (Table 21.2). Obviously in PrP^{ARQ}/PrP^{ARQ} sheep, other factors, for example the contaminated environment (infectious dose) or the scrapie source, play a more decisive role in inducing the disease than in sheep with a PrP^{VRQ} allele. This is consistent with recent conclusions of Westaway et al. (17), who found that scrapie-affected Suffolk sheep from the United States were homozygous A at codon position 136 and homozygous Q at codon position 171. However, they found this genotype also with a high frequency among phenotypically normal sheep at least 4 years old. In accordance with our conclusions, these authors indicated that the high percentage of healthy sheep with 136 A/A and 171 Q/Q formally excludes that this genotype itself, which probably resembles the PrP^{ARQ}/PrP^{ARQ} genotype described here, causes disease.

As indicated above, the PrP^{VRQ} and PrP^{ARR} alleles are antagonists in determining disease susceptibility with the PrP^{VRQ} allele associated with a high incidence of natural scrapie and PrP^{ARR} with a low incidence. PrP^{VRQ}/PrP^{ARR} heterozygotes were found in 3% of the scrapie-affected animals, in 4% of the healthy sheep from scrapie-free flocks, but in 13% of the healthy sheep from scrapie-affected flocks. We concluded that, at least in the scrapie-affected flocks, the PrP^{ARR} allele protects PrP^{VRQ}/PrP^{ARR} sheep from disease onset. This is consistent with the finding that the 171 R allele seems to be very rare among sheep affected by natural scrapie (13, 17). Laplanche and coworkers (13) showed that 3 Ile-de-France sheep with a 136 A/V, 154 R/R, 171 R/Q genotype (which is probably similar to the PrP^{VRQ}/PrP^{ARR} genotype) became scrapie-affected with a very late clinical onset of disease at 7 years of age. Apparently also in these sheep the PrP^{ARR} allele delays clinical manifestation of the disease. In this respect, it is interesting to note that we found PrP^{Sc} depositions in the brain, spleen, tonsil, and lymph nodes in 54 of the 55 scrapie-affected sheep that were investigated (18). In only one sheep we found PrP^{Sc} depositions in the brain but not in lymphoid tissue. This particular sheep turned out to be the animal with the PrP^{VRQ}/PrP^{ARR} genotype. Whether there is any correlation between this lack of PrP^{Sc} accumulation in lymphoid tissue and the relative protection attributed to the PrP^{ARR} allele remains to be established.

Studies on experimental scrapie with Cheviot sheep have shown that codon 136 V is associated with a high susceptibility to experimental infection (12, 15). The association of codon 136 V with natural scrapie was confirmed for Ile de France and Romanov sheep (13) and for Swaledales (16). The data described here confirm these findings. The association of codon 136 A with low susceptibility as predicted in several reports, however, needs refinement. The data presented here clearly indicate that only the PrP^{ARR} allele is associated with a low incidence of natural scrapie and not the PrP^{ARQ} and PrP^{ARH} alleles. Therefore, the detection of codon 136 A alone is not sufficient to predict scrapie resistance or susceptibility.

From the experiments described here, it cannot be concluded that the association of the PrP^{VRQ} and the PrP^{ARR} alleles with high and low scrapie susceptibility accounts for all scrapie "strains." The data suggest that the association at least accounts for most of the "strains" (if present) that circulate in the Netherlands. In addition recent data of Goldmann et al. (14) indicate that our observations have a more general significance. They showed that all sheep homozygous for codon 136 A and either homo- or heterozygous for codon 171 R are resistant to experimental challenge with scrapie isolates from the A-group (SSBP/1) and the C-group (CH1641), and to a BSE isolate. They also showed that all homozygotes or heterozygotes for codon 136 V that do not have a codon 171 R allele, became affected after experimental challenge with SSBP/1 and the BSE isolate. These results are completely consistent with the findings presented here. In contrast, the same study indicated that sheep, heterozygous for codons 136 (V/A) and 171 (Q/R) were very susceptible to experimental infection with SSBP/1 but not with the BSE isolate. These sheep are probably comparable with the PrP^{VRQ}/PrP^{ARR} sheep described here. In these sheep disease development might be restricted to the A-group agents, although the species barrier for the BSE isolate could have been the reason for the observed resistance as well.

The data presented in this chapter suggest that a positive selection for the PrP^{ARR} allele and/or a negative selection for the PrP^{VRQ} allele in breeding programs could help to control natural scrapie. We cannot exclude yet whether such breeding programs will result in the selection and circulation of other scrapie "strains." We also do not know whether sheep with the PrP^{ARR} allele are able to replicate the scrapie agent without showing clinical signs of disease and therefore can serve as a reservoir for infectious agent. Further research is needed to address these questions.

Acknowledgments. We thank E. de Haan for help in the collection of sheep material, L. van Keulen and coworkers for confirmation of the scrapie cases, I. Muileman and D. Prins for excellent technical assistance, and Drs. J. van Bekkum and J. Langeveld for critical reading of the manuscript.

References

1. Parry HB. Scrapie: a transmissible and hereditary disease of sheep. Heredity 1962;17:75–105.
2. Dickinson AG, Young GB, Stamp JT, Renwick CC. An analysis of natural scrapie in Suffolk sheep. Heredity 1965;20:485–503.
3. Brotherston JG, Renwick CC, Stamp JT, Zlotnick I. Spread of scrapie by contact to goats and sheep. J Comp Pathol 1968;78:9–17
4. Dickinson AG, Stamp JT, Renwick CC. Maternal and lateral transmission of scrapie in sheep. J Comp Pathol 1974;84:19–25.
5. Dickinson AG, Outram GW. Genetic aspects of unconventional virus infections: the basis of the virino hypothesis. In: Bock G, Marsh J, eds. Novel infectious agents of the central nervous system. Ciba Foundation Symposium. London: Wiley-Interscience, 1988;135:63–83.

6. Prusiner SB. Molecular biology of prion diseases. Science 1991;252:1515–22.

7. Prusiner SB. Genetic and infectious prion's diseases. Arch Neurol 1993; 50:1129–53.

8. Collinge J, Palmer MS, Dryden AJ. Genetic predisposition to iatrogenic Creutzfeldt-Jakob disease. Lancet 1991;337:1441–2.

9. Palmer MS, Dryden AJ, Hughes JT, Collinge J. Homozygous prion protein genotype predisposes to sporadic Creutzfeldt-Jakob disease. Nature 1991; 352:340–2.

10. Goldfarb LG, Petersen RB, Tabaton M, et al. Fatal familial insomnia and familial Creutzfeldt-Jakob disease: disease phenotype determined by a DNA polymorphism. Science 1992;258:806–8.

11. Goldmann W, Hunter N, Foster JD, Salbaum JM, Beyreuther K, Hope J. Two alleles of a neural protein gene linked to scrapie in sheep. Proc Natl Acad Sci USA 1990;87:2476–80.

12. Goldmann W, Hunter N, Benson G, Foster JD, Hope J. Different scrapie-associated fibril proteins (PrP) are encoded by lines of sheep selected for different alleles of the Sip gene. J Gen Virol 1991;72:2411–7.

13. Laplanche JL, Chatelain J, Westaway D, et al. PrP polymorphisms associated with natural scrapie discovered by denaturing gradient gel electrophoresis. Genomics 1993;15:30–7.

14. Goldmann W, Hunter N, Smith G, Foster J, Hope J. PrP genotype and agent effects in scrapie: change in allelic interaction with different isolates of agent in sheep, a natural host of scrapie. J Gen Virol 1994;75:989–95.

15. Maciulis A, Hunter N, Wang S; Goldmann W, Hope J, Foote W. Polymorphisms of a scrapie-associated fibril protein (PrP) gene and their association with susceptibility to experimentally induced scrapie in Cheviot sheep in the United States. Am J Vet Res 1992;53:1957–60.

16. Hunter N, Goldmann W, Benson G, Foster JD, Hope J. Swaledale sheep affected by natural scrapie differ significantly in PrP genotype frequencies from healthy sheep and those selected for reduced incidence of scrapie. J Gen Virol 1993;74:1025–31.

17. Westaway D, Zuliani V, Cooper CM, et al. Homozygosity for prion protein alleles encoding glutamine-171 renders sheep susceptible to natural scrapie. Genes Dev 1994;8:959–69.

18. Van Keulen LJM, Schreuder BEC, Meloen RH, et al. Immunohistochemical detection and localization of prion protein in brain tissue of sheep with natural scrapie. Vet Pathol 1995;32:299–308.

19. Schreuder BEC, De Jong MCM, Pekelder JJ, Vellema P, Bröker AJM, Betcke H. Prevalence and incidence of scrapie in the Netherlands: a questionnaire survey. Vet Rec 1993;133:211–4.

20. Sambrook J, Fritsch EF, Maniatis T. Molecular cloning: a laboratory manual. 2nd ed. Cold Spring Harbor, NY: Cold Spring Harbor Laboratory, 1989.

21. Belt PBGM, Muileman IH, Schreuder BEC, Bos- de Ruijter J, Gielkens ALJ, Smits MA. Identification of five allelic variants of the sheep PrP gene and their association with natural scrapie. J Gen Virol 1995;76:509–17.

22. Myers RM, Maniatis T, Lerman LS. Detection and localization of single bases changes by denaturing gradient gel electrophoresis. In: Wu R, ed. Methods Enzymol 1987;155:501–27.

23. Sanger F, Niklen S, Coulson AR. DNA sequencing with chain-terminating inhibitors. Proc Natl Acad Sci USA 1977;74:5463-7.
24. Hsiao K, Scott M, Foster D, Groth D, Dearmond S, Prusiner S. Spontaneous neurodegeneration in transgenic mice with mutant prion protein of Gerstmann-Sträussler syndrome. Science 1990;250:1587-90.

22

The Formation of Scrapie-Associated Prion Protein In Vitro

Byron Caughey, Suzette Priola, David Kocisko,
Gregory Raymond, Jon Come, Santosh Nandan,
Bruce Chesebro, and Peter Lansbury

The formation of the abnormal protease-resistant, amyloidogenic isoform of PrP (PrP-res) appears to play a central role in the pathogenesis of scrapie and other transmissible spongiform encephalopathies (TSEs). Thus, it is important to understand the process by which the normal, protease-sensitive PrP (PrP-sen) is converted to the protease-resistant state. To define the cellular and molecular details of this process and how it might be inhibited, we have performed in vitro studies using both scrapie-infected tissue culture cells and cell-free reactions. This chapter summarizes the recent results from these studies.

PrP-res Biosynthesis in Scrapie-Infected Neuroblastoma Cells

The development of chronically scrapie-infected mouse neuroblastoma (sc⁺-MNB) cells (1) provided an experimental system in which PrP-res could be metabolically labeled and conveniently manipulated in intact cells. Studies with these cultures have identified dramatic differences in the metabolism of PrP-res and PrP-sen (reviewed in 2, 3). In these cells a minor proportion of the PrP synthesized slowly becomes protease resistant (4, 5). Unlike normal PrP, PrP-res is resistant to phospholipase and protease treatments of intact scrapie-infected cells of their membranes (6–8) and shows no sign of turnover within the cells (4, 5). The formation of PrP-res occurs after its apparently normal (phospholipase- and protease-sensitive) PrP precursor reaches the cell surface (5). Soon after its formation, PrP-res in MNB cells is exposed to lysosomal or endosomal proteases and truncated at the N-terminus (9, 10). Thus, the conversion of PrP to the protease-resistant state occurs at the plasma membrane or along an endocytic pathway to the lysosomes (5, 9, 11). Ultimately, PrP-res appears to accumulate in lysosomes in these cells (5, 9, 12). This differs from the situation in vivo where much of the PrP-res accumulates extracellularly.

Chemical Inhibition of PrP-res Accumulation and Scrapie Agent Replication

Inhibitors of PrP-res accumulation have been identified using sc$^+$-MNB as a screening system. Congo red, a dye that has long been used as a diagnostic stain for amyloids (13) including those composed of PrP-res (14), was found to potently inhibit PrP-res accumulation (15) and scrapie agent replication (16). Congo red is a sulfonated molecule and we wondered if the sulfated glycans with prophylactic antiscrapie activity in mice and/or hamsters (17–22) might have anti–PrP-res activity similar to that of Congo red. Pentosan polysulfate, iota-carrageenan and dextran sulfate were as effective as Congo red with nearly complete inhibition of PrP-res accumulation observed at ~10 ng/mL (23). Other sulfated glycans with no antiscrapie effect in animals such as heparin and chondroitin sulfate were orders of magnitude less effective as inhibitors of PrP-res accumulation. Thus, there appeared to be a correlation between the therapeutic efficacies of these sulfated glycans and their potencies as inhibitors of PrP-res accumulation.

A mechanism of action for these compounds is suggested by their similarity to sulfated glycosaminoglycans (GAGs). Amyloid plaques, including those composed of PrP-res (24–26), contain highly sulfated glycosaminoglycans (summarized in 27). This observation led to proposals that endogenous sulfated proteoglycans are involved in the polymerization of proteins into amyloid filaments (24–26, 28). We have hypothesized that these inhibitors bind to PrP-res or PrP-sen and competitively inhibit an interaction of PrP with a specific cellular sulfated GAG that is essential for PrP-res formation or stabilization (23, 29). This model is supported by recent experiments indicating that PrP-sen can bind to both immobilized heparin and Congo red, and that either interaction can be blocked with free Congo red or sulfated glycan inhibitors of PrP-res accumulation (30). Thus, these inhibitors may bind PrP, but they lack features required to facilitate PrP-res accumulation within cells by themselves. The fact that PrP-res accumulation is blocked by these *exogenous* GAGs or GAG analogues is consistent with the idea that *endogenous* GAGs are important in amyloid formation.

The chemoprophylactic effects of sulfated glycans, Congo red, and other polyanions against scrapie have already been demonstrated (17–22, 31, 32). Since all naturally derived amyloids contain GAGs, it is tempting to speculate that the interaction of GAGs with a variety of amyloidogenic proteins might also be blocked by sulfated glycans, Congo red, or related compounds. Treatment with such inhibitors of GAG-(pre)amyloid interactions in vivo might reduce accumulation of amyloids associated with other diseases such as Alzheimer's disease or systemic amyloidoses (15, 29, 33, 34).

Interference with PrP-res Accumulation by Heterologous PrP-sen Molecules

As noted above, sc⁺-MNB cells express MoPrP-sen, accumulate MoPrP-res, and replicate mouse scrapie infectivity (1, 4, 5). These cells provide a controlled system in which to study the molecular interactions that are important in the formation of PrP-res. We have expressed recombinant PrP molecules in mouse scrapie-infected murine neuroblastoma cells and assayed the effect these heterologous PrP genes have on the formation and accumulation of PrP-res. Expression of the recombinant PrP molecules was assayed by using monoclonal antibody 3F4, which recognizes only PrP derived from the recombinant PrP genes and not PrP derived from the endogenous mouse PrP expressed in sc⁺-MNB cells. Expression of heterologous PrP molecules, which differ from the endogenous PrP by as little as one amino acid, profoundly interfered with the overall accumulation of PrP-res. This interference was mapped to residue 111 (35). Despite the ability of these recombinant PrP molecules to interfere with the conversion of the endogenous MoPrP-sen to MoPrP-res, some of the recombinants were themselves converted into PrP-res (35, and unpublished data). Recombinant PrP molecules between HaPrP-sen, which does not convert to PrP-res in sc⁺-MNB cells, and MoPrP-sen, which does convert to PrP-res in these cells, showed that homology in the region of MoPrP-sen from amino acid residues 108–189 was required for conversion of the recombinant PrP-sen to PrP-res (Fig. 22.1). These data strongly suggested that precise interactions between homologous PrP molecules were important in PrP-res accumulation, but did not provide definitive proof that in the absence of cellular factors direct PrP-PrP interactions were sufficient for the conversion of PrP-sen into PrP-res.

Hamster PrP-sen Dimers with Properties of Both PrP-sen and PrP-res

Several hypotheses predict that dimeric or other oligomeric forms of either PrP-sen or PrP-res may act as intermediates in the conversion process (36–39). We have recently identified a 60-kd PrP derived from hamster Prp expressed in murine neuroblastoma cells (40). Peptide mapping studies provided evidence that the 60-kd PrP was composed solely of PrP, and based on its molecular mass appeared to be a PrP dimer. It was not dissociated under several harsh denaturing conditions, which indicated that it

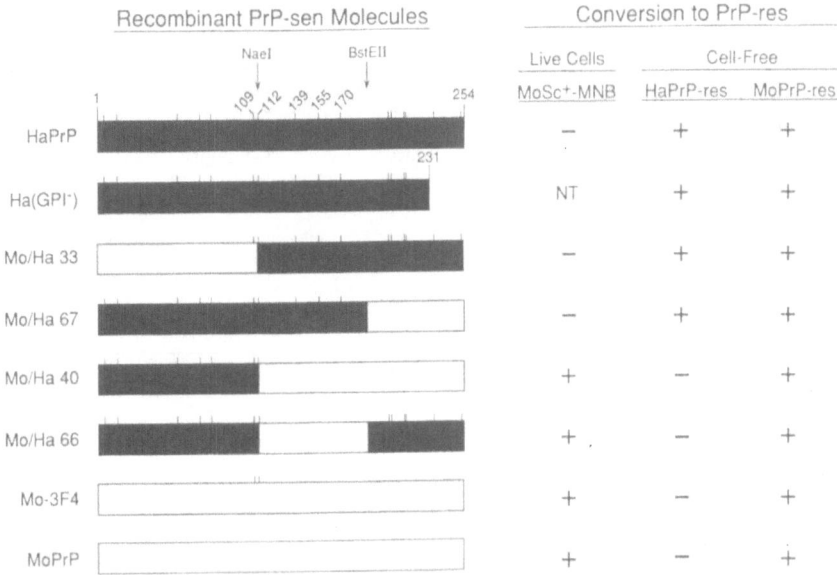

FIGURE 22.1. Conversion of recombinant PrP-sen molecules to PrP-res. Shaded areas indicate regions that were derived from hamster PrP DNA, and open areas were derived from mouse PrP DNA. Tick marks indicate where amino acid residues differ between mouse and hamster PrP and the ticks over the bars designate the presence of a hamster residue in that position. The five numbered residues were derived from hamster PrP in all constructs that coverted to resistant forms with hamster PrP-res (HaPrP-res) in the cell-free system, but were derived from mouse PrP in all constructs that converted to resistant forms in live mouse scrapie-infected cells (MoSc⁺-MNB). A plus signifies that the construct converted to a resistant form after being expressed in live scrapie-infected cells (MoSc⁺-MNB) or in the cell-free system after incubation with the indicated PrP-res. Notice that all constructs that converted with HaPrP-res had hamster residues (methionines) at residues 109 and 112 [defining the 3F4 epitope(43)] but these residues were not sufficient for conversion with HaPrP-res in the cell-free system as indicated by the lack of conversion of Mo-3F4. The wild-type mouse PrP gene (MoPrP) does not require the methionines at residues 109 and 112 to convert to PrP-res in either live cells (35) or in the cell-free system.

310   B. Caughey et al.

TABLE 22.1. Comparison of the hamster PrP dimer to PrP-sen and PrP-res.

	PrP-sen	HaPrP Dimer	PrP-res
Expressed in MNB[a] cells	+	+	−
Expressed in sc+-MNB cells	+	+	+
Proteinase K resistant	−	−	+
Half-life in tissue culture[b]	3–6 hours	3–6 hours	>24 hours
Aggregation	−	+	+
PIPLC sensitive (live cells)[c]	+	−	−
Protease resistant (live cells)[d]	−	+	+
Expressed on cell surface	+	−	−
Converted to PrP-res[e]	+	+	NA[f]
Molecular mass (kd)	25, 30, 35–41	60	19, 23–28

[a] Uninfected mouse neuroblastoma cells.
[b] Half-life of PrP molecule in sc+-MNB or MNB cells (5, 40).
[c] MNB or sc+-MNB cells monolayers were treated with PIPLC and both cell media and cell lysates were tested for PrP (5, 40).
[d] Same as footnote b except that cells were treated with proteinase K.
[e] Conversion into PrP-res was assayed using the cell-free conversion assay (41).
[f] Not applicable.

was covalently linked. Interestingly, the 60-kd PrP has characteristics of both PrP-sen and PrP-res (Table 22.1). For example, while the 60-kd PrP was similar to PrP-res in that it formed large aggregates, it resembled PrP-sen in that it was sensitive to proteinase K, had a short metabolic half-life in cell culture, and was converted to PrP-res in a cell-free conversion assay (41). In fact, the hamster PrP dimer has most of the characteristics of the PrP-res that accumulates in sc+-MNB cells (Table 22.1). However, formation and aggregation of the 60-kd hamster PrP occurs in uninfected mouse neuroblastoma cells, which suggests that hamster PrP has a predisposition to aggregate even in the absence of scrapie infectivity. Similar 60-kd PrP bands have been identified in scrapie-infected hamster brain but not in uninfected brain (40). Thus, the 60-kd hamster PrP fulfills many of the characteristics that might be expected of a dimeric intermediate in PrP-res formation: (1) it forms large aggregates, (2) it can be converted into PrP-res, and (3) it is present in scrapie-infected hamster brains.

Cell-Free Formation of PrP-res

One of the greatest difficulties in studying the PrP-sen–to–PrP-res conversion has been the fact that the simplest system able to perform the reaction was an intact scrapie-infected cell. To simplify the analysis of the conversion mechanism, we sought to establish a defined cell-free system that reconstitutes the PrP-sen–to–PrP-res conversion. These studies indicated that the incubation of purified PrP-sen with PrP-res under certain conditions led to the conversion of PrP-sen to a protease-resistant form resembling PrP-res from scrapie-infected sources (41).

The major components of the reaction were PrP-res purified from scrapie infected hamster brain and recombinant hamster ^{35}S-PrP-sen (^{35}S-HaPrP-sen) expressed in uninfected mouse neuroblastoma cells and immunoprecipitated. Typically, the PrP-res was pretreated with 3 M guanidine-HCl (Gdn HCl) overnight at 37°C and then mixed with labeled PrP-sen, which was also pretreated with 3 to 7.5 M Gdn HCl. The mixture was then diluted to 0.75 M Gdn HCl, incubated at 37°C for 2 days, and then treated with PK to digest any remaining ^{35}S-PrP-sen. When products of this digestion were analyzed using sodium dodecyl sulfate–polyacrylamide gel electrophoresis (SDS-PAGE) and fluorography, radiolabeled PK-resistant bands were observed that were the expected 6 to 7 kd lower in molecular weight than the ^{35}S-PrP-sen precursor. The conversion of ^{35}S-PrP-sen to these PK-resistant forms required the presence of the preexisting PrP-res and did not occur when the PrP-res was first fully denatured with high concentrations of Gdn HCl. The conversion reaction was specific for PrP-sen and PrP-res as indicated by observations that another amyloid aggregate, Alzheimer's beta-amyloid, could not be substituted for PrP-res and other ^{35}S-labeled could not be substituted for ^{35}S-PrP-sen (41).

This study has provided direct evidence that PrP-res is derived from specific PrP-res–PrP-sen interactions. The results are consistent with the protein-only hypothesis for scrapie agent replication; however, we have not yet been able to test whether new scrapie infectivity is generated with the cell-free conversion of PrP-sen to protease-resistant PrP, nor have we ruled out a role of non-PrP cofactors that might have copurified with PrP-res. This experimental system provides us with a unique opportunity to study the molecular mechanism of PrP-res formation and TSE agent replication under conditions that are much more defined than has been possible previously. For instance, we have used the cell-free conversion system to study the molecular basis of the TSE species barrier. Striking species specificity in the cell-free conversion system was observed that further indicated that this experimental system is a valuable and appropriate model of basic TSE biochemistry and biology.

Observation of In Vitro Species Barrier Effects

Mouse- or hamster-derived PrP-res was used in combination with mouse, hamster, and chimeric PrP-sen constructs to assess whether the TSE species barrier might be the result of incompatibilities in direct PrP-sen–PrP-res interactions (42). The PrP-sen constructs were expressed, metabolically labeled, and specifically immunoprecipitated from uninfected mouse neuroblastoma cells. The ^{35}S-PrP-sen and PrP-res preparations were combined, incubated, treated with PK, and analyzed for PK-resistant ^{35}S-PrP. When

^{35}S-labeled mouse PrP-sen (^{35}S-MoPrP-sen) was reacted with MoPrP-res, conversion of ^{35}S-MoPrP-sen to PK-resistant forms was observed. This conversion was dependent on the presence fo MoPrP-res and did not occur with HaPrP-res (Fig. 22.2). The potential influence of PrP-sen glycosylation in the species specificity of the conversion reaction was investigated using unglycosylated MoPrP-sen (labeled in the presence of tunicamycin) in reactions with MoPrP-res and HaPrP-res. Striking differences in the extent of conversion were observed, indicating the species specificity of the reactions was not dependent on N-linked glycosylation of the PrP-sen precursor.

In contrast to the lack of conversion of ^{35}S-MoPrP-sen by HaPrP-res, MoPrP-res was able to convert HaPrP-sen [with and without the glycophosphatidylinositol (GPI) anchor] to PK-resistant forms, but these were lower in molecular weight than those generated by incubation with HaPrP-res (Fig. 22.2) (42). The conversion of MoPrP-sen with MoPrP-res

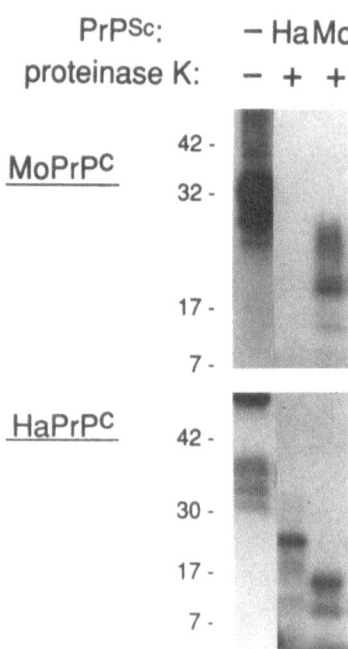

FIGURE 22.2. Species specificity of the cell-free conversion of PrP-sen to PrP-res. ^{35}S-labeled PrP-sen derived from mouse PrP (MoPrP^c) or hamster PrP (HaPrP^c) was mixed and incubated for 2 days with PrP-res (PrP^{sc}) derived from either scrapie-infected hamster brain (Ha) or scrapie-infected mouse brain (Mo) as described previously (42). Samples were then treated with proteinase K (+) followed by SDS-PAGE gel analysis and fluorography. Lane 1 shows the ^{35}S-labeled PrP-sen precursors in the absence of PrP-res (-) and without proteinase K treatment (-). Note the HaPrP-sen 60-kd dimer band in lane 1 of the lower panel. Molecular mass markers in kilodaltons are shown on the left.

was optimal at a somewhat different Gdn HCl concentration than the HaPrP-sen–HaPrP-res conversions. These results indicated that there was species specificity in the cell-free conversion reaction. Interestingly, the inability of HaPrP-res to convert MoPrP-sen to PK-resistant forms correlated with the inability of hamster scrapie strain 263K to infect mice. The contrasting ability of MoPrP-res to convert HaPrP-sen to PK-resistant forms correlated with the fact that the Chandler strain of mouse scrapie can be transmitted to hamsters, albeit with extended incubation periods. Thus, the species specificity of the cell-free conversion correlated with the corresponding scrapie species barriers observed in vivo. We conclude that there is species specificity in the cell-free interactions that lead to the conversion of PrP-sen to protease-resistant forms. This specificity could account for the barriers to interspecies transmission of scrapie and other TSEs in vivo.

To study which regions of the PrP-sen primary sequence influence the species specificity of the conversion to PK-resistant forms, we tested various mouse/hamster ^{35}S-PrP-sen chimeras in the conversion reaction with either MoPrP-res or HaPrP-res (Fig. 22.1) (42). All the constructs yielded PK-resistant products after incubation with MoPrP-res, with some variation in the size of the major bands produced. After incubation with HaPrP-res, only constructs with hamster residues at positions 109, 112, 139, 155, and 170 yielded obvious PK-resistant products up to 32 kd with a banding pattern typical of that seen when PrP-res is treated with PK. Thus, these residues are important in determining the species specificity of the reactions of PrP-sen constructs with HaPrP-res. This same region was shown to be ciritcal for influencing conversion of recombinant PrP-sen to PrP-res in live mouse sc$^+$-MNB cells (Fig. 22.1), but *mouse* PrP residues were required at positions 139, 155, and 170 for conversion in the presence of mouse PrP-res and mouse scrapie infectivity. Interestingly, these same residues were not required for conversion to PrP-res in the cell-free system in the presence of mouse PrP-res (Fig. 22.1). The reason for these differences is not known at present.

Since the cell-free conversion reaction has species specificity, it is possible to use it as an in vitro indicator of the likelihood that various TSE agent strains will be pathogenic in different species. Of particular concern is the question of whether various strains of TSE in animals can be transmitted to humans. The recent bovine spongiform encephalopathy (BSE) epidemic in Great Britain has raised fears that this apparently new strain of TSE might be infectious for humans. Accordingly, we are in the process of testing whether PrP-res derived from the brains of BSE-infected cattle can convert human PrP-sen to PrP-res.

References

1. Race RE, Fadness LH, Chesebro B. Characterization of scrapie infection in mouse neuroblastoma cells. J Gen Virol 1987;68:1391–9.

2. Caughey B. Cellular metabolism of mormal and scrapie-associated forms of PrP. Semin Virol 1991;2:189–96.

3. Caughey B. In vitro expression and biosynthesis of prion protein. Curr Top Microbiol Immunol 1991;172:931–107.

4. Borchelt DR, Scott M, Taraboulos A, Stahl N, Prusiner SB. Scrapie and cellular prion porteins differ in the kinetics of synthesis and topology in cultured cells. J Cell Biol 1990;110:743–52.

5. Caughey B, Raymond GJ. The scrapie-associated form of PrP is made from a cell surface precursor that is both protease- and phospholipase-sensitive. J Biol Chem 1991;266:18217–23.

6. Caughey B, Neary K, Buller R, Ernst D, Perry L, Chesebro B. Normal and scrapie-associated forms of prion protein differ in their sensitivities to phospholipase and proteases in intact neuroblastoma cells. J Virol 1990;64:1093–101.

7. Stahl N, Borchelt DR, Prusiner SB. Differential release of cellular and scrapie prion proteins from cellular membranes by phosphatidylinositol-specific phospholipase C. Biochemistry 1990;29:5405–12.

8. Safar J, Ceroni M, Gajdusek DC, Gibbs CJ Jr. Differences in the membrane interaction of scrapie amyloid precursor proteins in normal and scrapie- or Creutzfeldt-Jakob disease-infected brains. J Infect Dis 1991;163:488–94.

9. Caughey B, Raymond GJ, Ernst D, Race RE. N-terminal truncation of the scrapie-associated form of PrP by lysosomal protease(s). Implications regarding the site of conversion of PrP to the protease-resistant state. J Virol 1991;65:6597–603.

10. Taraboulos A, Raeber AJ, Borchelt DR, Serban D, Prusiner SB. Synthesis and trafficking of prion proteins in cultured cells. Mol Biol Cell 1992;3:851–63.

11. Borchelt DR, Taraboulos A, Prusiner SB. Evidence for synthesis of scrapie prion protein in the endocytic pathway. J Biol Chem 1992;267:16188–99.

12. McKinley MP, Taraboulos A, Kenaga L, et al. Ultrastructural localization of scrapie prion proteins in cytoplasmic vesicles of infected cultured cells. Lab Invest 1991;65:622–30.

13. Glenner GG. Amyloid deposits and amyloidosis. The Beta-fibrillosa (second of two prats). N Engl J Med 1980;302:1333–43.

14. Prusiner SB, McKinley MP, Bowman KA, et al. Scrapie prions aggregate to form amyloid-like birefringent rods. Cell 1983;35:349–58.

15. Caughey B, Race RE. Potent inhibition of scrapie-associated PrP accumulation by Congo red. J Neurochem 1992;59:768–71.

16. Caughey B, Ernst D, Race RE. Congo red inhibition of scrapie agent replication. J Virol 1993;67:6270–2.

17. Ehlers B, Rudolf R, Diringer H. The reticuloendothelial system in scrapie pathogenesis. J Gen Virol 1984;65:423–8.

18. Ehlers B, Diringer H. Dextran sulphate 500 delays and prevents mouse scrapie by impairment of agent replication in spleen. J Gen Virol 1984;65:1325–30.

19. Farquhar CF, Dickinson AG. Prolongation of scrapie incubation period by an injection of dextran sulphate 500 within the month before or after infection. J Gen Virol 1986;67:463–73.

20. Kimberlin RH, Walker CA. Suppression of scrapie infection in mice by heteropolyanion 23, dextran sulfate, and some other polyanions. Antimicrob Agents Chemother 1986;30:409–13.

21. Diringer H, Ehlers B. Chemoprophylaxis of scrapie in mice. J Gen Virol 1991;72:457–60.
22. Ladogana A, Casaccia P, Ingrosso L, et al. Suplphate polyanions prolong the incubation period of scrapie-infected hamsters. J Gen Virol 1992;73:661–5.
23. Caughey B, Raymond GJ. Sulfated polyanion inhibition of scrapie-associated PrP accumulation in cultured cells. J Virol 1993;67:643–50.
24. Snow AD, Kisilevsky R, Willmer J, Prusiner SB, DeArmond SJ. Sulfated glycosaminoglycans in amyloid plaques of prion diseases. Acta Neuropathol 1989;77:337–42.
25. Snow AD, Wight TN, Nochlin D, et al. Immunolocalization of heparan sulfate proteoglycans to the prion protein amyloid plaques of Gerstmann-Straussler syndrome, Creutzfeldt-Jakob disease and scrapie. Lab Invest 1990;63: 601–11.
26. Guiroy DC, Yanagihara R, Gajdusek DC. Localization of amyloidogenic proteins and sulfated glycosaminoglycans in nontransmissible and transmissible cerebral amyloidoses. Acta Neuropathol 1991;82:87–92.
27. Narindrasorasak S, Lowery D, Gonzalez-DeWhitt P, Poorman RA, Greengerg B, Kisilevsky R. High affinity interactions between the Alzheimer's beta-amyloid precursor proteins and the basement membrane form of heparan sulfate proteoglycan. J Biol Chem 1991;266:12878–83.
28. Guiroy DC, Gajdusek DC. Fibril-derived amyloid enhancing factor as nucleating agents in Alzheimer's disease and transmissible virus dementia. Disc Neurosci 1989;5:69–73.
29. Caughey B. Scrapie associated PrP accumulation and its prevention. Insights from cell culture. Br Med Bull 1993;49:860–72.
30. Caughey B, Brown K, Raymond GJ, Katzenstien GE, Thresher W. Binding of the protease-sensitive form of PrP (prion protein) to sulfated glycosaminoglycan and Congo red. J Virol 1994;68:2135–41.
31. Kimberlin RH, Walker CA. The antiviral compound HPA-23 can prevent scrapie when administered at the time of infection. Arch Virol 1983;78:9–18.
32. Ingrosso L, Ladogana A, Pocchiari M. Congo red prolongs the incubation period in scrapie-infected hamsters. J Virol 1995;69:506–8.
33. Snow AD, Sekiguchi R, Nochlin D, et al. An important role of heparan sulfate proteoglycan (perlecan) in a model system for the depostion and persistence of fibrillar A-beta-amyloid in rat brain. Neuron 1994;12:219–34.
34. Kisilevsky R, Lemieux LJ, Fraser PE, Kong X, Hultin PG, Szarek WA. Arresting amyloidosis in vivo using small-molecule anionic sulphonates or sulphates. implications for Alzheimer's disease. Nature Med 1995;1:143–8.
35. Priola SA, Caughey B, Race RE, Chesebro B. Heterologous PrP molecules interfere with accumulation of protease-resistant PrP in scrapie-infected murine neuroblastoma cells. J Virol 1994;68:4873–8.
36. Dickinson AG, Outram GW. The scrapie replication-site hypothesis and its implications for pathogenesis. In: Prusiner SB, Hadlow WJ, eds. Slow transmissible diseases of the nervous system. New York: Academic Press, 1979:13–31.
37. Hope J, Morton LJD, Farquhar CF, Multhaup G, Beyreuther K, Kimberlin RH. The major polypeptide of scrapie-associated fibrils (SAF) has the same size, charge distribution and N-terminal protein sequence as predicted for the normal brain protein (PrP). EMBO J 1986;5:2591–7.

38. Jarrett JT, Lansbury PT Jr. Seeding "one-dimensional crystallization" of amyloid. A pathogenic mechanism in Alzheimer's disease and scrapie? Cell 1993;73:1055–8.
39. Bolton DC, Bendheim PE. A modified host protein model of scrapie. In Bock G, Marsh J, eds. Novel infectious agents and the central nervous system. Chichester, UK: John Wiley & Sons, 1988:164–81.
40. Priola SA, Caughey B, Wehrly K, Chesebro B. A 60-kDa pion potein (PrP) with properties of both the normal and scrapie-associated forms of PrP. J Biol Chem 1995;270:3299–305.
41. Kocisko DA, Come JH, Priola SA, et al. Cell-free formation of protease-resistant prion protein. Nature 1994;370:471–4.
42. Kocisko DA, Priola SA, Raymond GJ, Chesebro B, Lansbury PT Jr, Caughey B. Species specificity in the cell-free conversion of prion protein to protease-resistant forms. A model for the scrapie species barrier. Proc Natl Acad Sci USA 1995;92:3923–7.
43. Bolton DC, Seligman SJ, Bablanian G, et al. Molecular location of a species specific epitope on the hamster scrapie agent protein. J Virol 1991;65:3667–75.

23

Proteinase K–Resistant Prion Protein Detection in Animal Tissues and In Vitro

RICHARD E. RACE AND DARWIN ERNST

Transmissible spongiform encephalopathies (TSEs) of man and animals are caused by unique transmissible agents that are unusually resistant to chemical inactivation and treatments that destroy or modify nucleic acids (1–4). In preparations enriched for infectivity, a predominant proteinase K–resistant form (PrP-res) of an endogenous protein (PrP-sen) has been identified (5–7). A great deal of controversy exists about the relationship between PrP-res and the infectious scrapie agent. In fact, some investigators believe that PrP-res and the infectious scrapie agent are the same (5, 8). Regardless of the true relationship, virtually all workers in the field recognize that a close association exists between the detection of PrP-res and the presence of spongiform encephalopathy. This association has led in recent years to new diagnostic tests for scrapie based on PrP-res detection by either immunoblotting (9, 10) or immunohistochemistry (11, 12). Both techniques offer distinct advantages over diagnosis based on microscopic evaluation of brain. We summarize here several of our studies that utilized PrP-res detection as a diagnostic test for scrapie. In addition, we report initial results of a study in which PrP-res was sought in brain of cattle presented with clinical symptoms suggestive of bovine spongiform encephalopathy (BSE).

Results

Comparison Between PrP-res Detection and Infectivity in Mouse Spleen and Brain

Scrapie infectivity in mouse spleen is known to increase rapidly following experimental inoculation of mouse-adapted scrapie agent. Several months after the initial replication of the scrapie agent in spleen, the infectivity titer in brain increases, finally reaching a titer approximately 100-fold higher than that in spleen (13). The increase in scrapie infectivity titer in spleen is not accompanied by any microscopic evidence of disease while the increase in brain is associated with pathologic changes. Therefore, it has been unclear whether PrP-res accumulates as a result of tissue damage or rather as an independent phenomenon. We wondered further if the accumulation of PrP-res in mouse spleen and brain would directly parallel the increase of scrapie infectivity in these tissues or if instead a divergence of the two parameters might be realized. If PrP-res and infectivity do not increase in parallel it would be unlikely for PrP-res and the infectious scrapie agent to be a single entity.

To make these determinations, we analyzed the spleens and brains from groups of mice killed at various times after inoculation with Chandler strain scrapie agent for the presence of PrP-res and infectivity. De novo synthesis of PrP-res was shown in mouse spleen by 2 weeks postinoculation. By 3 weeks postinoculation, the amount of PrP-res was approximately 65-fold higher than at 1 week after inoculation but then increased only an additional 15-fold during the next 17 weeks at which time the animals were beginning to die of scrapie. Infectivity in the spleen reached a maximum plateau level at approximately 3 weeks postinoculation (14). Therefore, the early substantial increase in PrP-res accumulation in spleen was accompanied by a corresponding increase in infectivity. However, after 3 weeks the increase in PrP-res did not correlate well with changes in infectivity.

In mouse brain, the scrapie agent titer reaches a level 100- to 1,000-fold higher than it does in mouse spleen (13). This level however, is not attained until late in the course of disease. We did not detect PrP-res in mouse brain at 3 weeks postinoculation. However, at 8 weeks postinoculation, PrP-res was easily detected by immunoblotting. Between 8 and 20 weeks postinoculation, the amount of PrP-res increased approximately 200-fold. Infectivity was low at 3 weeks after infection but increased 10,000-fold between 8 and 20 weeks after inoculation (14). Thus, in mouse brain between 8 and 20 weeks postinoculation there was approximately a 50-fold difference between the increase in PrP-res and the increase in infectivity. And, although the increases in PrP-res and infectivity seemed to occur in parallel, the relationship did not appear to be quantitative. Some of the difference could be the result of the innate margins of error associated with the mouse bioassay (5- to 10-fold) and immunoblot (2- to 4-fold). Therefore, the combination of potential error of the infectivity and PrP-res analyses could possibly account for the 15-fold increase in PrP-res between 3 and 20 weeks postinoculation in spleen, which was not accompanied by a corresponding increase in infectivity but does not fully account for the 50-fold difference between PrP-res and infectivity increases in brain between 8 and 20 weeks postinoculation. Therefore, we cannot conclude that PrP-res and the infectious scrapie agent are the same. It does appear that PrP-res accumulation is not dependent on tissue damage because substantial amounts of PrP-res were detected in mouse spleen, a tissue in which no scrapie-associated lesions have ever been described.

PrP-res Detection in Sheep Tissues

Before performing detailed analyses of sheep spleen and lymph nodes, we sought to verify that PrP-res could be detected by immunoblotting in scrapie-affected sheep brain. Initially, brain tissue was obtained from clinically positive sheep or sheep with no known exposure to scrapie. All the brains were analyzed by immunoblot for the presence of PrP-res. PrP-res was not detected in clinically normal sheep with no known exposure to scrapie. In sheep clinically positive for scrapie, PrP-res was detected in brain. PrP-res results were compared with diagnosis based on microscopic evaluation of brain. Every sheep that was considered positive or even suggestive for scrapie on the basis of microscopic evaluation of brain also had detectable PrP-res in brain. Several sheep that were not diagnosed as scrapie positive on the basis of microscopic evaluation also had PrP-res in their brains, and on this basis, we considered them to be scrapie-positive (9). Bioassay in mice of brain homogenates from several of these animals confirmed that brains that contained detectable PrP-res also had infectious scrapie agent (Race, unpublished). Thus, detection of PrP-res identified more scrapie-positive sheep than did microscopic examination, and provides a diagnostic approach for sheep scrapie that is superior to microscopic evaluation.

Because PrP-res was found in the spleens of scrapie-infected mice we wondered if the same situation might also be found in sheep. Detection of PrP-res in sheep spleen or lymph nodes would provide diagnostic advantages over the currently used method for diagnosis of sheep scrapie based on the microscopic evaluation of brain. This is so because spleen and lymph node are easier to collect than brain. Therefore, we next sought to detect PrP-res in sheep spleen and lymph nodes. Most, but not all, sheep that had PrP-res in brain also had PrP-res in spleen. In addition, all animals with PrP-res in spleen also had PrP-res in either the mesenteric or prescapular lymph nodes or both. In a few sheep, PrP-res was found in one of the lymph nodes but not the other, suggesting either quantitative differences in the amount of PrP-res accumulated or differences in scrapie pathogenesis in individual animals. Immunoblot analysis of sheep spleen or lymph nodes for PrP-res was not as sensitive at identifying scrapie-positive animals as analysis of brain (9). Nevertheless, evaluation of these peripheral tissues was superior to microscopic evaluation of brain in identifying scrapie-positive sheep and, considering the ease of collection, offers a reasonable alternative tissue for evaluation in those situations where brain cannot be obtained. In addition, under special circumstances, lymph nodes could be obtained surgically from living sheep and analyzed for the presence of PrP-res. However, in this regard we do not know how soon after infection PrP-res begins to accumulate in the spleen and lymph nodes and whether these tissues will allow routine antemortem diagnosis of scrapie. A systematic study in which experimentally inoculated sheep are killed at various intervals after inoculation and their tissues analyzed for the presence of PrP-res needs to be done to answer this important question.

In several instances we have analyzed spleen or lymph nodes by bioassay for the presence of scrapie agent. In every analysis, if PrP-res was detected by immunoblot, so was scrapie agent by bioassay even though to do so required crossing a species barrier (sheep to mouse). In a few situations, infectivity was found in the absence of detectable PrP-res, suggesting that immunoblot is less sensitive in detecting scrapie agent positive sheep than is bioassay. However, the number of scrapie-positive animals that would not be identified if only PrP-res and no bioassay were used for diagnosis would be a very small percentage (approximately 6%) of the total number analyzed (data not shown). In contrast, microscopic evaluation failed to identify a significant number of scrapie-positive sheep that were identified by PrP-res analysis (9). PrP-res can also be detected in severely autolyzed tissue that would be completely unsuitable for microscopic evaluation (10). In view of the many advantages that PrP-res analysis offers compared with microscopic evaluation it would seem reasonable to consider PrP-res analysis as a primary diagnostic test for sheep scrapie.

PrP-res Detection In Vitro

Several years ago we reported the establishment and characterization of a mouse neuroblastoma cell line infected with the Chandler scrapie agent (15, 16). We developed the cells because we reasoned that such cells would be free of many of the tissue-derived by-products encountered in animal tissues that have hampered studies of the biochemical characteristics of the scrapie agent. Cell cultures could provide an agent in a more suitable substrate. In the initial cultures that we developed, PrP-res was not detected by immunoblot even though high levels of infectivity were present (16). However, modifications of techniques then available for PrP-res detection eventually resulted in the routine demonstration of PrP-res in the scrapie-infected cultured cells (17–19). We have shown that PrP-res analysis can be useful to monitor the cells for the presence of scrapie agent. To verify that this is so, we obtained 30 clones by limiting cell dilution from a scrapie-positive parent culture. Each of the 30 subcultures was analyzed for the presence of scrapie agent by bioassay and was also analyzed for the presence of PrP-res by immunoblotting. Sixteen of the clones had both PrP-res and infectivity. Six of the cultures had no PrP-res but did have infectivity.

None of the cultures had PrP-res and no infectivity. Therefore, bioassay appears to be a more sensitive way to demonstrate the presence of the scrapie agent than does immunoblot for PrP-res. However, in no instance did PrP-res analysis result in the identification of any "false positives" nor has any uninfected culture ever given PrP-res–like protein bands on immunoblot. So, PrP-res analysis can be used to follow infectivity in cells and provides a powerful tool for studying biochemical interactions between the scrapie agent and its host cell.

PrP-res Detection in Experimentally Inoculated Cattle

PrP-res was detected in the brains of cattle experimentally inoculated with a brain homogenate derived from scrapie-positive sheep. The PrP-res analyses were done as a part of a larger study by Dr. Randall Cutlip et al. PrP-res data are presented in Chapter 7.

Search for PrP-res in Selected High-Risk Cattle

Because we detected PrP-res in the brains of animals affected with scrapie, we have considered the possibility that PrP-res analysis might be the preferred method to search for the presence of bovine spongiform encephalopathy (BSE) in American cattle. This is so because the microscopic pattern associated with the spongiform encephalopathies is variable among species and the lesions observed are often scarce or nonspecific. While BSE is unlikely to exist in the United States, it would nevertheless be reasonable to survey cattle for its presence. If the incidence is low or similar to that of sporadic Creutzfeldt-Jakob disease of humans (one per million) we would have virtually no chance of finding a positive animal as a part of this study. However, if the disease exists at a higher level and has not been recognized, then a positive animal might be found, which would allow control measures to be instituted before a major problem were to develop. In collaboration with Dr. Lyle Miller, Iowa State University, we are analyzing brain from cattle presented with clinical evidence of central nervous system disease of undetermined cause for the presence of PrP-res. To date, 62 brains have been analyzed. None of them were positive for PrP-res. While we recognize that the sample size is extremely small, we are nevertheless encouraged that no positive animals have been identified.

Conclusions

We detected PrP-res in brain, spleen, and lymph node of scrapie-infected sheep and goats, brain and spleen of scrapie-affected mice, and in vitro in scrapie-infected mouse neuroblastoma cells. In no study where we compared PrP-res detection with the presence of the scrapie agent did we find PrP-res by immunoblot and not detect scrapie agent by bioassay. However, the converse was not true. Some tissues that had detectable scrapie agent by bioassay did not have detectable PrP-res. We believe this observation relates to the greater sensitivity of bioassay compared with immunoblot. We estimate that bioassay is 100- to 1,000-fold more sensitive than PrP-res detection by immunoblot in identifying the presence of scrapie agent (data not shown). Nevertheless, we showed that more scrapie-infected sheep were identified by PrP-res analysis than by microscopic evaluation of brain, and that spleen or lymph nodes could be used when PrP-res analysis was used as the diagnostic test (9). PrP-res was also detected in scrapie infected tissue culture cells and correlated with the presence of the scrapie agent. So far, we have not detected PrP-res in brain from United States–derived cattle presenting with BSE-like clinical disease.

Even though we now have better techniques for diagnosis of TSE than before, the major obstacle to both control and understanding of TSE is the lack of a practical antemortem diagnostic assay. We have detected PrP-res in lymph node biopsies from clinically normal but scrapie-infected sheep. However, from a practical point of view, diagnosis based on analysis of lymph nodes derived surgically would have very limited application. Analysis of placenta offers a method for antemortem diagnosis of scrapie in sheep flocks. This is so because approximately one-third of scrapie-positive ewes had detectable PrP-res in placenta (data not shown). Serologic tests are needed but no serum-derived scrapie-specific antigen or molecule has been identified on which such tests might be based.

Surveillance of United States cattle for the presence of BSE presents a dilemma for regulatory agencies. If currently present in the United States, the incidence is likely to be so low as to make detection unlikely. In the absence of any positive BSE diagnoses, the cost of surveillance may be difficult to justify. If, however, surveillance programs are going to be utilized, they should be formulated to provide the best possible chance of detecting BSE-like disease. Choice of diagnostic method becomes a very important consideration. In Great Britain, diagnosis of BSE is based on the observation of typical clinical disease and the subsequent observation of typical microscopic lesions in brain of affected animals. As seen in Great Britain, BSE lesions tend to be present and uniform in nearly all affected individuals. However, in United States cattle inoculated intracerebrally with brain derived from scrapie-infected sheep, lesions were sparse and not diagnostic (20). Therefore, if one wishes to identify primary transmission of sheep scrapie from United States sheep to United States cattle, diagnosis based solely on microscopic examination of brain maybe inappropriate. On the other hand, cattle inoculated with sheep scrapie agent had easily detectable PrP-res in brain (20). It seems to us that any surveillance program for BSE in United States cattle must include analyses for PrP-res.

Summary

The proteinase K–resistant form (PrP-res) of the endogenous prion protein (PrP-sen) was detected in brain and spleen of experimentally inoculated mice; brain, spleen, and lymph node of naturally infected sheep; and in vitro in scrapie-infected mouse neuroblastoma cells. In these situations, detection of PrP-res by immunoblotting was compared to detection of infectious scrapie agent by bioassay in mice. In addition, we showed that detection of PrP-res in sheep brain, spleen, or lymph node identified more scrapie-positive sheep than could be identified using microscopic evaluation of sheep brain. PrP-res was also detected in the brains of cattle experimentally inoculated intracerebrally with a brain homogenate derived from scrapie-positive sheep. Our results suggested that surveillance for bovine spongiform encephalopathy-like disease in cattle should include diagnostic methods based on detection of PrP-res.

References

1. Alper T, Cramp WA, Haig DA, Clarke MC. Does the agent of scrapie replicate without nucleic acid? Nature 1967;214:764–6.
2. Bellinger-Kawahara C, Cleaver JE, Diener TO, Prusiner SB. Purified scrapie prions resist inactivation by UV irradiation. J Virol 1987;61:159–66.
3. Dees C, Wade WF, German TL, Marsh RF. Inactivation of the scrapie agent by ultraviolet irradiation in the presence of chlorpromazine. J Gen Virol 1985;66:845–9.
4. Latarjet R, Muel B, Haig DA, Clarke MC, Alper T. Inactivation of the scrapie agent by near monochromatic ultraviolet light. Nature 1970;227:1341–3.
5. Bolton DC, McKinley MP, Prusiner SB. Identification of a protein that purifies with the scrapie prion. Science 1982;218:1309–11.
6. Diringer H, Gelderblom H, Hilmert H, Ozel M, Edelbluth C, Kimberlin RH. Scrapie infectivity, fibrils and low molecular weight protein. Nature 1983; 306:476–8.
7. Hope J, Morton LJD, Farquhar CF, Multhaup G, Beyreuther K, Kimberlin RH. The major polypeptide of scrapie-associated fibrils (SAF) has the same size, charge distribution and N-terminal protein sequence as predicted for the normal brain protein (PrP). EMBO J 1986;5:2591–7.
8. Prusiner SB. Novel proteinaceous infectious particles cause scrapie. Science 1982;216:136–44.
9. Race R, Ernst D, Jenny A, Taylor W, Sutton D, Caughey B. Diagnostic implications of detection of proteinase K-resistant protein in spleen, lymph nodes, and brain of sheep. Am J Vet Res 1992;53:883–9.
10. Race RE, Ernst D, Sutton D. Severe autolysis does not prevent scrapie diagnosis in sheep. J Vet Diagn Invest 1994;6:486–9.
11. Miller JM, Jenny AL, Taylor WD, Marsh RF, Rubenstein R, Race RE. Immunohistochemical detection of prion protein in sheep with scrapie. J Vet Diagn Invest 1993;5:309–16.

12. Miller JM, Jenny AL, Taylor WD, et al. Detection of prion protein in formalin-fixed brain by hydrated autoclaving immunohistochemistry for the diagnosis of scrapie in sheep. J Vet Diagn Invest 1994;6:366–8.

13. Eklund CM, Kennedy RC, Hadlow WJ. Pathogenesis of scrapie virus infection in the mouse. J Infect Dis 1967;117:15–22.

14. Race RE, Ernst D. Detection of proteinase K-resistant prion protein and infectivity in mouse spleen by 2 weeks after scrapie agent inoculation. J Gen Virol 1992;73:3319–23.

15. Race RE, Fadness LH, Chesebro B. Characterization of scrapie infection in mouse neuroblastoma cells. J Gen Virol 1987;68:1391–9.

16. Race RE, Caughey B, Graham K, Ernst D, Chesebro B. Analyses of frequency of infection, specific infectivity, and prion protein biosynthesis in scrapie-infected neuroblastoma cell clones. J Virol 1988;62:2845–9.

17. Caughey B, Neary K, Buller R, et al. Normal and scrapie-associated forms of prion protein differ in their sensitivities to phospholipase and proteases in intact neuroblastoma cells. J Virol 1990;64:1093–101.

18. Caughey B, Raymond GJ, Ernst D, Race RE. N-terminal truncation of the scrapie-associated form of PrP by lysosomal protease(s): implications regarding the site of conversion of PrP to the protease-resistant state. J Virol 1991; 65:6597–603.

19. Caughey B, Race RE. Potent inhibition of scrapie-associated PrP accumulation by Congo red. J Neurochem 1992;59:768–71.

20. Cutlip RC, Miller JM, Race RE, et al. Intracerebral transmission of scrapie to cattle. J Infect Dis 1994;169:814–20.

24

Elimination of Scrapie-Agent Infectivity in Naturally Derived Biologics

ALESSANDRO DI MARTINO

It is now almost 10 years since bovine spongiform encephalopathy (BSE) was first recognized, and considerable effort has been made to answer questions raised by its unexpected and undesirable appearance (1, 2). Epidemiologic studies have confirmed that BSE on a large-scale is found exclusively in adult cattle of the United Kingdom, while rare cases have been confirmed in indigenous cattle of very few other European nations. The disease has also appeared sporadically in cattle exported from the United Kingdom to other countries in Europe and elsewhere. The probable cause of the outbreak in the United Kingdom has been associated with significant changes in the manufacturing procedures used to prepare the meat and bone meal subsequently fed to cattle (3).

The pathogenicity of the BSE agent for humans remains unknown and is still a matter of intense discussion (4–7). There are some suggestions that the risk of contracting Creutzfeldt-Jakob disease (CJD) from bovine-derived products might well be unimportant. Since humans have been exposed for many decades to ovine scrapie, the possibility that sporadic CJD in the human population might have originated from animals has been studied intensively. To date, all investigations have failed to show any epidemiologic relationship between the incidence of spongiform ence-phalopathies in animals and man (8–13). Moreover, transmission of scrapie-like agents between different species must always overcome the "species barrier" that naturally limits the spread of infection. Finally, precautions taken to contain BSE in the United Kingdom have led to a decline in number of recorded cases per year, and as no transmission between animals will ever be able to sustain the epidemic, the disease should fade out in the future (14, 15). As a consequence, the risk of a new BSE epidemic in other countries is virtually negligible.

However, even the most remote possibility of BSE constituting a risk to humans must not be overlooked, especially since the biology of all scrapie-like agents is poorly understood (16–19). The appearance of BSE coincided with the widening of the spectrum of naturally occurring transmissible spongiform encephalopathies (TSEs) in animals (20, 21). This suggests that the agent may have mutated its host range, and the widespread human use of bovine-derived products might mean increased exposure to infection. Last but not least, the unfortunate transmission of CJD via human-derived products or surgical procedures has shown that these agents are highly pathogenic (22–24). Thus, the uncertainty surrounding BSE represents a major concern for health authorities, regulatory agencies, and manufacturers of biologics and pharmaceuticals, and the issue seems to be discussed recurrently (25–27).

Minimization of the Risk

The problem of reducing contamination in materials extracted from natural sources is not new, since in the past a number of biologics administered to humans have been shown to be contaminated by various pathogens. All guidelines insist that careful selection of source material is the key point in securing product safety, as no risk can exist if raw materials are obtained from uninfected sources. In the case of contamination by BSE or any other scrapie-like agent, for which there is no valid preclinical diagnostic assay to formally prove absence of infectivity, special criteria must be applied to reduce any potential contamination. Since the scrapie-free status is recognized a priori only for those territories where the disease has been eradicated and strict restrictions are in force on animal and animal product importation (28), some national regulatory authorities have decided to keep surveillance programs active and to embark on formal risk assessment studies to exclude any risk of BSE in their own countries (29, 30).

The safety of products can be further assured by evaluating manufacturing procedures and testing their capacity to remove or inactivate pathogens. All infectious pathogens causing TSEs are known to be unlike any other known infectious entity and are extremely resistant to inactivation. Several conditions can affect agent viability: exposure to extreme alkaline pH, organic solvents, detergents, protein-denaturing agents, chaotropic salts, and phenol (31–37). However, only prolonged exposure to concentrated sodium hydroxide or sodium hypochlorite and autoclaving above 130°C can sterilize high-titered preparations, and are recommended for disinfection (38, 39). In addition, the only methodology that guarantees detection of infectivity is in vivo animal bioassay. Validation experiments should be carried out as required by each specific type of contamination and should allow the detection of minimal amounts of infectivity. As such, tests can be performed in laboratory animals using species-adapted strains whenever contamination by any scrapie-like agent is involved. Bioassay in hamster using the 263K strain of scrapie specifically represents a good "model virus" bioassay because it is well described and shows the highest titer and shortest incubation time of any natural or experimentally induced TSE (40).

Validation Experiments: Rationale, Design, and Experimental Procedure

Collaborative studies were promptly undertaken with the aim of developing and/or validating procedures to consistently assure sterility from scrapie-like agents as soon as the early United Kingdom cases were seen to becoming epidemic (41).

The distribution of scrapie infectivity in vivo is most evident in brain and central nervous tissue. It was therefore considered that, if procedures commonly used for extraction of products from brain could be shown to yield sterile material, similar procedures should have been applicable to extraction from other tissue. Gangliosides represented a good reference molecule or "product"; bovine brain tissue is the major source of these glycosphingolipids, which have also been used as therapeutics (42, 43), and large-scale preparations are performed by taking advantage of the resistance of gangliosides to strong alkali (44, 45), and hence, to procedures that would cause a substantial loss of scrapie infectivity.

On the other hand, other products of potential use might have been irreversibly affected by concentrated alkali, and there was a need to evaluate other procedures. The efficacy of a protocol for extraction and purification of phospholipids from brain tissue was therefore evaluated. This particular process was attractive since phospholipids were prepared by exclusive and consistent use of organic solvents throughout the procedure, thus allowing determination of the effect of repeated solvent extractions in scrapie decontamination. Experiments were based on original manufacturing processes deliberately scaled down to a laboratory bench scale while maintaining the conditions of original protocols. They were all made in compliance with the notes for guidance on virus removal and inactivation procedures as well as those for minimizing the risk of transmission of agents causing spongiform encephalopathies via medicinal products (46, 47). Experiments were performed in duplicate assuming unrealistic worst-case scenarios where the source material were pools of 263K-infected hamster brains obtained from terminally ill hamsters. The overall process validation experiments aimed to demonstrate the complete absence of infectivity in the final preparations and at which point in the procedure infectivity was lost. They were performed similarly for both gangliosides and phospholipids; the scaled-down manufacturing process was applied to a batch of infected hamster brains and residual infectivity was determined at several stages of the procedure and in final preparations (autoclaved product). Entire volumes of autoclaved products were administered to assay animals (more than 150 per experiment) to exclude the chance of missing even a single infectious particle, thus minimizing the most relevant limit of quantitative infectivity bioassay (48, 49).

The clearance factor for every separate key step of the ganglioside process could be assessed by applying the scaled-down procedure to a batch of defatted bovine brain powder spiked with a deliberately high amount of infected hamster brain homogenate. This could not be done for phospholipids because spiking a brain homogenate would have significantly altered the conditions with respect to the original protocol by introducing an aqueous component in an organic phase, and also because we did not succeed in spiking high titers of scrapie agent into the organic solvents used in the process conditions.

Infectivity was determined by in vivo end-point titrations in golden Syrian hamsters injected with undiluted and 10-fold serially diluted samples while the autoclaved products were injected into animals, intracerebrally or intraperitoneally, as undiluted, 10-fold, or 20-fold concentrated preparations with respect to the starting volume. Animal groups were identified by code numbers and were examined daily up to a 15-month period for the appearance of clinical signs of scrapie. At the end all surviving animals were sacrificed. Microscopic histopathologic analyses were carried out by an independent investigator who worked "blind" on coded brain slides to confirm the diagnosis in animals with clinically manifest disease and to exclude the possibility of incorrect evaluation in negative animals. Titers were expressed as $\log LD_{50}/mL$ (50).

Ganglioside Sample Preparation

Frozen scrapie-infected hamster brains were ground to a fine powder under liquid nitrogen. The powder was resuspended to obtain a 16.7% (w/v) homogenate, then processed through the entire purification procedure, which included one extraction with a mixture of methanol/methylene chloride/NaOH (key step 1), one treatment with a warm solution of methanol/NaOH (key step 2), one exposure of purified gangliosides to 1 N NaOH for 1 hour at room temperature (key step 3), and finally autoclaving of preparations at 121°C for 30 minutes (key step 4) until achievement of the autoclaved product. A similar experiment was also performed on a batch of healthy hamster brains as a normal control run. Aliquots were taken immediately after key step 1 to key step 3, namely raw material 1, raw material 2, and final product, to evaluate residual infectivity. Spiking experiments were performed by spiking the bovine brain powder with a volume of scrapie-infected brain homogenate that maintained at the same value the infectivity titer per volume unit for every key step. Samples obtained from the various steps of the procedure were assayed by immunoblot to determine where PrPSc, the only marker associated with infectivity, could no longer be detected (51). The detailed description of all experimental procedures is reported elsewhere (52, 53).

Phospholipid Sample Preparation

Frozen scrapie-infected hamster brains were ground to a fine powder under liquid nitrogen. The brain powder was dispersed in acetone/1,1,1-trichloroethane to obtain a 33% (w/v) suspension. The upper water/acetone phase and all the solid insoluble material were discarded while the phospholipid-containing organic phase was precipitated, washed, and resolubilized in 1,1,1-trichloroethane/absolute ethanol. The solution was partitioned with double-distilled water and the aqueous phase was discarded. The phospholipid-containing organic phase was recovered (raw phospholipids), resolubilized in chloroform/methanol, and partitioned again with a sodium chloride solution. The aqueous phase was discarded and purified phospholipids were recovered by precipitation and washing (final product). This was further subjected to steam autoclaving at 121°C for 15 minutes (autoclaved product). Immunoblot assay for determination of PrP^{Sc} were also made on samples recovered during the procedure (54).

Results and Discussion of Ganglioside Experiments

Results are summarized in Tables 24.1 and 24.2. The infectivity titer of the starting homogenate used in the overall process validation experiments was 9.2 log LD$_{50}$/mL (corresponding to 10.0 log LD$_{50}$/g of wet tissue). The infectivity titer of the scrapie-spiked bovine brain powder used in spiking experiments was 7.8 log LD$_{50}$/mL (9.8 log LD$_{50}$/g of wet tissue). These values are in excellent agreement with the average titer of the 263K strain of scrapie (40) and are also much higher than those found for the BSE agent (55, 56).

No residual infectivity was detected in the undiluted sample obtained after the extraction with methylene chloride/methanol/NaOH (key step 1).

TABLE 24.1. Infectivity determined at different steps of the scaled-down ganglioside process.

Sample	Infectivity (log LD$_{50}$/mL)	Infectivity in undiluted inoculum
Brain homogenate (source material)	9.2	Yes
Raw material 1	≤1.4	No
Raw material 2	≤1.0	No
Final product	≤1.0	No
Autoclaved product	0.0[a]	No

[a] Infectivity not detected in entire volumes undiluted, 10-fold, or 20-fold concentrated.
Data from Di Martino et al. (52, 53) with permission.

TABLE 24.2. Inactivation of the 263K strain of scrapie obtained for each key step after spiking of infectivity.

Key step	Treatment	Time (hrs)	Temperature (°C)	Decrease ($\log LD_{50}/mL$)	Infectivity in undiluted inoculum
1	Extraction with $CH_2Cl_2/CH_3OH/NaOH$	3	31–33	≥7.8	No
2	Treatment with $CH_3OH/NaOH$	4	40	≥7.8	No
3	Exposure to 1N NaOH	1	r.t.	6.0	Yes
4	Steam autoclaving	0.5	121	3.8	Yes

Data from Di Martino et al. (52, 53) with permission.

We observed one animal that had received a 10-fold diluted sample with suspected signs of neurologic disease but no histopathologic analysis could be made to confirm the clinical diagnosis. Residual infectivity was too low to be determined by end-point titration, and estimates indicated that any infectivity left was ≤1.4 $\log LD_{50}/mL$, with a minimum clearance factor ≥7.8 $\log LD_{50}/mL$. The same clearance factor was observed in spiking experiments, where no infectivity was present in the infectivity-spiked sample subjected to key step 1. This confirmed that the procedure performed in key step 1 was very effective in inactivating the scrapie agent.

Infectivity was assayed again after key step 2, but none of the animals that were injected with any dilution of the sample showed clinical or histopathologic signs of disease. The value reported in Table 24.1 (≤1.0 $\log LD_{50}/mL$) represents the limit of detectability of the bioassay. This prolonged treatment with warm methanol and NaOH followed by solvent partitioning also reduced infectivity to undetectable levels in the treated scrapie-spiked sample (see Table 24.2), and a clearance factor ≥7.8 $\log LD_{50}/mL$ was obtained again.

The high values of infectivity clearance obtained for both steps indicated that any infectivity was virtually eliminated from the process after key step 2, and none was expected to be present in any sample of overall process validation experiments from this point onward. This was confirmed when no infectivity was detected in the sample obtained after exposure of the ganglioside mixture to 1 N NaOH for 1 hour (key step 3). This also confirms that NaOH is the best sterilizing chemical for unconventional slow viruses.

Exposure of the scrapie agent to 1 N NaOH at room temperature for 1 hour is recommended for sample sterilization, and previous data had shown complete inactivation of the 263K agent in a 10% (w/v) crude brain homogenate (57). However, the specific evaluation of this step as performed in spiking experiments showed the presence of clinical cases of scrapie (confirmed histopathologically) in animals injected with the undiluted sample. Residual infectivity was 1.8 $\log LD_{50}/mL$ and the clearance factor was 6.0 $\log LD_{50}/mL$. Although the procedure was very effective and infectivity was decreased by at least one million–fold, we cannot exclude that the conditions of use play a very important role in inactivation. This has been further confirmed by recent studies (56).

Finally, none of the animals injected with 20-fold concentrated, 10-fold concentrated, or undiluted final ganglioside preparation (autoclaved product) showed any clinical or histopathologic signs of disease. As the entire sample was injected into animals, we concluded that final ganglioside preparations were noninfectious. No infectivity was detected in gangliosides extracted from noninfected hamster brains.

Infectivity was detected in spiking experiments after steam autoclaving (key step 4) in samples diluted down to 100-fold with respect to the original volume. All clinical diagnoses were confirmed histopathologically. Residual infectivity was 4.0 $\log LD_{50}$/mL, indicating that autoclaving at 121°C for 30 min achieved a clearance factor of 3.8 $\log LD_{50}$/mL.

Total decontamination was not expected from this step, since the 263K strain had previously been shown to be resistant at this temperature and time of exposure (57). We used these autoclaving conditions because gangliosides can withstand these high temperatures and extended times of sterilization without losing their chemical and biologic characteristics (52).

Immunoblot indicated that extraction with methylene chloride/methanol/NaOH (key step 1) reduced PrP^{Sc} below the limit of detection. PrP^{Sc} was detected only in the solid insoluble material that was eliminated from the process after the first extraction (Fig. 24.1).

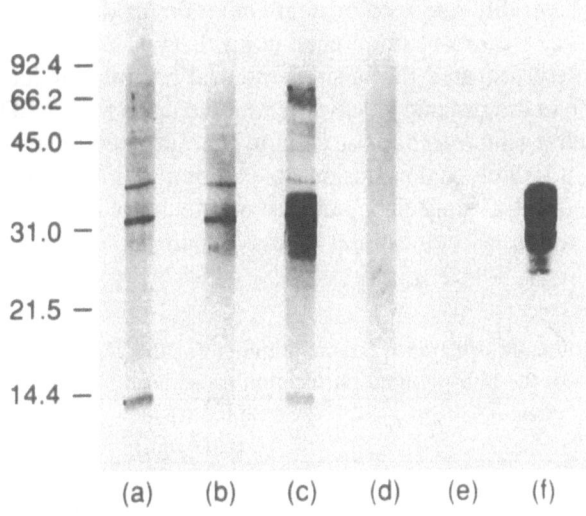

FIGURE 24.1. Immunoblot of PrP^{Sc} in (a) scrapie-infected hamster brain homogenate; (b) defatted scrapie-infected hamster brain powder; (c) insoluble proteinacious material removed by filtration and centrifugation at key step 1; (d) ganglioside-containing lipid fraction recovered after key step 1; (e) final sterilized ganglioside preparation. Lane (f) represents a purified preparation of hamster PrPSc. Reproduced with permission from Di Martino et al. (52, 53).

Result and Discussion of Phospholipids Experiments

Results are summarized in Table 24.3. The infectivity of the 33% w/v starting homogenate was 9.0 $\log LD_{50}$/mL, corresponding to 9.5 $\log LD_{50}$/g of wet tissue. No clinical or histopathologic signs of scrapie were observed in animals injected with a sample of raw phospholipids. The infectivity titer of the sample was ≤ 1.5 $\log LD_{50}$/mL, indicating an infectivity clearance ≥ 7.5 $\log LD_{50}$/mL. None of the animals injected with either undiluted or 10-fold concentrated final product and autoclaved product preparations showed any clinical signs of scrapie. All clinically negative animals in these groups were also confirmed by microscopic histopathologic evaluation. The total clearance obtained for the phospholipid process was ≥ 9.0 $\log LD_{50}$/mL.

These experiments confirm that extraction by organic solvents using the organic phase for product purification can also be considered a useful procedure for reduction and/or removal of infectivity. Cessation of solvent extraction for the production of animal feed has been indicated as one of the major concurrent causes of the sudden BSE epidemic (3), and, although single exposures to organic solvents are known to have a limited effect, continued and prolonged treatments would significantly reduce infectivity titers. Moreover, there is no doubt that the scrapie agent is poorly soluble in most organic solvents.

The scaled-down phospholipid protocol that was evaluated included an early step of extraction with acetone/1,1,1-trichloroethane. This procedure not only considerably removed protein contaminants but also reduced the presence of PrP^{Sc} down to undetectable levels (Fig. 24.2). Similar experiments have demonstrated that a single neutral partitioning leaving no detectable PrP^{Sc} in the organic phase may remove up to 5.3 $\log LD_{50}$/mL (58). Whereas such a single treatment cannot guarantee complete removal or inactivation, it is likely that a considerable amount of infectivity would have been removed at this stage. The absence of infectivity in the raw phospholipid preparations, as determined in our experiments, was the result of

TABLE 24.3. Infectivity clearance factors obtained for the 263K strain of scrapie at different steps of the phospholipid purification procedure.

Sample	Infectivity ($\log LD_{50}$/mL)	Clearance ($\log LD_{50}$/mL)
Brain homogenate (source material)	9.0	—
Raw phospholipids	≤ 1.5[a]	≥ 7.5
Final product	0.0[b]	≥ 9.0
Autoclaved product	0.0[b]	≥ 9.0

[a] Infectivity not detected in undiluted inoculum. The data represent the minimum detectable level by end-point titration in the sample injected.
[b] Infectivity not detected in the entire volumes, undiluted, or 10-fold concentrated.
Data from Di Martino et al. (54) with permission.

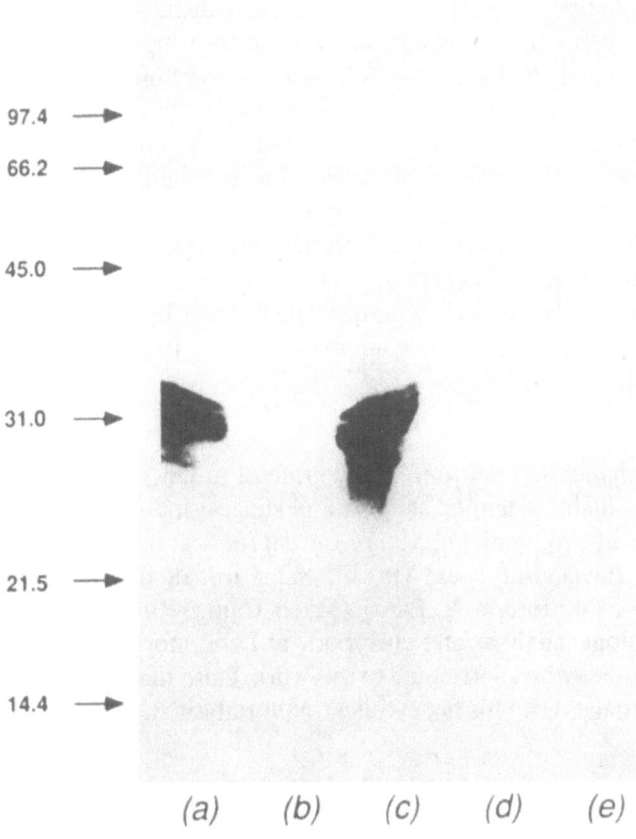

97.4 ⟶

66.2 ⟶

45.0 ⟶

31.0 ⟶

21.5 ⟶

14.4 ⟶

    (a)   *(b)*   *(c)*   *(d)*   *(e)*

FIGURE 24.2. Immunoblot of PrP^{Sc} in (a) scrapie-infected hamster brain homogenate; (b) aqueous acetonic phase and (c) insoluble proteinacious materials removed by centrifugation; (d) phospholipid-containing lipid fraction recovered after centrifugation; (e) raw phospholipids. Reproduced with permission from Di Martino et al. (54).

further differential solubilizations in organic solvents and several liquid-liquid partitionings yielding a minimum clearance factor $\geq 7.5 \log LD_{50}/mL$. Subsequent steps of liquid-liquid partitioning and further exposure to organic solvents achieved complete decontamination of both unautoclaved and autoclaved final preparations, undiluted or concentrated 10-fold, and yielded a total clearance factor $\geq 9.0 \log LD_{50}/mL$ even before steam autoclaving.

Conclusions

Despite existing uncertainties on all etiologic agents that cause TSEs, adequate collection policies of source materials may be sufficient to minimize the risk of potential contamination. The experiments of validation described above have in any case demonstrated that it is possible to produce sterile biologics, even in the case of unrealistic worst-case scenarios where the tissue source is contaminated with a high-titered preparation of scrapie agent. In fact, a positive value of both studies was the use of 263K infected hamster brain tissue as starting material, in order to obtain high clearance factors. There is little doubt that concentrated NaOH is one of the best sterilizing chemicals for scrapie-like agents, although the conditions of use may play a very important role in agent inactivation. Nevertheless, a combination of proper treatments known to affect or remove infectivity is always to be preferred whenever the ultimate purpose is the achievement of safety margins unaffected by failures in the good manufacturing procedure or contamination due to improper care of the equipment.

Acknowledgments. I am profoundly grateful to Dr. Clarence J. Gibbs, Jr. for his invaluable scientific support and supervision of the whole study. I would also like to thank Dr. Mauro Ceroni (now at Istituto Neurologico C. Mondino, Pavia, Italy) and Dr. Jiri Safar for all their help and useful discussions, Dr. Joseph E. Parisi (Mayo Clinic, Rochester, MN) for all histopathologic analyses and everybody at Laboratory of Central Nervous System Studies who contributed to this work. I also thank E.J. Hornby (OFI Ltd, Stevenage, UK) for his excellent editorial assistance.

References

1. Wells GAH, Scott AC, Johnson CT, et al. A novel progressive spongiform encephalopathy in cattle. Vet Rec 1987;121:419–20.
2. Wilesmith JW, Wells GAH, Cranwell JP, Ryan JBM. Bovine spongiform encephalopathy: epidemiological studies. Vet Rec 1988;123:638–44.
3. Kimberlin RH, Wilesmith JW. Bovine spongiform encephalopathy. Epidemiology, low dose exposure and risks. Ann NY Acad Sci 1994;724:210–20.
4. Dealler SF, Lacey RW. Transmissible spongiform encephalopathies: the threat of BSE to man. Food Microbiol 1990;7:253–79.
5. Coyle E, Harvey I. Bovine spongiform encephalopathy and risk to health. Br Med J 1992;304:1509.
6. Ridley RM, Baker HF. Occupational risk of Creutzfeldt-Jakob disease. Lancet 1993;341:641–2.
7. Kaaden OR. Unconventionelle erreger—eine gefahr fur mensch und tier? Berl Munch Tierarztl Wochenschr 1994;107:44–8.
8. Goldberg H, Alter M, Kahana E. The Libyan Jewish focus of Creutzfeldt-Jakob disease: a search for the mode of natural transmission. In: Prusiner SB, Hadlow WJ, eds. Slow transmissible diseases of the nervous system, vol 1. New York: Academic Press, 1979:451–60.

9. Masters CL, Harris JO, Gajdusek DC, Gibbs CJ Jr, Bernoulli C, Asher DM. Creutzfeldt-Jakob disease: patterns of world-wide occurrence and the significance of familial and sporadic clustering. Ann Neurol 1979;5:177–88.
10. Davainpour Z, Alter M, Sobel E, Callahan M. Sheep consumption: a possible source of spongiform encephalopathy in humans. Neuroepidemiology 1985; 4:240–9.
11. Brown P, Cathala F, Raubertras RF, Gajdusek DC, Castaigne P. The epidemiology of Creutzfeldt-Jakob disease: conclusion of a 15-year investigation in France and review of the world literature. Neurology 1987;37:895–904.
12. Cousens SN, Harries-Jones R, Knight R, Will RG, Smith PG, Matthews WB. Geographical distribution of cases of Creutzfeldt-Jakob disease in England and Wales 1970–84. J Neurol Neurosurg Psychiatry 1990;53:459–65.
13. Alperovitch A, Brown P, Weber T, Pocchiari M, Hofman A, Will RG. Incidence of Creutzfeldt-Jakob disease in Europe in 1993. Lancet 1994; 343:918.
14. Wilesmith JW, Ryan JBM. Bovine spongiform encephalopathy: observation of the incidence during 1992. Vet Rec 1993;132:300–1.
15. Hoinville LJ. Decline in the incidence of BSE in cattle born after the introduction of the "feed ban". Vet Rec 1994;134:274–5.
16. Gajdusek DC. Spontaneous generation of infectious nucleating amyloids in the transmissible and nontransmissible cerebral amyloidoses. Mol Neurobiol 1994;8:1–13.
17. Hope J. The nature of the scrapie agent: the evolution of the virino. Ann NY Acad Sci 1994;724:282–9.
18. Pocchiari M. Prions and related neurological diseases. Mol Aspects Med 1994;15:195–291.
19. Prusiner SB, Hsiao KK. Human prion diseases. Ann Neurol 1994;35:385–95.
20. Kirkwood JK, Wells GAH, Wilesmith JW, Cunningham AA, Jackson SI. Spongiform encephalopathy in an Arabian oryx (*Oryx leucoryx*) and a greater kudu (*Tragelaphus strepsiceros*). Vet Rec 1990;127:418–20.
21. Wyatt JM, Pearson GR, Smerdon TN, Gruffydd Jones TJ, Wells GAH, Wilesmith JW. Naturally occurring scrapie-like spongiform encephalopathy in five domestic cats. Vet Rec 1991;129:233–6.
22. Brown P, Gajdusek DC, Gibbs CJ Jr, Asher DM. Potential epidemic of Creutzfeldt-Jakob disease from human growth hormone therapy. N Engl J Med 1985;313:728–31.
23. Brown P, Preece MA, Will RG. "Friendly fire" in medicine: hormones, homografts, and Creutzfeldt-Jakob disease. Lancet 1992;340:24–7.
24. Gibbs CJ Jr, Asher DM, Brown P, Fradkin JE, Gajdusek DC. Creutzfeldt-Jakob disease infectivity of growth hormone derived from human pituitary glands. N Engl J Med 1993;328:358–9.
25. Minor P. Bovine spongiform encephalopathy and biological products for human use. Biologicals 1990;18:77–80.
26. Baron H. Bovine spongiform encephalopathy and related agents: risk factors of biological products. An informal consultation/working meeting. Toxicol Pathol 1991;19:293–7.
27. Di Martino A. Transmissible spongiform encephalopathies and the safety of naturally-derived biologicals. Biologicals 1993;21:61–6.

28. MacDiarmid SC. Risk analysis and the importation of animals and animal products. Rev Sci Tech 1993;12:1093–107.

29. Bleem AM, Crom RL, Francy DB, Hueston WD, Kopral C, Walker KD. Risk factors and surveillance for bovine spongiform encephalopathy in the United States. J Am Vet Med Assoc 1994;204:644–51.

30. Schudel AA, Carrillo BJ, Gimeno EJ, et al. Bovine spongiform encephalopathy surveillance in Argentina. Rev Sci Tech 1994;13:801–36.

31. Millson GC, Hunter GD, Kimberlin RH. The physicochemical nature of the scrapie-agent. In: Kimberlin RH, ed. Slow virus diseases of animals and man. Amsterdam: North Holland, 1976:243–66.

32. Dickinson AG, Taylor DM. Resistance of scrapie agent to decontamination. N Engl J Med 1978;229:1413–4.

33. Gibbs CJ Jr, Gajdusek DC, Latarjet R. Unusual resistance to ionizing radiation of the viruses of kuru, Creutzfeldt-Jakob disease and scrapie. Proc Natl Acad Sci USA 1978;75:6268–70.

34. Millson GC, Manning EJ. The effect of selected detergents on scrapie infectivity. In: Prusiner SB, Hadlow WJ, eds. Slow transmissible diseases of the nervous system, vol 2. New York: Academic Press, 1979:409–24.

35. Prusiner SB, Groth DF, McKinley MP, Cochran SP, Bowman KA, Casper KC. Thiocyanate and hydroxyl ions inactivate the scrapie agent. Proc Natl Acad Sci USA 1981;78:4606–10.

36. Bellinger-Kawahara CG, Diener TO, McKinley MP, Groth DF, Smith DT, Prusiner SB. Purified prions resist inactivation by procedures that hydrolyze, modify or shear nucleic acids. Virology 1987;160:271–4.

37. Ernst DR, Race RE. Comparative analysis of scrapie agent inactivation methods. J Virol Methods 1993;41:193–201.

38. Department of Health and Social Security. The management of patients with spongiform encephalopathy (Creutzfeldt-Jakob disease). DHSS Circular 16, 1984.

39. Rosenberg RN, White CL III, Brown P, et al. Precautions in handling tissues, fluids, and other contaminated material from patients with documented or suspected Creutzfeldt-Jakob disease. Ann Neurol 1986;19:75–7.

40. Kimberlin RH, Walker CA. Transport, targeting and replication of scrapie in CNS. In: Court LA, Dormont D, Brown P, Kingsbury DT, eds. Unconventional virus diseases of the central nervous system. Fontenay-aux-Roses, France: Commissariat à l'Énergie Atomique, 1989:547–62.

41. Di Martino A. Attempts to detect evidence of scrapie-associated protein in bovine tissue-derived products. J Am Vet Med Assoc 1990;196:1685–6.

42. Geisler FH, Dorsey FC, Coleman WP. Recovery of motor function after spinal cord injury. A randomized, placebo-controlled trial with GM1 ganglioside. N Engl J Med 1991;324:1829–38.

43. Schneider JS, Pope A, Simpson K, Taggart J, Smith MG, Di Stefano L. Recovery from experimental parkinsonism in primates with GM1 ganglioside treatment. Science 1992;256:843–6.

44. Ariga T, Sekine M, Yu RK, Miyatake T. Disialogangliosides in bovine adrenal medulla. J Biol Chem 1982;257:2230–5.

45. Ledeen RW, Yu RK. Gangliosides: structure, isolation and analysis. Meth Enzymol 1982;83:139–91.

46. Committee for Proprietary Medicinal Products: Ad Hoc Working Party on Biotechnology/Pharmacy and Working Party on Safety of Medicines. EEC regulatory document. Note for guidance. Validation of virus removal and inactivation procedures. Biologicals 1991;19:247–51.
47. Committee for Proprietary Medicinal Products: Ad Hoc Working Party on Biotechnology/Pharmacy and Working Party on Safety of Medicines. EEC regulatory document. Note for guidance. Guidelines for minimizing the risk of transmitting agents causing bovine spongiform encephalopathy via medicinal products. Biologicals 1992;20:155–8.
48. Pocchiari M, Peano S, Conz A, et al. Combination ultrafiltration and 6M urea treatment of human growth hormone minimizes risk from potential CJD contamination. Horm Res 1991;35:161–6.
49. Pocchiari M. Methodological aspects of the validation of purification procedures of human/animal-derived products to remove unconventional slow viruses. Dev Biol Stand 1991;75:87–95.
50. Reed LJ, Muench H. A simple method of estimating fifty per cent endpoints. Am J Hyg 1938;27:493–7.
51. Safar J, Ceroni M, Piccardo P, Gajdusek DC, Gibbs CJ Jr. Scrapie-associated precursor proteins: antigenic relationship between species and immunochemical localization in normal, scrapie and Creutzfeldt-Jakob disease brain. Neurology 1990;40:513–7.
52. Di Martino A, Safar J, Ceroni M, Gibbs CJ Jr. Purification of non-infectious gangliosides from scrapie-infected hamster brains. Arch Virol 1992;124:111–21.
53. Di Martino A, Safar J, Ceroni M, Gibbs C, Jr. Inactivation of the scrapie agent in a scaled-down procedure for the purification of gangliosides from brain tissue. Dev Biol Stand 1993;80:187–94.
54. Di Martino A, Safar J, Gibbs CJ Jr. The consistent use of organic solvents for purification of phospholipids from brain tissue effectively removes scrapie infectivity. Biologicals 1994;22:221–5.
55. Fraser H, Bruce ME, Chree A, McConnell I, Wells GAH. Transmission of bovine spongiform encephalopathy and scrapie to mice. J Gen Virol 1992; 73:1891–7.
56. Taylor DM, Fraser H, McConnell I, et al. Decontamination studies with the agents of bovine spongiform encephalopathy and scrapie. Arch Virol 1994; 139:313–26.
57. Brown P, Rohwer RG, Gajdusek DC. Newer data on the inactivation of scrapie virus or Creutzfeldt-Jakob disease virus in brain tissue. J Infect Dis 1986; 153:1145–8.
58. Safar J, Ceroni M, Piccardo P, et al. Subcellular distribution and physicochemical properties of scrapie-associated precursor protein and relationship with the scrapie agent. Neurology 1990;40:503–8.

25

Cellular and Scrapie Prion Protein Immunolocalization and In Vitro Amyloid Formation

MAURO CERONI, JIRI SAFAR, PEDRO PICCARDO, PAUL P. LIBERSKI, PAOLA PERGAMI, AND CLARENCE J. GIBBS JR.

Spongiform encephalopathies in humans and animals include fatal neurodegenerative disorders that can be sporadic, infectious, or inherited (1). Regardless of their origin, spongiform encephalopathies are all transmissible to an appropriate animal species (2), with the exception of fatal familial insomnia (FFI) (3), whose transmission is still unknown. The unique features of the infectious agent that causes spongiform encephalopathies prompted Prusiner (4) to coin the term *prion* to define a new category of pathogens. In Prusiner's terminology, the etiologic definition is substituted for the neuropathologic one, and spongiform encephalopathies are called prion diseases. Considerable neuropathologic variability among this group of diseases has been described (5). Some forms are characterized by widespread spongiform changes, while others show mostly prion protein (PrP) amyloid deposition (5). Scrapie, a disease naturally occurring in sheep and goats, is the prototype of prion diseases and has been adapted to various experimental rodent species.

The Prion Protein

Detergent extracts from scrapie-infected mouse and hamster brain homogenates contain disease-specific fibrillar structures that retain infectivity and have amyloid properties (6, 7). Sodium dodecyl sulfate–polyacrylamide gel electrophoresis (SDS-PAGE) analysis of these preparations, following proteinase K (PK) digestion, shows a single molecular species, a 27- to 30-kd protein designated prion protein (PrP27-30) (8, 9). Fibrillar amyloid structures and PrP27-30 are specific markers for scrapie (and the other prion diseases), and are found exclusively in infected brains. PrP27-30 is a hydrolysis product derived from PK digestion of a 33- to 35-

kd protein (PrP^{Sc}). PrP27-30 and/or PrP^{Sc} are the main constituents of the fibrillar amyloid-like structures (10, 11). PrP^{Sc} and its isoform PrP^{C}, found in normal brain, have the same molecular weight and are coded by a host gene (12). There are no differences in the cDNA sequence of the PrP gene obtained from scrapie-infected and normal brains (13). The PrP messenger RNA (mRNA) level remains constant during the course of the disease and is equivalent to that of matched controls (14, 15). In spite of these similarities, PrP^{Sc} and PrP^{C} have distinct physicochemical behaviors; PrP^{Sc} copurifies with infectivity, is PK-resistant, and forms amyloid-like structures, whereas PrP^{C} does not (16). The lack of differences in posttranslational modifications (17), such as glycosylation (18, 19), acylation (12), and the glycolipid anchor (20, 21), has led to the conclusion that PrP conformation is responsible for the biochemical dissimilarities. Recent data on the conformation of PrP^{C} and PrP^{Sc} strongly support this conclusion (22–24). However, the precise mechanism of amyloid production is unknown.

PrP^{C} and PrP^{Sc} Immunodetection in the Brain

There are several reports of immunostained amyloid plaques in the brains of humans and animals with prion diseases using antibodies raised against purified PrP27-30 and also against synthetic peptides representing various portions of the PrP protein (11, 25–31). However, there are few data about reactivity with PrP^{C} and PrP^{Sc} in neurons. The majority of these antibodies were produced against unfolded PrP27-30 following denaturation under harsh conditions using SDS, formic acid, acetone, or methanol, as discussed in a previous paper (32). Thus, the apparent little or no reactivity with PrP^{Sc} and PrP^{C}, which are membrane-bound proteins, may be explained by a molecular conformation different from that occurring in plaques or after SDS electrophoresis. Furthermore, antibodies produced against epitopes of the unfolded protein may not recognize the native form of the protein if these epitopes are buried in the complex tertiary structure (32). DeArmond et al. (33) reported localization of both PrP^{Sc} and PrP^{C} in hamster brain neurons using a special fixation technique, proteinase K treatment, and long incubation time with monoclonal antibodies.

Much work has been done to improve the sensitivity of immunohistochemical techniques to assess the regional distribution of PrP^{Sc} and/or PrP27-30 in the brains of various host species affected by different scrapie strains, and to be used as a parameter for the diagnosis of prion diseases (34). PrP^{Sc} and PrP27-30 detection is greatly enhanced by pretreatment with 80% to 100% formic acid for 5 to 60 min (35–37) 4 M guanidine thiocyanate for 2 hours (38), proteinase K (PK) 10 µg/mL (33), hydrolytic autoclaving 2.5 mM HCl (39), or a combination of them. All the treatments obtain two major goals: (a) they greatly diminish the amount of protein competing with PrP during the immunologic detection, exploiting PrP^{Sc} and/or

PrP27-30 resistance to such treatments; and (b) they reduce or abolish an immunologic reaction due to PrP^C, so that it is unlikely PrP^{Sc} is sensitive to these treatments. Results from these techniques are useful in studying PrP^{Sc} and/or PrP27-30 distribution, but provide no information on PrP^C localization.

A new technique called histoblot has been developed (40). Cryostat sections of unfixed brains are transferred to nitrocellulose membranes and subjected to treatments with PK and guanidine cyanate. In histoblots performed without PK treatment, PrP^C can be determined in different brain regions; however, this methodology does not allow morphologic studies (40).

Recently a number of authors (41–43) were able to show detection of PrP^C in normal hamster brain. The staining was limited to distinct neuronal systems and was localized in the cytoplasm and on the cytoplasmic membrane of neurons. Additionally PrP^C was detected in a variety of other extraneural tissues (41). There has been a report that PrP^{Sc} and/or PrP27-30 may be found in astrocytes (42). However, immunostaining at the ultrastructural level demonstrated association of PrP with axonal plasmalemma, amyloid plaques, and neuropil (44, 45), but not with astrocytes.

PrP and Amyloid Deposits In Vivo

Prion diseases are not always characterized by brain amyloid deposition composed of abnormal PrP. Periodic acid-Schiff–positive birefringent amyloid-containing plaques are found in 75% of kuru cases (2). PrP amyloid plaques are regularly present in all Gerstmann-Sträussler-Scheinker (GSS) syndrome cases (46, 47). However, PrP amyloid plaques are found in only 5% to 14% of Creutzfeldt-Jakob disease (CJD) cases (2, 48). The presence of amyloid plaques is rare in sheep and goat natural scrapie (2). The appearance of amyloid plaques in mice depends on the scrapie strain and breed of the infected mice (49).

When amyloid plaques are present they are intensely stained by antibodies raised against purified PrP27-30. In fact, GSS cases were incorrectly diagnosed as familial Alzheimer disease, because histochemistry showed an abundance of neurofibrillary tangles (47) and amyloid plaques morphologically similar to those in Alzheimer's which are readily recognized when anti-PrP antibodies are used (48).

Because of the presence of amyloid plaques, prion diseases are often classified as cerebral amyloidoses (50). However, it must be stressed that only prion diseases have been proven to be transmissible, whereas the other cerebral amyloidoses are not (51).

PrP amyloid composition may be studied in brain sections by immunohistochemistry using antibodies against a variety of synthetic peptides mimicking different parts of the PrP sequence. These studies have shown that in

GSS the amyloid plaques exhibiting typical tinctorial and optical properties were exclusively labeled by the antibody to the PrP27-30 N-terminus peptide, whereas PrP deposits lacking typical amyloid properties were stained by all anti-synthetic antibodies including the one raised against the PrP^{Sc} N-terminus peptide that does not label PrP27-30 (52).

Aim of the Work

Pathogenesis of the prion diseases remains largely undetermined. The relative distribution of PrP^C and PrP^{Sc} in scrapie brains and the manner in which neuropathologic damage is produced are not yet understood. To study PrP^C and PrP^{Sc} distribution in brain sections, we raised polyclonal antibodies against nondenatured PrP27-30 and against a 15 amino acid synthetic peptide with the sequence of the hamster PrP27-30 N-terminus able to stain both PrP^C and PrP^{Sc} (32). Here we present and discuss results obtained from immunohistochemical studies in brain sections from normal and scrapie-infected animals. Golden Syrian hamsters (LVG/LAK) were infected with the 263K scrapie strain and NIH random-bred Swiss-Webster mice were inoculated with the C506 scrapie strain isolated in our laboratory. Results from these studies have been reported in part previously (53). Moreover, as the pathway of formation and the composition of amyloid deposits in scrapie brains remain unclear, we decided to analyze amyloid formation in vitro starting from membrane- or liposome-bound PrP^{Sc}.

Results and Discussion

PrP^C and PrP^{Sc} Immunohistochemistry

Polyclonal antibodies against nondenatured PrP27-30 and a polyclonal antibody against a 15 amino acid synthetic peptide with the sequence of the hamster PrP27-30 N-terminus give similar results when used on sections of normal and scrapie-infected mouse brains (53). Positive PrP immunoreactivity is mainly found in the cytoplasm of neurons throughout the brain of normal (Figs. 25.1 and 25.2) and scrapie-infected (Figs. 25.3 and 25.4) brains. The staining reveals spherical structures with nonreactive cores or appearing as dots. There is no staining in the white matter, the blood vessel walls, meninges, or choroid plexus. Positive staining is present in the neocortex, hippocampus, basal ganglia, and in variable amounts in the Purkinje and granule cell layers of the cerebellum. The staining pattern in normal controls is similar to that seen in infected brains (compare Figs. 25.1 and 25.2 with Figs. 25.3 and 25.4). A faint brown staining in the neuropil is present in scrapie-infected brains and to a milder degree even in normal brains. Such diffuse staining is not present with antibodies adsorbed with

342 M. Ceroni et al.

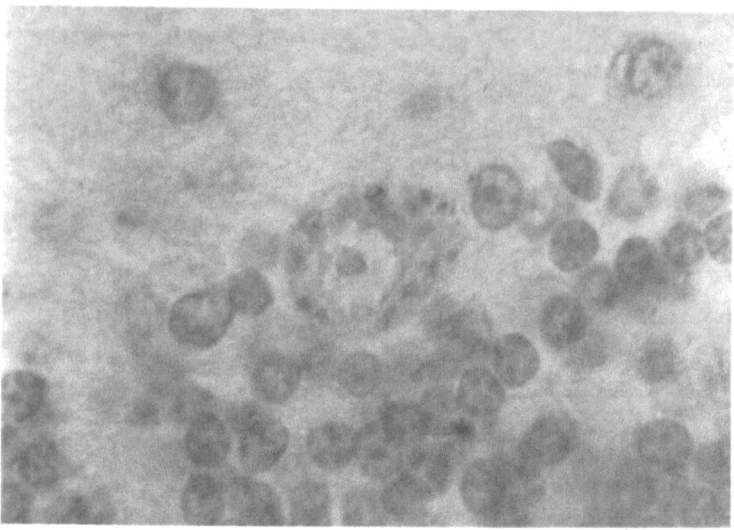

FIGURE 25.1. Intracytoplasmic PrP staining in Purkinje cell neurons of a normal mouse, following reaction with rabbit polyclonal antibody raised against a 15 amino acid synthetic peptide homologous to the hamster PrP27-30 N-terminus; the antiserum was used at dilution 1:2,000 and amplified by avidin-biotin peroxidase ($\times$1,200 before 34% reduction).

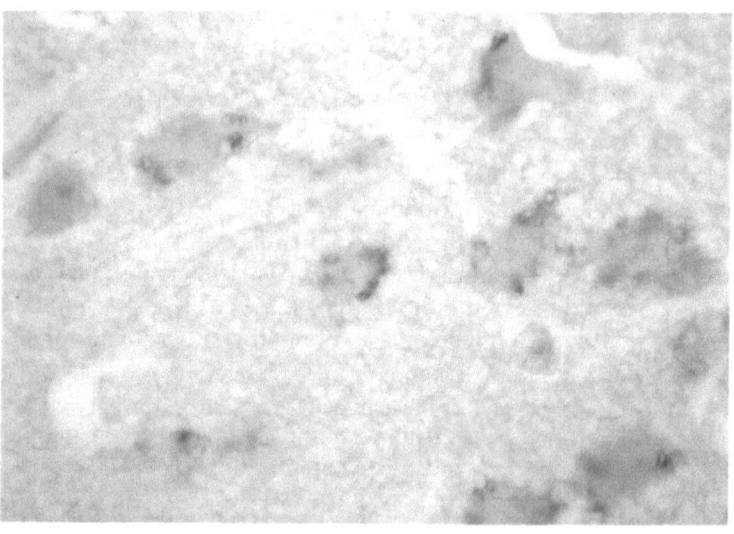

FIGURE 25.2. Intracytoplasmic PrP staining in cortical neurons of a normal mouse, following reaction with rabbit polyclonal antibody raised against purified nondenatured hamster PrP27-30; the antiserum was used at dilution 1:100 and amplified by avidin-biotin peroxidase ($\times$1,200 before 34% reduction).

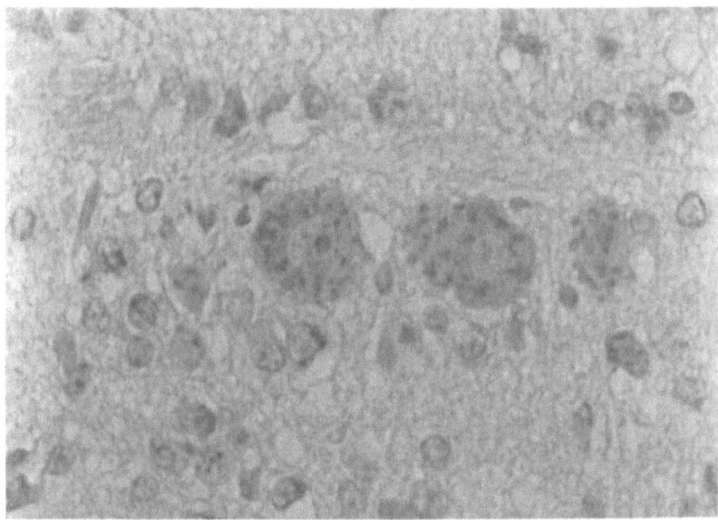

FIGURE 25.3. Intracytoplasmic PrP staining in neurons of the subcortical gray matter of a scrapie-infected mouse, following reaction with rabbit polyclonal antibody raised against a 15 amino acid synthetic peptide homologous to the hamster PrP27-30 N-terminus; the antiserum was used at dilution 1:2,000 and amplified by avidin-biotin peroxidase (×1,200 before 34% reduction).

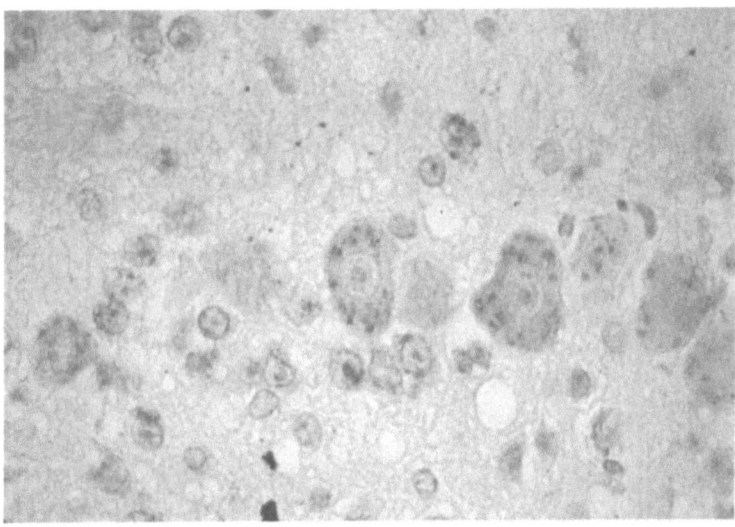

FIGURE 25.4. Intracytoplasmic PrP staining in cortical neurons of a scrapie-infected mouse, following reaction with rabbit polyclonal antibody raised against a 15 amino acid synthetic peptide homologous to the hamster PrP27-30 N-terminus; the antiserum was used at dilution 1:2,000 and amplified by avidin-biotin peroxidase (×1,200 before 34% reduction).

purified PrP27-30 or the N-terminus of PrP27-30 synthetic peptide or with preimmune sera. This immunostaining pattern has been reported by various authors and attributed to PrP^{Sc} accumulation (33, 40, 44). In scrapie-infected brains it is greatly enhanced by a variety of brain section pretreatments (summarized in 34). In our experimental conditions this diffuse specific immunostaining may represent neuropil localization of PrP^{C} in normal brain and of PrP^{C} and PrP^{Sc} in infected brain, respectively.

Polyclonal antibodies to PrP27-30 strongly labeled PrP amyloid plaques with typical tinctorial properties (Fig. 25.5) and other PrP deposits without amyloid tinctorial properties, exclusively in infected brains. Typical amyloid plaques are found only in mice infected with the C506 scrapie strain, whereas in hamsters infected with the 263K scrapie strain only Congo red negative PrP deposits are found, particularly at the subependymal level. Extracellular PrP deposits are accompanied by slightly more intense neuropil staining. Interestingly the antibody raised against the synthetic peptide homologous to the PrP27-30 N-terminus does not stain PrP plaques.

The intracytoplasmic staining pattern similar in normal and in scrapie brains suggests that only PrP^{C} is immunolabeled in large amounts within some cytoplasmic organelle (probably the endoplasmic reticulum and Golgi apparatus) in normal as well as in scrapie-infected neurons. No apparent increase in intracellular PrP staining was observed in scrapie-infected brain sections compared with normal brains, even though PrP^{Sc} accumulates in

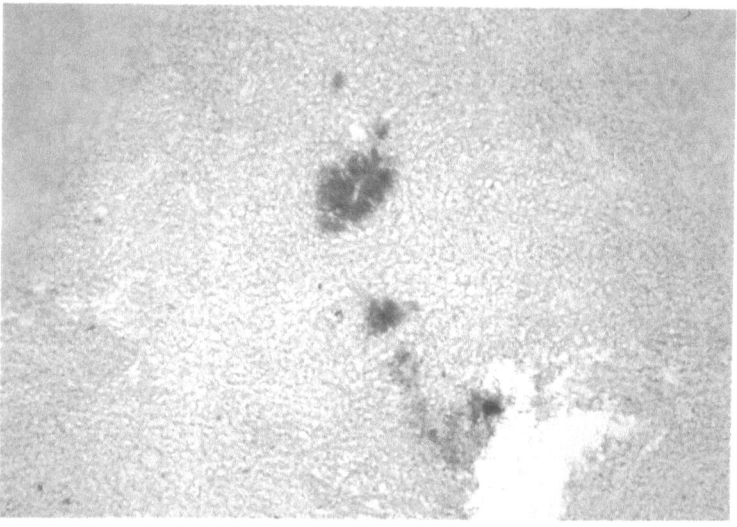

FIGURE 25.5. PrP amyloid plaques intensely stained by a rabbit polyclonal antibody raised against purified nondenatured hamster PrP27-30; part of the tissue is out of focus because it is detached from the glass.

scrapie-infected brains. These findings are supported by studies on PrP mRNA expression in the brain. In fact, PrP mRNA level is highest in neurons (54), remains constant during the course of the disease, and is equivalent to that of matched controls (14, 15). The similarity of intracytoplasmic staining in normal and scrapie-infected brains might be attributed to a lack of exposed epitopes of PrP^{Sc} with respect to PrP^C. In contrast with this is the fact that polyclonal antibodies against purified PrP27-30 readily detect PrP amyloid plaques and other PrP extracellular deposits that are probably composed of PrP^{Sc}. A lower affinity to PrP^{Sc} has been suggested for these antibodies, but this contrasts with the fact that the antibodies were raised against purified PrP27-30 (a hydrolytic product of PrP^{Sc}) and not against PrP^C. Our findings and the data on PrP mRNA expression in the brain suggest that PrP^C is present in neurons from normal and scrapie brains at about the same concentration and that PrP^{Sc} does not accumulate in the cytoplasm of neurons.

It has been estimated from PrP27-30 purification procedures and immunoblot results that PrP^{Sc} is 10 to 100 times more concentrated than PrP^C in scrapie brains (16, 33). Our findings suggest that the amount of PrP^C and PrP^{Sc} in scrapie brains is not so dramatically different. Our immunohistochemical observations are further supported by experiments with trypsin-limited proteolysis of normal and scrapie brain membrane fractions. Unlike other proteases that readily and completely digest PrP^C, limited proteolysis of normal brain membrane fractions with trypsin consistently yields a PrP^C hydrolysis product of 21 to 23 kd (21). When scrapie brain membrane fractions are incubated with trypsin and then subjected to SDS-PAGE and immunobloting, two PrP bands with similar intensity are found representing PrP27-30 and the 21- to 23-kd hydrolysis product of PrP^C respectively (21).

Where and how PrP^C is transformed into PrP^{Sc} remains to be established. There is evidence in scrapie-infected cell cultures that the event takes place in the endocytic pathway (55), but other authors believe that the conversion may occur on the plasmalemma (44). It seems unlikely that in scrapie-infected brains PrP^{Sc} accumulates in lysosomes. In fact, PrP purification procedures that include protease inhibitors yield almost exclusively PrP^{Sc} (56), whereas lysosomal proteases would readily hydrolyse PrP^{Sc} to PrP27-30.

In Vitro PrP Amyloid Formation

It has already been mentioned that detergent extracts from scrapie-infected mouse brain homogenates contain disease-specific fibrillar structures that have amyloid properties (6). Indeed, this was the first hallmark of the disease to be discovered. Moreover PrP^{Sc} and PrP27-30 purification procedures are largely based on conditions that selectively favor PrP aggregation.

The use of protease inhibitors during the purification procedure from mouse and hamster scrapie-infected brain homogenate yields a great majority of PrP^{Sc} compared with PrP27-30 (56). During scrapie in the brain, PrP^C is converted into PrP^{Sc}. PrP^{Sc} is exquisitely susceptible to hydrolysis by various proteases, such as trypsin, papain, PK, and pronase, yielding PrP27-30 (Ceroni, personal observation). However, only negligible hydrolysis of PrP^{Sc} occurs in vivo and the protein accumulates as such in infected-brains. A simplified purification procedure for PrP^{Sc} or PrP27-30 is shown in Table 25.1. Omission of PK and inclusion of protease inhibitors during the procedure yields PrP^{Sc}, whereas the use of PK yields, PrP27-30 as a final product. In fractions with variable degrees of PrP purity, amyloid formation can be detected using Congo red and negative staining electron microscopy.

As can be seen in Table 25.1, the procedure step that consists of incubation with 1% sarcosyl and 10% NaCl was varied (1 hour versus overnight), because amyloid formation is considered a slow process requiring time to be completed. Table 25.2 summarizes our results. Both the synaptosomal-microsomal membrane fraction and phospholipid-rich phase from Triton X-114 extraction (Fig. 25.6) with or without PK never show fibrils with electron microscope (EM) detection. Congo red staining was not done on these crude preparations because various structures, such as collagen fibrils, have amyloid properties and are indistinguishable from PrP fibrils under these conditions.

We observed various preparations of 30% to 40% pure PrP^{Sc}, as judged by SDS-PAGE and silver staining, and found no fibrillar structures (negative data not shown). Congo red results substantially confirmed the EM findings. When purified PrP27-30 preparations (PK was added to the initial homogenate and to the final step, and Triton X-114 was omitted) were

TABLE 25.1. PrP^{Sc} and PrP27-30 purification procedure.

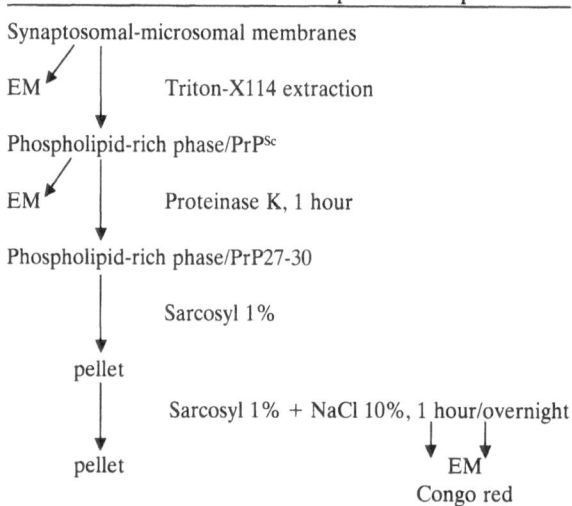

TABLE 25.2. EM fibril detection and Congo red staining in various PrP preparations.

Preparation	Fibrils on EM	Congo red
Synaptosomal-microsomal membrane fraction	Neg	ND
Phospholipid-rich phase (PRP) from Triton X-114	Neg	ND
Purified PrP^{Sc}	Neg	$--/+$
Purified Prp27–30 (Diringer)	$+++$	$+++$
PRP + proteinase K (PK) → sarcosyl → sarcosyl + NaCl 1 hour	$+$	$++$
PRP + PK → sarcosyl → sarcosyl + NaCl overnight	$++$	$++$

ND, not done.

subjected to EM examination, fibrils were readily detected in all grids (Fig. 25.7). Additionally, Congo red staining was intensely positive for birefringent material. The addition of PK to the phospholipid-rich phase readily converted all of PrP^{Sc} to PrP27-30 (data not shown). Under these conditions, at the end of the purification procedure, fibrils may be detected by EM after both the brief and long extractions in sarcosyl and NaCl (Fig. 25.8). However, fibrils are much less numerous than in a classic PrP27-30 preparation, and the short incubation further diminishes fibril formation. These findings suggest that Triton X-114 extraction negatively interferes with amyloid formation and that fibril formation is a time-requiring process. Our data confirm that the PrP^{Sc} N-terminus hydrolysis to PrP27-30 is essential to allow amyloid formation, as reported by others (57). Also, these data are supported by studies on conformational

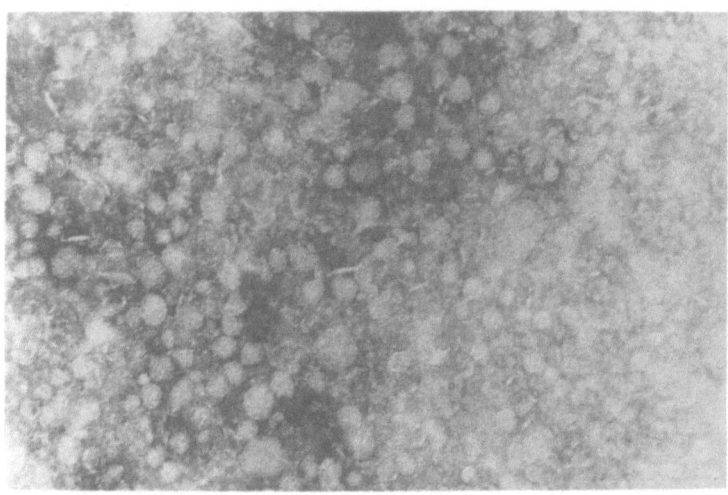

FIGURE 25.6. EM negative staining ($\times 70,000$) of phospholipid-rich phase Triton X-114 extraction from scrapie-infected mouse brain. No scrapie fibrils or rods were ever found in this kind of preparation (see also Table 25.2).

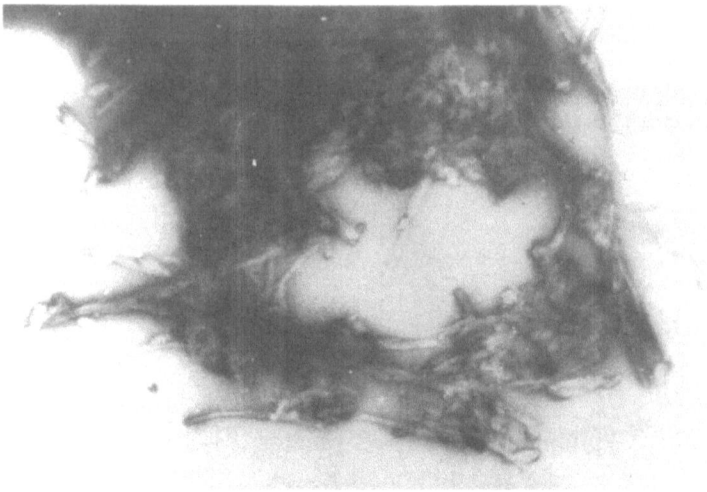

FIGURE 25.7. EM negative staining (×40,000) of typical scrapie fibrils in a preparation of PrP27-30 purified according to Diringer's method.

transitions, dissociation, and unfolding of PrP^{Sc} and PrP27-30 that demonstrate the importance of the amino-terminal cleavage of PrP^{Sc} to induce stability and alignment of the amyloid-forming PrP27-30 monomers (23). In fact, in all human amyloidoses, with the exception of the one caused by β_2-microglobulin in patients with renal insufficiency (58), amyloid formation results from peptide hydrolysis from a precursor protein.

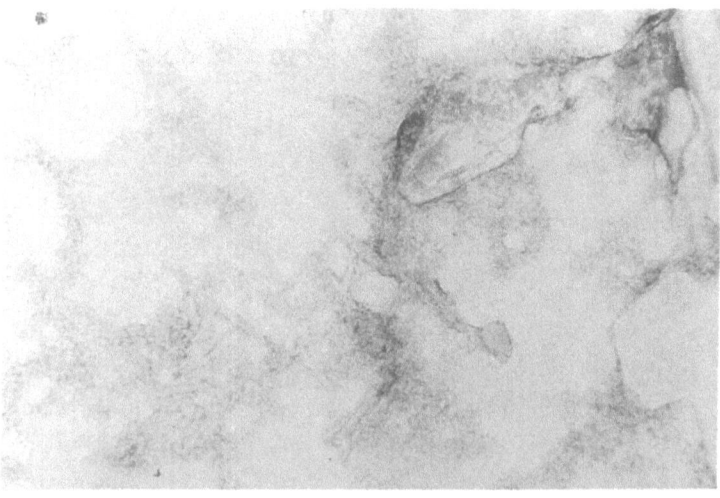

FIGURE 25.8. EM negative staining (×33,000) of atypical scrapie fibrils obtained after overnight extraction in sarcosyl and NaCl of a scrapie brain phospholipid-rich phase with the addition of PK (see also Table 25.2).

Infectivity and Amyloid Deposition

It has been suggested that scrapie agent proliferation may occur as a crystallization process of PrP27-30 (50, 59). According to this "unifying" hypothesis, prion diseases would represent one of many cerebral amyloidoses and share the same pathogenesis with them. Amyloidogenesis would require stagnation, pooling, and concentrating a precursor protein that through a nucleation process initiates crystal formation (2, 59). The whole process would then perpetuate itself and produce the disease.

However, the relevance of amyloid formation in the pathogenesis of cerebral amyloidoses and particularly of prion diseases remains to be established. The difference between transmissible and nontransmissible cerebral amyloidoses must be stressed (51). Indeed, transmissibility has important potential implications on pathogenesis. In prion diseases, the infectious agent may travel from cell to cell causing neurodegeneration; such a mechanism cannot be supposed in nontransmissible amyloidoses.

Moreover, it has been proven that infectivity is retained by a preparation of PrP^{Sc} monomers (56), and dispersion of prion rods composed of amyloid material by sonication into mixed micelles results in a 10- to 100-fold increase in infectivity, in spite of the rods' disappearance (60). Therefore, infectivity seems to depend on an intrinsic property of PrP^{Sc} rather than PrP^{Sc} aggregation by itself, even though we cannot exclude that PrP^{Sc} activity depends on dimer or polymer formation. Moreover, in prion diseases most of the abnormal PrP accumulates as PrP^{Sc} and can be purified as such by adding protease inhibitors (56), which does not form amyloid (57). Finally, amyloid plaques are not a constant finding in various prion diseases, such as sporadic CJD (61), and no amyloid material can be detected in cell cultures permanently infected with scrapie (however, no pathological changes can be demonstrated in these cells) (62).

Synthetic peptides homologous to the hydrophobic core of PrP27-30 have been shown under appropriate physicochemical conditions to form amyloid fibers and to produce neuronal death in cell cultures (63). However, it is not yet clear whether the peptide neurotoxicity depends on a specific receptor-mediated activity or is due to amyloid toxicity by itself.

We think that extracellular amyloid deposition by itself cannot account for neurotoxicity in prion diseases, but rather represents a late event in the pathogenesis of these diseases. We cannot exclude that amyloid formation can contribute to the production of neuronal damage in the last stages of the disease or in a more relevant way in those forms that show diffuse and intense amyloid formation.

The tendency of PrP^{Sc} to tightly aggregate and to form amyloid following hydrolysis to PrP27-30 is an intrinsic property of PrP^{Sc} monomers. Unlike other amyloidoses in which amyloid formation is the result of increased concentration of normal proteins, e.g., β_2-microglobulin in chronic hemodialysis-related amyloidosis (58), or of some of their hydrolysis products (for amyloidosis of the nervous system see 64), PrP^C never precipitates as amyloid, whereas PrP^{Sc} forms amyloid plaques when it is hydrolyzed to PrP27-30.

Conclusions

Our findings contribute to better understanding the complex relations among PrP^{C}, PrP^{Sc}, and PrP27-30 in scrapie infected-brains. PrP^{C} is synthesized in large amount and immunolabeled within some cytoplasmic organelles (probably endoplasmic reticulum and Golgi apparatus) in neurons of different regions of the cerebrum, the brain stem, and the cerbellum. PrP neuronal staining is observed in all brain regions that were studied, but whether PrP^{C} synthesis is restricted to particular subsets of neurons or is promoted in a particular neuronal system at any particular time remains to be established. Our data are in agreement with the fact that PrP mRNA expression is 100-fold greater in neurons than in any other cell type (65), and that its concentration remains constant during the course of the disease and is equivalent to that of matched controls (14, 15).

The inability to detect PrP^{C} in normal brains by various antibodies raised against purified PrP27-30 may be due to the fact that these antibodies have been produced against denatured PrP27-30 and do not recognize PrP^{C} epitopes (32).

There appears to be no significant difference in cytoplasmic immunostaining between normal and scrapie brains. Even though in some instances an intense staining may be found in the cytoplasm of neurons in scrapie-infected brains, as reported by others (33), we do not think that such a finding can be attributed to PrP^{Sc} deposits in the cytoplasm. Neuronal cytoplasmic immunostaining might represent PrP^{C} actively synthesized in the rough reticulum and posttransationally processed in the endoplasmic reticulum and the Golgi apparatus.

It is well known that PrP^{C} synthesis is necessary for prion replication and PrP^{Sc} and/or PrP27-30 accumulation. In fact, mice with complete ablation of the wild-type PrP gene are resistant to scrapie and do not produce prions (66). It can be speculated that at least in some neurons in scrapie-infected brains, PrP^{C} synthesis may be accelerated in an attempt to respond to the lack of protein function due to conversion of PrP^{C} to PrP^{Sc}.

Permanently infected cell cultures have shown that PrP^{Sc} accumulates in cytoplasmic vesicles where it is hydrolyzed to PrP27-30 (67). It seems highly unlikely that this occurs in infected brains. In fact, scrapie PrP accumulates in the brain mostly as PrP^{Sc} and not as PrP27-30. Therefore, in vivo PrP^{Sc} does not seem to enter the lysosomal pathway that would readily convert PrP^{Sc} into PrP27-30.

In what cell compartment PrP^{C} is transformed into PrP^{Sc} remains to be elucidated. Infected cell culture experiments show that PrP^{C} must first reach the outer membrane before being converted to PrP^{Sc} (55, 67). It has been suggested that the transformation may occur in the endocytic pathway (55). Other authors have produced evidence that PrP^C converts into PrP^{Sc} on the plasmalemma (44). Whatever the case, both PrP^C and PrP^{Sc} may be transported along the axons and the dendrites to reach the neuronal plasmalemma everywhere on the cell surface (68).

The mechanism of production of the neuropathologic damage in prion diseases remains obscure. It seems unlikely that accumulation of PrP^{Sc} by itself in the cells can account for neuron degeneration. In fact, we were unable to find intracellular vacuoles containing large amounts of PrP^{Sc}. Our data suggest that in scrapie brains, PrP^C is partially replaced by PrP^{Sc} in the loci where its physiologic function is accomplished. At this level, PrP^{Sc} may disrupt some basic mechanism of cellular metabolism or of membrane turnover, causing neuronal degeneration and/or pathologic events resulting from loss of PrP^C function.

Following membrane disruption and formation of multilaminated membranes (69) and/or axonal distrophy (70) and/or complete neuronal degeneration and/or neuronal apoptosis (63), PrP^{Sc} seems to accumulate in the extracellular space. The existence of PrP deposits exclusively in scrapie brains with no Congophylic properties argues that PrP^{Sc} forms aggregates in the extracellular space (especially subependymal in some scrapie models) and that extracellular processing leads to formation of mature PrP amyloid plaques (52). Our findings suggest that PrP^{Sc} hydrolysis to PrP27-30 is a necessary step for the formation of mature PrP amyloid plaques. The absence of PrP amyloid plaques in various human and animal prion diseases and their late appearance in the course of the disease in most experimental models suggest that amyloid formation is a late event in the pathogenesis of the disease and that it is not necessarily required for the production of the neuropathologic damage.

Table 25.3 summarizes our view of the complex relations among PrP^C, PrP^{Sc}, and PrP27-30 and of their characteristics. PrP^C is probably converted into PrP^{Sc} through a conformational change driven by the interaction with one or more PrP^{Sc} molecules. At this moment we cannot exclude that another cooperative molecule may be necessary for the transformation process. In the presence of various types of proteases, in the extracellular space in vivo, PrP^{Sc} is hydrolyzed to PrP27-30, which forms amyloid material through an ordered aggregation that can be defined as a nucleation-

TABLE 25.3. Relations and characteristics of PrP^C, PrP^{Sc}, and PrP27-30.

	Mechanism	Infectivity	SAF	Congo red
PrP^c		0	0	0
↓				
PrP^{Sc}	Conformational	+	0	0
↓				
PrP27-30	Proteolysis nucleation-polymerization	+	+	+

dependent polymerization process (59). PrP^{Sc} exhibits a strong tendency to aggregate and forms deposits in the extracellular space in the brain, but without birefringent optical properties. The PrP^{Sc} purification procedure exploits its tendency to aggregate. Prp^{Sc} hydrophobicity is higher than that of PrP^{C} (71), but produces neither fibrils detectable by electron microscope nor Congo red–positive material at the light microscope. The key step in prion disease pathogenesis appears to be the transformation of PrP^{C} into PrP^{Sc}. Prion disease pathogenesis depends on some direct or indirect toxic effect of PrP^{Sc} rather than on a nucleation-dependent polymerization process with amyloid formation. Studies on the conformation of PrP^{C} and PrP^{Sc}, and on the processing of PrP^{C} to PrP^{Sc}, appear to have the potential to solve the fascinating biologic problems posed by prion diseases.

References

1. Prusiner SB. Molecular biology of prion diseases. Science 1991;252:1515–22.
2. Gajdusek DC. Subacute spongiform encephalopathies: transmissible cerebral amyloidoses caused by unconventional viruses. In: Fields BN, Knipe DM, eds. Virology. New York: Raven Press, 1990:2289–324.
3. Medori R, Tritshler HJ, Leblanc A, et al. Fatal familial insomnia, a prion disease with a mutation at codon 178 of the prion protein gene. N Engl J Med 1992;326:444–9.
4. Prusiner SB. Novel proteinaceous infectious particles cause scrapie. Science 1982;216:136–44.
5. Kitamoto T, Tateishi J. Human prion diseases with variant prion protein. In: Molecular biology of prion diseases. Philos Trans R Soc Lond (B 1.1) 1994;343:391–8.
6. Merz PA, Somerville RA, Wisniewski HM, Iqbal K. Abnormal fibrils from scrapie infected brain. Acta Neuropathol (Berl) 1981;54:63–74.
7. Prusiner SB, McKinley MP, Bowman KA, et al. Scrapie prions aggregate to form amyloid-like birefringent rods. Cell 1983;35:349–58.
8. Prusiner SB, Bolton DC, Groth DF, Bowman KA, Martinez HM. Further purification and characterization of scrapie prions. Biochemistry 1982;21:6942–50.
9. Bolton DC, McKinley MP, Prusiner SB. Identification of a protein that purifies with the scrapie prion. Science 1982;218:1309–10.
10. McKinley MP, Bolton DC, Prusiner SB. A protease-resistent protein is a structural component of the scrapie prion. Cell 1983;35:57–62.
11. DeArmond SJ, McKinley MP, Barry RA, Braunfeld MB, McColloch JR, Prusiner SB. Identification of prion amyloid filaments in scrapie-infected brain. Cell 1985;41:221–35.
12. Oesch B, Westaway D, Wälchli M, et al. A cellular gene encodes scrapie PrP27-30 protein. Cell 1985;40:735–46.
13. Basler K, Oesch B, Scott M, et al. Scrapie and cellular PrP isoforms are encoded by the same chromosomal gene. Cell 1986;46:417–28.
14. Kretschmar HA, Stowring LE, Westaway D, Stublebine WH, Prusiner SB, DeArmond SJ. Molecular cloning of human prion protein cDNA. DNA 1986;4:315–24.

15. Caughey B, Race RE, Ernst D, Buchmeier MJ, Chesebro B. Prion protein biosynthesis in scrapie infected and uninfected neuroblastoma cells. J Virol 1989;63:175–81.
16. Meyer RK, McKinley MP, Bowman KA, Braunfeld MB, Barry RA, Prusiner SB. Separation and properties of cellular and scrapie prion proteins. Proc Natl Acad Sci USA 1986;83:2310–4.
17. Stahl N, Baldwin MA, Teplow DB, et al. Sructural studies of the scrapie prion protein using mass spectrometry and amino acid sequencing. Biochemistry 1993;32:1991–2002.
18. Haraguchi T, Fisher S, Olofsson S, et al. Asparagine-linked glycosylation of the scrapie and cellular prion proteins. Arch Biochem Biophys 1989;274:1–13.
19. Endo T, Groth D, Prusiner SB, Kobata A. Diversity of oligosaccharide structures linked to asparagines of the scrapie prion protein. Biochemistry 1989;28:8380–8.
20. Stahl N, Borchelt DR, Hsiao K, Prusiner SB. Scrapie prion protein contains a phosphatidylinositol glycolipid. Cell 1987;51:229–40.
21. Safar J, Ceroni M, Gajdusek DC, Gibbs CJ Jr. Differences in the membrane interaction of scrapie amyloid precursor proteins in normal and scrapie- or Creutzfeldt-Jakob disease-infected brains. J Infect Dis 1991;163:488–94.
22. Pan K-M, Baldwin M, Nguyen J, et al. Conversion of α-helices into β-sheets features in the formation of the scrapie prion proteins. Proc Natl Acad Sci USA 1993;90:10962–6.
23. Safar J, Peter PR, Gajdusek DC, Gibbs CJ Jr. Conformational transitions dissociation, and unfolding of scrapie amyloid (prion) protein. J Biol Chem 1993;268:20276–84.
24. Safar J, Roller PP, Gajdusek DC, Gibbs CJ Jr. Scrapie amyloid (prion) protein has the conformational characteristics of an aggregated molten globule folding intermediate. Biochemistry 1994;33:8375–83.
25. Bendheim PE, Barry RA, DeArmond SJ, Stites DP, Prusiner SB. Antibodies to a scrapie prion protein. Nature 1984;310:417–28.
26. Bode L, Pocchiari M, Gelderblom H, Diringer H. Characterisation of antisera against scrapie associated fibrils (SAF) from affected hamster and cross-reactivity with SAF from scrapie affected mice and from patients with Creutzfeldt-Jakob disease. J Gen Virol 1985;66:2471–8.
27. McBride PA, Bruce ME, Fraser M. Immunostaining of scrapie cerebral amyloid plaques with antisera raised to scrapie associated fibrils (SAF). Neuropathol Appl Neurobiol 1988;14:325–36.
28. Kitamoto T, Tateishi J, Tashima T, et al. Amyloid plaques in Creutzfeldt-Jakob disease stain with prion protein antibodies. Ann Neurol 1986;20:204–8.
29. Barry RA, McKinley MP, Bendheim PE, Lewis GK, DeArmond SJ, Prusiner SB. Antibodies to the scrapie prion decorate scrapie rods. J Immunol 1985; 135:603–13.
30. Barry RA, Vincent MT, Kent SBH, Hood LE, Prusiner SB. Characterization of prion proteins with monospecific antisera to synthetic peptides. J Immmunol 1988;140:1188–93.
31. Rubenstein R, Kascsak RJ, Merz PA, et al. Detection of scrapie associated fibrils (SAF) proteins using anti-SAF antibody in non-purified tissue preparations. J Gen Virol 1986;67:671–8.
32. Safar J, Ceroni M, Piccardo P, Gajdusek DC, Gibbs CJ Jr. Scrapie-associated precursor proteins: antigenic relationship between species and immunocy-

tochemical localization in normal, scrapie and Creutzfeldt-Jakob disease brains. Neurology 1990;40:513–7.

33. DeArmond SJ, Mobley WC, Demott DL, Barry RA, Beckstead JH, Prusiner SB. Changes in the localization of brain prion proteins during scrapie infection. Neurology 1987;37:1271–80.

34. Bell JE, Ironside JW. Neuropathology of spongiform encephalopathies in humans. In: Allen IV, ed. British Medical Bulletin. London: Churchill Livingstone, 1993:738–77.

35. Kitamoto T, Tateishi J, Sato Y. Immunohistochemical verification of senile and kuru plaques in Creutzfeldt-Jakob disease and the allied disease. Ann Neurol 1988;24:537–42.

36. Powers JM, Liu Y, Hair LS, Kascsack RJ, Lewis LD, Levy LA. Concomitant Creutzfeldt-Jakob and Alzheimer diseases. Acta Neuropathol 1991;83:95–8.

37. Lantos PL. From slow virus to prion: a review of transmissible spongiform encephalopathies. Histopathology 1992;21(4):397.

38. Serban D, Taraboulos A, DeArmond SJ, Prusiner SB. Rapid detection of Creutzfeldt-Jakob disease and scrapie prion proteins. Neurology 1990;40:110–7.

39. Kitamoto T, Muramoto T, Mohri S, Doh-Ura K, Tateishi J. Abnormal isoform of prion protein accumulates in follicular dendritic cells in mice with Creutzfeldt-Jakob disease. J Virol 1991;65:6292–5.

40. Taraboulos A, Jendroska K, Serban D, Yang SL, Dearmond SJ, Prusiner SB. Regional mapping of prion proteins in brain. Proc Natl Acad Sci USA 1992;89:7620–4.

41. Bendheim PE, Brown HR, Rudelli RD, Scala LJ, Goller NL, Wen GY, et al. Nearly ubiquitous tissue distribution of the scrapie agent precursor protein. Neurology 1992;42:149–56.

42. Diedrich JF, Bendheim PE, Kim KS, Carp RI, Haase AT. Scrapie-associated prion protein accumulates in astrocytes during scrapie infection. Proc Natl Acad Sci USA 1991;88:375–9.

43. Bruce ME, McBride PA, Farquhar CF. Precise targeting of the pathology of the sialoglycoprotein, PrP, and vacuolar degeneration in mouse scrapie. Neurosci Lett 1989;102:1–6.

44. Jeffrey M, Goodsir CM, Bruce ME, McBride PA, Scott JR, Halliday WG. Infection specific prion protein (PrP) accumulates in neuronal plasmalemma. Neurosci Lett 1992;147:106–9.

45. Bruce ME, McBride PA, Jeffrey M, Scott JR. PrP in pathology and pathogenesis in scrapie-infected mice. Mol Neurobiol 1994;8:105–12.

46. Master CL, Gajdusek DC, Gibbs CJ Jr. Creutzfeldt-Jakob disease virus isolations from the Gerstmann-Sträussler syndrome, with an analysis of the various forms of amyloid plaque deposition in the virus-induced spongiform encephalopathies. Brain 1981;104:559–88.

47. Ghetti B, Tagliavini F, Masters CL, et al. Gerstmann-Straussler-Scheinker disease. II. Neurofibrillary tangles and plaques with PrP-amyloid coexist in an affected family. Neurology 1989;39:1453–61.

48. Prusiner SB. Genetic and infectiuos prion diseases. Arch Neurol 1993;50:1129–53.

49. Bruce ME, Fraser H. Effect of rate of infection on the frequency and distribution of cerebral amyloid plaques in scrapie mice. Neuropathol Appl Neurobiol 1981;1:289–98.

50. Gajdusek DC. Transmissible and non-transmissible amyloidoses: autocatalytic post-translational conversion of host precursor proteins to beta-pleated sheet configuration. J Neuroimmunol 1988;20:95–110.
51. Gajdusek DC. Transmissible and nontransmissible dementias: distinction between primary cause and pathogenetic mechanism in Alzheimer's disease and aging. Mt Sinai J Med 1988;55:3–5.
52. Giaccone G, Verga L, Bugiani O, Frangione B, Serban D, Prusiner SB. Prion protein preamyloid and amyloid deposits in Gerstmann-Straussler-Scheinker disease, Indiana kindred. Proc Natl Acad Sci USA 1992;89:9349–53.
53. Piccardo P, Safar J, Ceroni M, Gajdusek DC, Gibbs CJ Jr. Immunohistochemical localization of prion protein in spongiform encephalopathies and normal brain tissue. Neurology 1990;40:518–22.
54. McKinley MP, Hay B, Lingappa VR, Lieberburg I, Prusiner SB. Developmental expression of prion protein gene in brain. Dev Biol 1987;121:105–10.
55. Borchelt DR, Taraboulos A, Prusiner SB. Evidence for synthesis of scrapie prion proteins in the endocytic pathway. J Biol Chem 1992;267:16188–99.
56. Safar J, Wang W, Padgett MP, et al. Molecular mass, biochemical composition, and physicochemical behavior of the infectious form of the scrapie precursor protein monomer. Proc Natl Acad Sci USA 1990;87:6373–7.
57. McKinley MP, Taraboulos A, Kenaga L, et al. Ultrastructural localization of scrapie prion proteins in cytoplasmic vesicles of infected cultured cells. Lab Invest 1991;65:622–30.
58. Cohen AS. Amyloidosis. In: Wilson JD, Braunwald E, Isselbacher KJ, et al., eds. Harrison's principles of internal medicine. New York: McGraw-Hill, 1991:1417–21.
59. Jarret JT, Lansbury PT. Seeding "one-dimensional crystallization" of amyloid: a pathogenetic mechanism in Alzheimer's disease and scrapie. Cell 1993; 73:1055–8.
60. Gabizon R, McKinley MP, Prusiner SB. Purified prion proteins and scrapie infectivity copartition into liposomes. Proc Natl Acad Sci USA 1987;84:4017–21.
61. Masters CL, Gajdusek DC. The spectrum of Creutzfeldt-Jakob disease and the virus-induced subacute spongiform encephalopathies. In: Smith WT, Cavanagh JB, eds. Recent advances in neuropathology. Edinburgh-London-Melbourne-New York: Churchill Livingstone, 1982:139–63.
62. Race RE, Fadness LH, Chesebro B. Characterization of scrapie infection in mouse neuroblastoma cells. J Gen Virol 1987;68:1391–9.
63. Forloni G, Angeretti N, Chisa R, et al. Neurotoxicity of a prion protein fragment. Nature 1993;362:543–6.
64. Glenner GG, Murphy MA. Amyloidosis of the nervous system. J Neurol Sci 1989;94:1–28.
65. Kretzschmar HA, Prusiner SB, Stowring LE, DeArmond SJ. Scrapie prion proteins are synthesized in neurons. Am J Pathol 1986;122:1–5.
66. Bueler H, Aguzzi A, Sailer A, et al. Mice devoid of PrP are resistant to scrapie. Cell 1993;73:1339–47.
67. Taraboulos A, Raeber AJ, Borchelt DR, Serban D, Prusiner SB. Synthesis and trafficking of prion proteins in cultured cells. Mol Biol Cell 1992;3:851–63.
68. Borchelt DR, Koliatsos VE, Guarnieri M, Pardo CA, Sisodia SS, Price DL. Rapid anterograde axonal transport of the cellular prion glycoprotein in the peripheral and central nervous systems. J Biol Chem 1994;269:14711–4.

69. Beck E, Daniel PM, Davey A, Gajdusek DC, Gibbs CJ Jr. The pathogenesis of spongiform encephalopathies: an ultrastructural study. Brain 1982;104:755–86.
70. Liberski PP, Yanagihara R, Gibbs CJ Jr, Gajdusek DC. Scrapie as a model for neuroaxonal dystrophy: ultrastructural studies. Exp Neurol 1989;106:133–41.
71. Ceroni M, Safar J, Pergami P, Camana C, Gibbs CJ Jr. Glycosilation and membrane interaction of normal and scrapie prion proteins and the role in scrapie infectivity. In: Nicolini M, Zatta PF, Corain B, eds. Alzheimer disease and related disorders. London: Pergamon Press, 1993:277–8.

Part V

Public Health Considerations of Human and Animal Spongiform Encephalopathies

26

Real and Theoretical Threats to Human Health Posed by the Epidemic of Bovine Spongiform Encephalopathy

RICHARD T. JOHNSON

This chapter discusses the question of whether or not food, pharmaceuticals, and cosmetics derived from animal tissues may pose a danger to human health. More specifically, can any of these products transmit Creutzfeldt-Jakob disease (CJD), or a variant thereof, to humans? The initial conclusions are straightforward: there is no solid evidence that they do, but there is theoretical evidence that they could. Because the question involves a form of human CJD, I will define the disease and the problems in diagnosis, review the evidence that CJD can be transmitted from human to human by different tissues and by different routes of inoculation, and note the absence of evidence of transmission from animal to man despite the formidable experience with intraspecies transmission of other spongiform encephalopathies among nonhuman species.

Clinical Diagnosis of CJD

CJD is a presenile dementia found throughout the world. The incidence is 0.5 to 1.0 cases per million population per year but may be more frequent in urban than rural areas (1). The mean age of onset is about 60 years. In its classical form the disease includes cognitive, motor, and sensory abnormalities. The cognitive deficits may begin insidiously, but progress from week to week and even day to day rendering the patient severely demented within a few months. Multifocal neurologic signs may be present at the outset or evolve after cognitive deficits. Motor deficits, visual impairment, abnormal movements, cerebral ataxia, and/or spinal cord signs with amyotrophy may be found. The most characteristic and consistent neurologic abnormality, however, is myoclonus, which usually increases in severity during the course of the disease from small choreic movements to symmetric body myoclonic jerks that are often stimulus sensitive. The ingravescent course with dissolution of intellect within 6 months of onset is characteristic. Over 80% of patients die within 12 months of onset, although a small group, particularly those with the spinal cord form of the disease, survive over 2 years (2, 3).

Diagnosis may be obscured under several circumstances. First, the onset of the disease may be abrupt with aphasia, hemiparesis, or other signs suggesting occlusive cerebrovascular disease. Progressive hemiparesis, movement disorder, cerebellar signs, or blindness may suggest other diagnoses (4). However, over time the rapid course of the dementia and the development of myoclonus usually clarify the diagnosis. In some cases, ataxia or other movement disorders may be sufficiently intense that myoclonus is obscured. In 5% to 10% of patients with more chronic disease and prominent spinal cord signs, the diagnosis is more frequently missed.

Clinical diagnosis is based on (a) the characteristic rapid course of dementia with development of myoclonus; (b) a spinal fluid examination that is normal except for possible modest protein elevation; (c) an abnormal EEG that may not be present initially but develops typical, triphasic sharp wave complexes that are characteristic but not pathognomonic of CJD (5); and (d) a brain biopsy that shows the pathognomonic spongiform changes with cytoplasmic vacuolarizations in neurons and glia.

Diagnosis may be missed in CJD, but greater awareness and familiarity with the classic disease among neurologists and even internists would, in general, make correct diagnosis more likely now than it was 20 years ago. On the other hand, problems have evolved that may interfere with the diagnosis: (a) Fewer biopsies are done in suspected cases both because of the physician's confidence that the clinical course and the EEG can secure the diagnosis and because of the reticence of neurosurgeons and pathologists to perform or process biopsies. (b) Fewer autopsies are done to detect unsuspected cases. Autopsy rates in the United States have fallen from 50% to 11%, so atypical cases remain undetected. (c) New health management systems now discourage consultations and may even block tertiary consultation, decreasing the likelihood of correct diagnosis.

CJD is not a reportable disease in North America. Reasonable estimates of the number of typical cases of CJD can be made from death certificates (6), and unusual temporal or geographic clustering of cases or occurrences of disease in children would have a good chance of detection by alert physicians. Many of the transmitted cases of CJD have been atypical, however, and the reliance on a passive reporting system and death certificates combined with diminished pathologic studies and expert consultation make us vulnerable to missing small outbreaks or to delayed recognition of a major outbreak.

Human to Human Transmission

CJD can be transmitted from human to subhuman primates and other laboratory animals, from human to human. The first possible human-to-human transmission was reported in 1974. CJD developed in a 55-year-old woman 18 months after receiving a corneal transplant from a man who died after a 2-month history of incoordination, memory loss, involuntary movements, and myoclonus. The recipient's fulminant disease was so dominated by abnormal movements that the diagnosis of CJD, although entertained, was not made until autopsy. Twenty years later, a subsequent suspected case of CJD related to a corneal transplant has been reported, so it does not appear to be a highly efficient mode of transmission. In contrast, 20 cases of transmission have occurred with dural grafts; transplantation directly into the brain appears to be an effective procedure to transmit disease. Injection of growth hormone and gonadotropin derived from human cadavers have now been blamed for over 60 cases of CJD. Over 10,000 children in the United States received pituitary human growth hormone between 1963 and 1985, but thus far fewer than 10 have developed CJD.

A strange aspect of the growth hormone transmissions has been the atypical presentation of CJD in recipients. Those young patients have presented with progressive cerebellar ataxia without cognitive dysfunction until later in the disease. The clinical syndrome resembles kuru (7), where conjunctival and subcutaneous exposure may have been the mode of transmission during ritual cannibalism (Table 26.1).

Transmission has not been associated with blood transfusions from donors who developed CJD. Nevertheless, the Food and Drug Administration (FDA) concluded that blood products given by a person who subsequently develops CJD should be withdrawn, but they excluded stable products such as albumin. A recent, incomplete report has appeared of CJD developing in a 60-year-old liver transplant recipient who received albumin derived from the blood of a donor who subsequently developed CJD (8). The scrutiny

TABLE 26.1. Transmission of spongiform encephalopathy from human to human.

Transmission mode	Example
Intracranial transplant or inoculation	Dural grafts Inadequately sterilized surgical instruments Stereotactic electrodes
Extracranial transplant	Corneal graft
Extraneural inoculation of neural tissue	Human growth hormone
Extraneural inoculation or oral exposure to neural and extraneural tissue	Ritual cannibalism (kuru)

previously directed at transfusions would argue that these occurrences were unrelated, but continued surveillance is needed.

Lack of Transmission from Nonhuman Species to Humans

Despite the transmission of spongiform encephalopathies from sheep or cattle to mink, from sheep to cows, and from bovine and/or ovine bone meal to a variety of exotic ruminants and ominously to the omnivorous house cat, there is no evidence of transmission from animals to humans. Many anecdotal reports of clustering of cases and possible common sources such as the eating of sheep's eyeballs, have been reported. Evaluations of the reports have shown probably occurrences of chance clustering or familial aggregates of CJD. The most convincing evidence against transmission to humans, particularly by the oral route, is the worldwide consumption of lamb and mutton; yet in scrapie-infested countries the rate of CJD appears to be the same as in countries that are scrapie-free.

Even though there is no evidence of transmission from animals to humans, the risk cannot be totally denied. The theoretical risk probably is increased by the source of the inoculum as well as by the route of inoculation. Animal brain would appear more dangerous than animal lymphatic organs or cells that, in turn, would compose a greater hazard than muscle. Route of inoculation is a second factor; intracerebral inoculation would be of greater risk than subcutaneous inoculation, which would be a greater risk than oral exposure. Both source and route of inoculation may, in part, determine dose. For example, on the assumption that no aspect of life is risk free, one might assume that gel caps made from bone of a scrapie- or BSE-infected animal taken orally would be less dangerous than a cosmetic containing infected rendered products that was put on cracked lips; this, in turn, would be less dangerous than the injection of a pharmaceutical derived from animal hematopoietic cells, which would be less dangerous than the intracerebral transplantation of dural grafts of animal origin. There is no evidence of transmission from animals to humans supported by extensive epidemiologic studies, yet there is no biologic reason to question the possibility of transmission from animals to man based on cross-species transmission in other species and the transmission of human disease to subhuman species.

The Broader Threat

The epidemic of BSE poses a theoretical threat of greater magnitude. It demonstrates not only the complexity of the food chain, but of the medical profession's lack of knowledge and the government's limited regulation of some aspects of the food chain. BSE, theoretically, could have become a human epidemic if the transmission material had been a supplement added to infant formula. Rather than simply denying such potential dangers, it is better that we develop a greater understanding of the nature of these agents and their modes of transmission, so that we can anticipate problems rather than respond in haste after they have become established (9).

References

1. Brown P, Cathala F. Creutzfeldt-Jakob disease in France: I. Retrospective study of the Paris area during the ten-year period 1968–1977. Ann Neurol 1979;5:189–92.
2. Harries-Jones R, Knight R, Will RG, Cousens S, Smith PG, Matthews WB. Creutzfeldt-Jakob disease in England and Wales, 1980–1984: a case-control study of potential risk factors. J Neurol Neurosurg Psychiatry 1988;51:1113–9.
3. Brown P, Gibbs CJ Jr, Rodgers-Johnson P, et al. Human spongiform encephalopathy: the National Institutes of Health series of 300 cases of experimentally transmitted disease. Ann Neurol 1994;35:513–29.
4. Bernoulli CC, Masters CL, Gajdusek DC, Gibbs CJ Jr, Harris JO. Early clinical features of Creutzfeldt-Jakob disease (subacute spongiform encephalopathy). In: Pruisner SB, Hadlow WJ, eds. Slow transmissible diseases of the nervous system. New York: Academic Press, 1979:229–51.
5. Burger LJ, Rowan AJ, Goldensohn ES. Creutzfeldt-Jakob disease. Arch Neurol 1972;26:428–33.
6. Davanipour Z, Smoak C, Bohr T, Sobel E, Liwnicz B, Chang S. Death certificates: an efficient source for ascertainment of Creutzfeldt-Jakob disease cases. Neuroepidemiology 1995;14:1–6.
7. Nau J-Y. CJD and albumin? Lancet 1995;345:9442.
8. Brown P. The decline and fall of Creutzfeldt-Jakob disease associated with human growth hormone therapy. Neurology 1988;38:1135–7.
9. Johnson RT. Emerging infections of the nervous system. J Neurol Sci 1994;124:3–14.

27

Incidence of Creutzfeldt-Jakob Disease in the European Community

R.G. WILL

Epidemiologic surveillance of Creutzfeldt-Jakob disease (CJD) was reinstituted in the United Kingdom in 1990 following a recommendation by the Southwood Committee (1). The aim of the surveillance program is to determine whether there is a change in the epidemiologic or clinicopathologic features of CJD as a result of transmission of bovine spongiform encephalopathy (BSE) to the human population, should this indeed occur. National surveillance programs for CJD have been established in France, Germany, Italy, and the Netherlands, and all these European projects now share common methodologies, assisted by a grant awarded through the BIOMED1 program. The importance of conducting epidemiologic research in CJD has been heightened by the occurrence of BSE in several member states of the European Community (EC) and an increased level of controversy about the perceived risk posed by BSE. The risk to the human population has been judged to be remote by a range of official bodies including the Southwood Committee, the United Kingdom Agricultural Select Committee, EC committees, and the World Health Organization (2). This consensus has been challenged, often on the basis of assertion rather than scientific evidence. An assessment of the risk posed by BSE must inevitably be based on an assessment of a body of scientific information including evidence from transmission studies, molecular biology, and protein chemistry. However, currently the only direct evidence on whether or not BSE has caused disease in the human population is from epidemiologic surveillance.

The incidence of CJD in the United Kingdom has increased from 0.09 (1964–1973) to 0.71 (1993–1994), with an incidence figure in 1992 of 0.87 representing the highest annual incidence since systematic surveillance of CJD was initiated in the United Kingdom in 1970 (3). This increase is in part related to changes in research methodology, the occurrence of iatrogenic cases after 1985, and the improved identification of familial/

genetic cases of CJD through DNA analysis (Table 27.1). The number of iatrogenic cases in the United Kingdom has increased in recent years due to the occurrence of CJD in human growth hormone (HGH) recipients ($n = 11$) and human dura mater recipients ($n = 5$). With the exception of the first case of CJD in an HGH recipient in the United Kingdom in 1985 (4) and one dura mater recipient in 1978 (5), all the iatrogenic cases have occurred since 1988 and the majority since 1990. Although there are only small numbers of iatrogenic cases, the identification of these patients has had a significant effect on the overall incidence because of the rarity of CJD.

Between 1980 and 1984 only one definite familial case of CJD was identified (6) in comparison to nine between 1990 and 1994, despite the use of similar mechanisms of case identification and diagnostic classification. The introduction in 1990 of screening DNA from incident cases for mutations of the prion protein (PrP) gene has resulted in improved identification of familial/genetic cases of CJD. Although a proportion of cases of mutation-associated CJD, notably those associated with codon 200 mutations, would have been identified as CJD without genetic analysis, other cases may have been missed using clinical criteria alone. For example, one referred case of suspect CJD was classified as "not CJD" following clinical and neuropathologic assessment, but was reclassified as fatal familial insomnia (FFI) following the discovery of a mutation at codon 178 of the PrP gene and review of the histology, which showed thalamic gliosis. Improved identification of familial cases of CJD influences the overall incidence figures in addition to the effect on incidence of the iatrogenic cases.

However, there has been an increase in the number of sporadic cases from 26 in 1990 to 44 in 1992 and 34 in 1993. This increase in the numbers of sporadic cases is the main reason for the increased overall incidence of CJD in the United Kingdom. One possible explanation for this increase is improved ascertainment of cases of CJD related to the extraordinary increase in public and professional awareness of the disease in the United

TABLE 27.1. Deaths from Creutzfeldt-Jakob disease by calendar year (1985–1993).

Year	Sporadic	Iatrogenic	Familial	GSS
1985	26	1	1	0
1986	26	0	0	0
1987	23	0	0	1
1988	21	1	1	0
1989	28	1	1	0
1990	26	5	0	0
1991	32	1	3	0
1992	44	2	4	1
1993	34	3	2	1
1994 (up to April)	7	1	0	0

GSS, Gerstmann-Sträussler-Scheinker syndrome.

TABLE 27.2. Sporadic CJD deaths from 1980–1993 under 75 years of age.

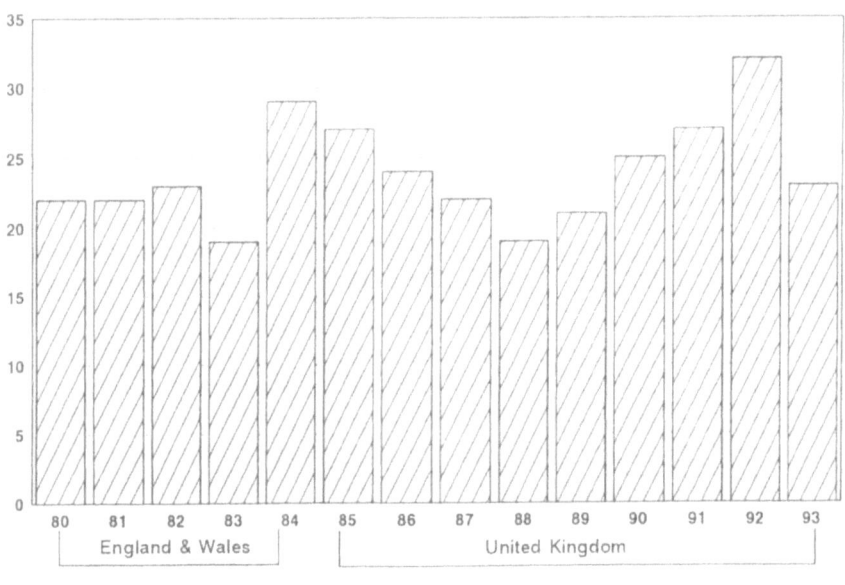

Kingdom. The possibility that cases of CJD might be missed in epidemiologic surveillance programs has been discussed in the past, (7) and there has been concern that the diagnosis of CJD may be missed in the elderly. Analysis of age-specific incidence rates of CJD demonstrates an increased incidence in patients over age 75 in the last 2 years in the United Kingdom. Furthermore, the total number of cases of CJD in patients younger than 75 has been relatively constant over the past 13 years (Table 27.2), while the numbers of cases of CJD in patients 75 and over has increased dramatically in the last 5 years (Table 27.3). The likeliest explanation for this change is improved diagnosis or referral of CJD in the elderly, and it is of note that the clinical features of CJD in the elderly are indistinguishable from those in younger patients and that such cases have been identified from throughout the United Kingdom.

If there were a new risk factor for human prion disease in the United Kingdom this might result in an increased overall incidence of human spongiform encephalopathy in relation to other countries. Through a grant from the European Union (EU), systematic national surveillance of CJD has been coordinated in five European countries, and these projects now share common methodologies including mechanism of case ascertainment and diagnostic criteria. In 1993 the incidence of CJD in participating countries was remarkably consistent, indicating no increased relative risk of CJD in the United Kingdom despite a marked variation in the incidence of BSE in Europe (8) (Table 27.4).

TABLE 27.3. Sporadic CJD deaths from 1980–1993 75 years of age and over.

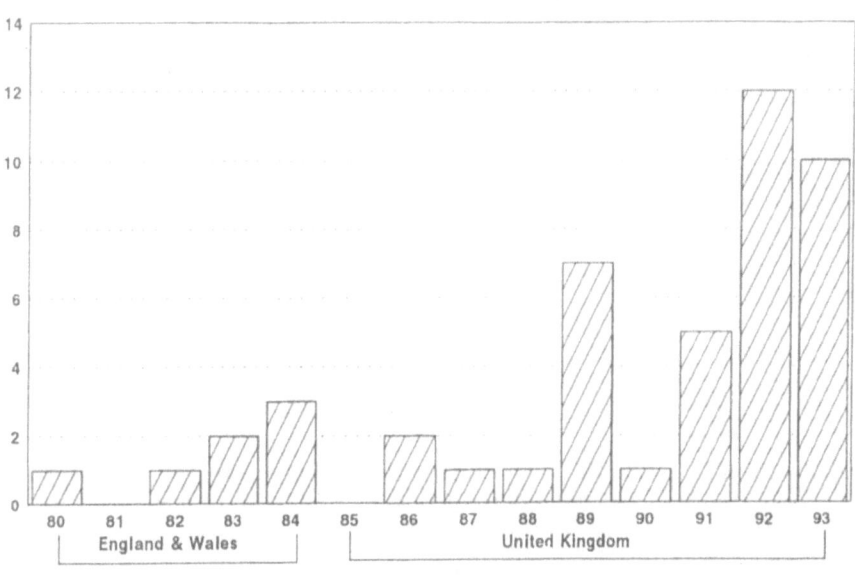

The incidence of CJD in a number of countries worldwide is listed in Table 27.5. The national incidence figures in this table are derived largely from published sources but some figures are currently unpublished and the methodologies of case ascertainment vary; for example, some are prospective studies, some retrospective, and some studies include only pathologically confirmed cases, while others use unverified death certificate data. Despite these caveats there are two apparent conclusions. First, the incidence of CJD in the more recent surveys is consistent worldwide, providing further evidence that there is no new or distinct risk factor for human prion diseases in the United Kingdom yet identifiable. The exception is the very low incidence figure for CJD in India (9), but this is likely to be an underestimate since the published study was not part of a systematic analysis of CJD and the geographic distribution of cases demonstrates that the majority of cases were identified from two centers, Bombay and Bangalore.

TABLE 27.4. Incidence of Creutzfeldt-Jakob disease in Europe in 1993.

	France	Germany	Italy	Netherlands	UK
Definite CJD	6	6	11	2	24
Probable CJD	22	13	20	8	8
Total	28	19	31	10	32
Incidence/million person-years	0.50	0.47[a]	0.54	0.68	0.56

[a] Extrapolated from figures available from June to December, 1993.

TABLE 27.5. Annual incidence of Creutzfeldt-Jakob disease.

Country	Period	Incidence (cases/million)
Australia	1979–1992	0.75
Chile	1955–1972	0.10
	1973–1977	0.31
	1978–1983	0.69
Czechoslovakia	1972–1986	0.66
France	1968–1977	0.34
	1978–1982	0.58
	1992–1994	0.91
Germany	1979–1990	0.31
	1993–1994	0.59
Iceland	1960–1990	0.27
India	1971–1990	0.002
Israel	1963–1987	0.91
Italy	1958–1971	0.05
	1972–1985	0.11
	1993–1994	0.51
Japan	1975–1977	0.45
Netherlands	1993–1994	0.92
New Zealand	1980–1989	0.88
Switzerland	1988–1990	0.80
United Kingdom	1964–1973	0.09
	1970–1979	0.31
	1980–1984	0.47
	1985–1989	0.46
	1990–1994	0.70
United States	1973–1977	0.26
	1986–1988	0.83

Moreover, the authors indicated that this was a pilot study and deliberately provided no analysis of prevalence or mortality.

Second, the incidence of CJD has risen with each survey in all countries in which serial surveillance of CJD has been undertaken. This is likely to reflect improving awareness and diagnosis of CJD and underlines the importance of not overinterpreting an apparent rise in the incidence of CJD within any individual country and the necessity for contemporary data in comparisons of the incidence of CJD between countries.

The poor correlation between the worldwide incidence of CJD and scrapie does not suggest that CJD is a zoonosis (10). Although Australia and New Zealand are scrapie free, CJD occurs with the expected incidence in these countries. The scrapie status of some other countries is difficult to establish because of uncertainties about the adequacy of veterinary surveillance. There is also tantalizing evidence that BSE might occur as a random event in any cattle population (11), and could be undetected because of the low frequency of such an occurrence. In one risk assessment, the hypothetical frequency of the spontaneous occurrence of BSE has been calculated at a maximum of 1 in 900,000 (12), which is similar to the incidence of CJD. If either sheep or cattle spongiform encephalopathy were occurring undetected at a very low rate, it could be hypothesized that such a zoonotic source of CJD might be undetected. To address this possibility, it is of interest to compare the incidence of CJD with the relative exposure of individuals to sheep or cattle as judged by the number of sheep or cattle per head of human population (Table 27.6).

Table 27.6 demonstrates no relationship between the incidence of CJD and the relative exposure of populations of various countries to either sheep or cattle. There is, for example, at least a 50-fold difference between the number of sheep or cattle per head of population in Japan and New Zealand but no significant difference in the incidence of CJD in these two countries. These data do not support the hypothesis that CJD is causally related to the occurrence of a rare, spontaneous spongiform encephalopathy in sheep or cattle.

In the search for an environmental source of infection, the geographic distribution of cases of CJD has been analyzed within individual countries in order to look for areas of high local incidence, implying an increased localized risk of infection. The geographic distribution of CJD in systematic surveys, however, has demonstrated no spatiotemporal clustering of cases (13), although in some surveys an increased incidence in urban areas has been described (14, 15). A study of the geographic distribution of CJD in the United Kingdom over the past 20 years strongly suggests that the

TABLE 27.6. Incidence of CJD and sheep/cattle population per country.

Country	CJD incidence	Cattle population[a] (1,000 head)	No./person	Sheep population[a] (1,000 head)	No./person
Australia	0.75	23,000	1.4	135,000	8
Chile	0.69	3,500	0.3	31,000	2.4
France	0.91	23,000	0.4	11,000	0.2
Germany	0.87	21,000	0.3	3,000	0.04
Iceland	0.27	70	0.3	800	3.2
Israel	0.91	300	0.07	250	0.06
Italy	0.49	9,000	0.2	9,000	0.2
Japan	0.45	4,500	0.04	10	~0.0
Netherlands	0.92	5,000	0.34	800	0.05
New Zealand	0.88	8,000	2.4	67,000	20.3
Switzerland	0.32	2,000	0.31	300	0.05
United Kingdom	0.70	12,500	0.22	22,000	0.39
United States	0.83	102,000	0.41	12,000	0.05

[a] Based on FAO data provided Dr. B. Hornlimann.

disease occurs randomly in space and time both before and after the occurrence of BSE (3). This evidence virtually precludes case-to-case transmission as an etiologic mechanism in sporadic CJD and also suggests that an environmental source of infection, should this exist, must be widely and uniformly distributed. Although not formally analyzed, the geographic distribution of CJD cases in other European countries shows a similar pattern.

Preliminary analysis has demonstrated that the age distribution of cases, sex ratio, and duration of illness in CJD in Europe is consistent in all participating countries. The proportion of cases of CJD in the United Kingdom presenting with a cerebellar syndrome is similar to the previous study of CJD in the United Kingdom and is not increasing (3). This type of clinical presentation is potentially of importance in relation to BSE as peripheral inoculation in human growth hormone recipients who develop CJD and in kuru results in a cerebellar presentation rather than the usual multifocal cerebral dysfunction in iatrogenic cases due to CNS inoculation and in sporadic cases. The implication is that a peripheral route of inoculation may be a major determinant to the clinical presentation in CJD. No case of sporadic CJD has been identified in the United Kingdom with a clinical syndrome similar to HGH recipients.

The greatest risk of transmission of BSE to the human population may potentially be through inoculation of tissues containing high titers of infectivity. The Southwood Committee identified occupational exposure as a potential area of risk, for example in abattoir workers or butchers through occupational cross-contamination. No individuals employed in these occupations since 1985 have been identified in the United Kingdom. However, three dairy farmers with CJD who had BSE affected animals in their herds have been identified in the United Kingdom in the past 3 years (16, 17). No mechanism of cross-contamination has been identified in these cases, and the clinical and pathologic features are consistent with sporadic CJD. Statistical analysis indicates that the likelihood of identifying three such cases since the reinstitution of surveillance in 1990 is low. Statistical analysis, however, is complicated by the range of potential denominators and the possibility of an apparently increased occupational risk occurring by chance. The case-control study of occupation in the United Kingdom has shown no evidence of an increased risk in relation to any specific occupation, including farmers, and comparison of the incidence of CJD in farmers in the European Community has shown no relative increase in the United Kingdom. CJD has previously been described in a neurosurgeon (18), a neuropathologist (19), two laboratory technicians (20, 21) and an individual who handled dura mater (22). Although these cases are disquieting, no conclusion can be reached on a causative link in any of these case reports. The systematic study of CJD in an increasing number of countries will inevitably result in the identification of a broadening range of occupations in CJD patients, including occupations that are theoretically associated with an increased occupational risk of CJD. Table 27.7 lists occupations through-

TABLE 27.7. Lifetime occupations in 50 cases of definite and probable sporadic CJD.

Accountant	Estate agent	Radio wirer
Accountant	Factory packer	RAF member
Air hostess	Factory worker	Receptionist
Air raid warden	Factory worker	Sales
Ammunition worker	Farmer	Sales representative
Apprentice optician	Fat melting business	School cook
Army corporal	worker	School dinner lady
Army officer	Fireman	Schoolteacher
Baker	Fireman	Secretary
Bank clerk	Governess	Shop assistant
Bellboy	GP	Shop assistant
Biscuit factory worker	Greengrocer	Shop assistant
Bomb disposal worker	Grocer	Shop assistant
Bomb disposal (Army)	Guest house worker	Shop assistant
Bookkeeper	Home help worker	Shop assistant
Builder	Hotel receptionist	Bridal shop assistant
Bus driver	Housewife	Shop assistant (conf/pharm)
Butcher	Housewife	Show card assembler
Canteen worker	Housewife	Sweet packer
Caretaker	Journalist	Sweet packer
Carpenter	Laborer	Tank components worker
Cashier	Laundry assistant	Telecomm traffic
Child minder	Lecturer	superintendent
Children/elderly person	Machine operator	Telephonist
home	Machine tool engineer	Time keeper
Civil servant	Machinist	Toy maker
Civil servant	Managing director	Train driver
Cleaner	Managing director	Trainee nurse
Cleaner and dyer	Medical secretary	Transport depot manager
Clerical worker	Milkman	Tutor
Clerical worker	Milkman	Typist
Clerical worker	Miner	Typist
Clerk	Jehovah's Witness minister	Upholsterer
Clothing industry worker	Navy (apprentice)	Voluntary services worker
Colonial service worker	Packer	Waitress
Contractor (metal storage)	Packer for dairy firm	Warehouseman
Contracts manager	Pattern technician	Water board worker
Cook	Post office clerk	Weaver
Cook (ham)	Priest	Weaver
Crofter	Private secretary	Welder
Diplomat	Process worker	Window cleaner
Domestic	(metalworks)	Wire Netting manufacturer
Engineer	Processing worker	Wood yard worker
Engineer	(vegetables)	

out life of 50 recent cases of CJD in the United Kingdom illustrating the difficulty of assessing a potential causative occupational link in any individual patient. It is also difficult to obtain meaningful information on relative risk in specific occupational groups because of the rarity of CJD and the small subset populations (23). This is illustrated in Table 27.8 which, although showing an increased apparent risk of CJD in farmers and paramedical staff, also shows a higher incidence in professional drivers and vicars.

Analysis of the dietary case-control study of CJD in the United Kingdom has demonstrated an apparent increase in the risk of developing CJD in relation to consumption of a variety of food products, most notably an increased relative risk of developing CJD of 13.3 in relation to consumption of veal at least once a year (3) and an increased relative risk of 9.3 from the consumption of venison at least once a year. These odds ratios are based on small numbers and are therefore susceptible to change from year to year. An apparently increased risk of developing CJD in relation to the consumption of "black puddings" identified in 1992 was no longer apparent in the analysis of 1993. Furthermore, the level of media coverage with regard to BSE has resulted in the great majority of relatives of patients being aware of the hypotheses being tested in the case control study. This raises the possibility of recall bias, and analysis of the frequency of consumption of veal and venison in referred cases with suspect CJD subsequently classified as "possible" cases or "not CJD" showed a similar frequency of exposure to these particular foodstuffs. This provides good evidence that bias is a major confounding factor in the dietary case-control study.

Molecular biologic analysis has been carried out in a proportion of all cases of CJD identified since 1990 in the United Kingdom and in participating European countries. A mutation of the PrP gene has been identified in approximately 12.5% of cases of definite or probable CJD in the United Kingdom, with a similar proportion of "genetic" cases in the aggregated results from three participating countries in Europe. This represents an

TABLE 27.8. Occupation at diagnosis in sporadic cases of CJD in the United Kingdom (1990–1995).[a]

Occupation	Number of cases	Incidence/million[b]
Manager	14	0.3
Secretary/Clerical	10	0.5
Shop Workers	7	1.7
Medical/Paramedical	6	5.7
Farmer	3	4.1
Teacher	3	0.7
Driver	3	8.2
Vicar	2	11.8
Journalist	1	2.5
Abbatoir Worker/Butcher/Vet	0	0

[a] $n = 169$.

[b] Incidence figures do not take into account proportion of total population in employment (46%) or proportion of patients retired from employment (43%).

approximate doubling of the frequency of familial/genetic CJD in relation to the frequency of familial cases identified from previous systematic national surveys (7). This is likely to be related to the atypical clinical features that often occur in familial CJD and also to the identification of individual cases with no apparent family history of a neurodegenerative disorder. Analysis of the distribution of genotypes at codon 129 of the prion protein gene has confirmed that approximately 80% of cases of sporadic CJD are methionine homozygotes.

Evidence from the combined studies of CJD in the European Community has demonstrated no significant difference in the incidence of CJD in countries with or without BSE. There is no conclusive evidence of any change in CJD in the United Kingdom attributable to BSE, although it is clearly important to continue the study with particular reference to occupation and diet. The potentially long incubation period in this group of diseases, however, indicates that these findings should be interpreted with caution. If, for example, exposure to cattle or cattle products carries an increased risk of CJD since 1985, it may be too soon to see evidence of that increased risk.

References

1. Southwood Committee. Report of the Working Party on Bovine Spongiform Encephalopathy. Department of Health and Ministry of Agriculture, Fisheries and Food. ISBN 1989;185197 405 9.
2. World Health Organization. Report of a WHO consultation on public health issues related to animal and human spongiform encephalopathies. Geneva, 12–14 November, 1991.
3. Creutzfeldt-Jakob Disease Surveillance in the United Kingdom—Third Annual Report. Edinburgh: CJD Surveillance Unit, 1994.
4. Powell-Jackson J, Weller RO, Kennedy P, Preece MA, Whitcombe EM, Newsom-Davis J. Creutzfeldt-Jakob disease after administration of human growth hormone. Lancet 1985;2:244–6.
5. Esmonde TFG, Lueck CJ, Symon L, Duchen LW, Will RG. Creutzfeldt-Jakob disease and lyophilised dura mater grafts: report of two cases. J Neurol Neurosurg Psychiatry 1993;56:999–1000.
6. Harries-Jones R, Knight R, Will RG, Cousens S, Smith PG, Matthews WB. Creutzfeldt-Jakob disease in England and Wales, 1980–1984: a case-control study of potential risk factors. J Neurol Neurosurg Psychiatry 1988;51:1113–9.
7. Will RG, Matthews WB, Smith PG, Hudson C. A retrospective study of Creutzfeldt-Jakob disease in England and Wales 1970–1979. II: Epidemiology. J Neurol Neurosurg Psychiatry 1986;49:749–55.
8. Alperovitch A, Brown P, Weber T, Pocchiari M, Hofman A, Will RG. Incidence of Creutzfeldt-Jakob disease in Europe 1993. Lancet 1994;343:918.
9. Satishchandra P, Shankar SK. Creutzfeldt-Jakob disease in India (1971–1990). Neuroepidemiology 1991;10:27–32.
10. Will RG. Is there a potential risk of transmission of BSE to the human population and how may this be assessed? In: Bradley R, Savey M, Marchant B, eds.

Sub-acute spongiform encephalopathies. Dordrecht: Kluwer Academic Publishers for the EEC, 1991:179–86.

11. Marsh RF. Transmissible mink encephalopathy. In: Prusiner SB, Collinge J, Powell J. Anderton B, eds. Prion diseases of humans and animals. Proceedings of Prion Diseases of Humans and Animals, London 2–4 September, 1992. Chichester: Ellis Horwood, 1992:300–7.

12. Robinson MM. Transmissible encephalopathy research in the United States. Transmissible spongiform encephalopathies. Proceedings of a consultation on BSE with the Scientific Veterinary Committee of the European Communities held in Brussels, 14–15 September 1993, Brussels: European Commission, 1993:261–9.

13. Cousens SN, Harries-Jones R, Knight R, Will RG, Smith PG, Matthews WB. Geographical distribution of cases of Creutzfeldt-Jakob disease in England and Wales 1970–84. J Neurol Neurosurg Psychiatry 1990;53:459–65.

14. Brown P, Cathala F, Sadowsky D. Correlation between population density and the frequency of Creutzfeldt-Jakob disease in France. J Neurol Sci 1983;60: 169–76.

15. Galvez S, Masters C, Gajdusek DC. Descriptive epidemiology of Creutzfeldt-Jakob disease in Chile. Arch Neurol 1980;37:11–4.

16. Sawcer SJ, Yuill GM, Esmonde TFG, et al. Creutzfeldt-Jakob disease in an individual occupationally exposed to BSE. Lancet 1993;341:642.

17. Davies PTG, Jahfar S, Ferguson IT. Creutzfeldt-Jakob disease in individual occupationally exposed to BSE. Lancet 1993;342:680.

18. Schoene WC, Masters CL, Gibbs CJ Jr, et al. Transmissible spongiform encephalopathy (Creutzfeldt-Jakob disease). Atypical clinical and pathological findings. Arch Neurol 1981;38 (8):473–7.

19. Gorman DG, Benson F, Vogel DG,Vinters HV. Creutzfeldt-Jakob disease in a pathologist. Neurology 1992;42:463.

20. Miller DC. Creutzfeldt-Jakob disease in histopathology technicians. N Engl J Med 1988;318:853–4.

21. Sitwell L, Lach B, Atack E, Atack D. Creutzfeldt-Jakob disease in histopathology technicians. N Engl J Med 1988;318:854.

22. Weber T, Tumani H, Holdorff B, Collinge J, Palmer M, Kretzschmar HA, Felgenhauer K. Transmission of Creutzfeldt-Jakob disease by handling of dura mater. Lancet 1993;341:123–4.

23. Brown P, Cathala F, Raubertas RF, Gajdusek DC, Castaigne P. The epidemiology of Creutzfeldt-Jakob disease: conclusion of a 15-year investigation in France and review of the world literature. Neurology 1987;37:895–904.

28

Problems in the Evaluation of Theoretical Risks for Humans to Become Infected with BSE-Contaminated Bovine-Derived Pharmaceutical Products

Maurizio Pocchiari

The safety of bovine-derived pharmaceutical products with respect to bovine spongiform encephalopathy (BSE) has been the topic of several scientific meetings in the last few years (1–13), and although a consensus on the factors that must be considered for their evaluations has been reached, it is still difficult for manufacturers and authorities to defermine an objective estimate of the potential level of BSE contamination for each drug or to allocate precise figures to each risk factor. This chapter weighs each risk factor in order to facilitate uniform application by competent authorities. However, this chapter does not give a magic formula with which to decide whether a product should be kept on, or removed from, the market. The decision regarding the usefulness of a medicinal product should take into consideration not only the estimated potential risk but also, and most important, its inherent benefit.

As stated above, the risk factors of a product concerning BSE have been identified, and can be summarized as follows: the source of animals, the tissue taken, the production process, the route of administration, the total weight of tissue used to manufacture a batch of product, and the proportion of the batch used in the maximum therapeutic regime. Evidently, apart from the last two items which are easily and precisely measurable, the allocation of figures for the other factors is debatable and depends on the available scientific and epidemiologic evidence, which is subject to modification as a result of new data.

Source of Animals

It is easily understandable that the highest assurance of safety for a medicinal product comes from sourcing from healthy, BSE-free animals. Both the European Economic Community (EEC) (8) and the World Health Organization (WHO) (9) guidelines on the safety of medicinal products with respect to BSE strongly recommend sourcing from countries where there is no reported BSE (or from closed herds), and where animals are not fed by ruminant derived protein. Sourcing from infected animals is unacceptable. It is therefore of great importance to produce reliable documentation on sourcing. This should include an active surveillance program able to recognize clinically legitimate suspicious cases directly on farms or in slaughterhouses. Article 3.2.13.2.(1) of the Office of International des Epizooties (OIE) International Animal Health Code concludes by saying that "in the absence of surveillance data, the status of a country must be considered as unknown." This implies that sourcing from such countries should be interdicted and that medicinal products derived from tissues taken from animals of uncertain origin should not receive the approval from health authorities. Thus, there would be no estimated figure when the sourcing is from countries where the BSE status is unknown because, if this occurs, the safety of the medicinal product cannot be guaranteed regardless of the tissue taken, the validation studies, and the route of drug administration.

On the other hand, even when sourcing is from BSE-free countries or closed herds, the assumption that random and sporadic cases of BSE do not occur cannot be completely ruled out (see Chapter 8).

This unlikely event may be estimated to occur at the rate of less than 1 case in a million animals. This figure is computed by considering that the identification of a few cases of BSE in a cattle population of several million has been proven to be feasible in countries with an active surveillance system (14, 15). Thus, the incidence of BSE in countries or closed herds with a similar active surveillance system, but where no cases have occurred, can be estimated to be, in the worst possible scenario (i.e., not taking into consideration that the animals are not fed ruminant-derived protein), of one case per million animals. The estimated correction factor is therefore 10^{-6}.

Tissue Taken

The level of infectivity in different tissues, organs, excretions, or body fluids of BSE-affected animals has given, so far, unexpected results. In fact, only the brain and the spinal cord appear to have a measurable infectivity titer [measured as the lethal dose that kills 50% (LD_{50}) of injected mice] similar to that reported for natural scrapie in sheep, while all other tissues tested so far had an infectivity titer under the level of detection in the mouse system

(16, 17). However, to keep a more conservative and restrictive view regarding the distribution of BSE infectivity in different tissues, it is wise to maintain the EEC-proposed classification of tissues into four categories (8): category I consists of tissue with high infectivity (brain and spinal cord); category II includes tissues with medium infectivity (the lymphoreticular organs, the upper part of the intestinal tract, and the tonsils); category III includes all those tissues with a minimal level of infectivity; and category IV includes those in which no infectivity has ever been measurable (8). Precautions must be taken to avoid contamination of low-risk tissue (e.g., lung) with high-risk tissue (e.g., brain, lymph nodes, spleen) during their removal. This division is based on the amount of infectivity measured in natural scrapie of sheep and goats (18, 19). Thus, the estimated levels of BSE infectivity for categories I, II, III, and IV is 10^8, 10^6, 10^4, and 10^2 LD_{50} per gram of tissue or milliliter of excretion or body fluids, respectively. For tissues that fall into category IV (no measurable infectivity), the value of 10^2 LD_{50} per gram of tissue is estimated considering that the limit of detectability of the bioassay system in mice is 1 LD_{50} per inoculum, i.e., 0.03 mL of a 1% tissue homogenate. Although the level of infectivity in tissues of BSE-affected animals may not be the same as those mentioned above, for the purpose of risk assessment this classification is, so far, maintained.

Production Process

Each step in the extraction and purification procedure of biologic products needs to be carefully evaluated to verify whether it removes and/or inactivates unconventional agents. In some instances it is even possible to introduce some extra steps (ultrafiltration, chemical treatments) into the production that are known to increase the safety of the product from any possible risk of slow virus contamination (20–26). Although an approximate estimate can be made by simply examining the method of production, it is always worthwhile to validate the entire process using an experimental procedure (27). Such a study consists of introducing a known amount of the infectious agent at the beginning of the procedure to validate and measure the residual infectivity at the end of the process. The infectious agent is usually added as a clarified suspension of a 10% to 20% homogenate of a pool of brains taken from clinically affected animals. A variety of BSE/scrapie/CJD rodent-adapted strains can be used for these experiments, provided they are well characterized in terms of incubation periods and brain lesion profiles (28).

Two distinctive experiments are usually performed to prove the safety of the product (27). The first (experiment 1) consists of spiking the raw material, running the procedure using a scaled-down manufacturing process, and measuring the residual infectivity by injecting intracerebrally (i.c.) the entire amount of the end product into weanling animals. The purpose of experiment 1 is to prove the absence, rather than the presence, of residual infectivity. The absence of virus can only be proven by inoculating the entire sample under examination, rather than an aliquot as is usually done in virus titration studies (29). The limit of this procedure is the relatively low viral clearance measurable by this assay, which is about 10^8 LD_{50}. This is the maximum amount available in the volume (0.05 mL in hamster) of the i.c. inoculum.

The aim of experiment 2 is to overcome the limit described above, demonstrating the maximum viral clearance factor of the extraction and purification procedure of the bovine-derived products. This is accomplished by measuring the residual infectivity of those stages of the process believed to be effective in the removal/inactivation of unconventional slow virus(es) and then summating the amount of virus removed at each step. The study is designed so that the purification procedure is divided into three or four steps. The virus is added at the beginning of each step as described above, the step is performed and the residual infectivity is then measured by end-point dilution (30). The criticism of experiment 2 is that each step might remove/inactivate less resistant viral particles, leaving the rest of the viral population unaffected (31, 32).

Several alternatives may occur:

1. Both experiments are performed, none of the animals injected with undiluted material (with respect to the initial spike) of the end product of experiment 1 have developed the disease, and the value obtained by summing the amount of infectivity removed at each step of the process (experiment 2) is greater than 10^8 LD_{50} units. In this case, the correct estimate of the removal and/or inactivation of the BSE agent by the extraction/purification procedures is somewhere between 10^{-8} LD_{50} (experiment 1, see above) and the value obtained in experiment 2. However, to keep this estimate within a safety margin, I suggest using the most restrictive figure of 10^{-8}.

2. As in 1, both experiments are performed, but half or fewer than half of the animals injected with the undiluted end product of experiment 1 have developed the disease, indicating that 1 or less than 1 LD_{50} of virus infectivity is still present in the inoculum. This denotes that at least 10^7 LD_{50} of viral infectivity has been removed. However, the safety figure of 10^{-6} is suggested.

3. As in 1 or 2, but the value obtained by summing the amount of infectivity removed at each step of the process (experiment 2) is equal to or less than 10^8 LD_{50} units. In this unlikely, but theoretically possible, event, I suggest using as an estimated figure the best single value (but no greater than 10^6) obtained in experiment 2. Assuming, for example, that the entire process has been divided into three major stages that had removed, respectively, 2, 3, and 2 logs of infectivity, then the estimated figure would be 10^{-3}.

4. The same rule for computing the estimated figure described in 3 is applied when both experiments are performed, but more than 50% of the animals injected with the undiluted end product of experiment 1 develop the disease or when experiment 2 alone has been performed.

5. If experiment 1 alone has been done, the most restrictive figures of 10^{-6} and 10^{-4} are applied, respectively, when none of the animals and 50% or fewer animals injected with undiluted material with respect to the initial spike develop the disease. Finally, there are insufficient data to estimate a correct figure when more than 50% of the animals develop the disease.

Route of Administration

The notion that some routes of inoculation are much more efficient in producing the disease than others is a very well known and generally accepted phenomenon. From experimental studies in rodents, it has been shown that the i.c. route is about 400 to 40,000 times more efficient (depending on the scrapie strain) than the intraperitoneal (i.p.) route (33, 34), about 25,000 times more efficient than the subcutaneous route (33), and 10^5 to 10^9 times more effective than the oral route (35, 36). However, the success of oral transmission of BSE in mice (37) and the findings that the i.c. and i.p. routes have the same efficiency in establishing the infection of the BSE agent in mice (38) call for an increased caution in the computation of the estimated figure. Thus, I propose a correction factor of 10^{-1} for the intramuscular, ophthalmic, and rectal routes, and 10^{-2} for the subcutaneous and oral routes. No correction factors are applied for the i.c. and the intravenous routes of drug administration.

Amount of Tissue and Maximum Therapeutic Regime

These are precise figures and need no further comment. The ratio between these two values gives the measure of the concentration or dilution of the pharmaceutical product with respect to the original amount of raw tissue.

Conclusion

The overall estimate of the amount of BSE infectivity that a patient will receive during a maximum therapeutic regime with a given medicinal product is obtained by summing the exponential values of each factor (Table 28.1). The value obtained from this analysis indicates the LD_{50} of BSE infectivity given during therapy. A patient receiving 1 LD_{50} has, therefore, a 50% chance of being infected with the BSE agent. It follows that the administration of 10^{-1} (0.1) LD_{50} gives 5% odds, that of 10^{-2} (0.01) LD_{50}

TABLE 28.1. Risk factors related to BSE and their estimated figures.

Risk factors	Figures
Sourcing	
Status of BSE is known	10^{-6}
Status of BSE is unknown	Not applicable; product should not be approved
Tissue	
Cat. I	10^{+8}
Cat. II	10^{+6}
Cat. III	10^{+4}
Cat. IV	10^{+2}
Amount of tissue (kg)	$kg \times 10^{+3}$
Production process	10^{-n*}
Route	
Intravenous, intrarachis	10^{0}
Intramuscular, ophthalmic, rectal	10^{-1}
Subcutaneous, oral	10^{-2}
Therapeutic regime	10^{t**}

*See text for computing the correct estimated figure.
**This figure is computed as follows: number of maximum doses of drug given to the patient divided by the number of doses obtained per batch of product.

0.5%, and so on. Because sporadic CJD has a worldwide incidence of about 1 case per million people per year (39, 40), then the maximum overall 'natural' risk of developing CJD during the entire lifetime for the general population is somewhere between 1 in 10,000 (1 in 1,000,000 per year $\times$ 100 years) and 1 in 100,000 (1 in 1,000,000 per year $\times$ 10 years). Thus, supposing that the BSE agent is as pathogenic for humans as it is for cattle, a patient receiving an estimated infectious dose of 10^{-5} LD_{50} during his treatment with a bovine-derived medicinal product will have roughly doubled his probability of developing CJD compared with the general population. However, if the infectious dose is 10- (10^{-6}) or 100- (10^{-7}) fold lower, the patient increases his risk by only one tenth or one hundredth.

In conclusion, a medicinal product with an overall figure equal to or less than 10^{-6} LD_{50} per therapeutic regime may be considered at low risk for humans concerning BSE. However, this value should not be considered per se a passe-partout through which every product with an estimated figure within that limit is accepted for human therapy and those falling out are rejected.

I believe that everyone is willing to accept a drug with its own intrinsic risk, provided, however, that its benefit is clearly discernible.

Acknowledgments. I thank Prof. Giuseppe Vicari and Dr. Anna Ladogana for their helpful discussion and critical reading of the manuscript, and Ms. Deborah Wool for editorial assistance.

References

1. Bovine Spongiform Encephalopathies (BSE) Roundtable. Bethesda, MD: National Institutes of Health, 1989.
2. BSE and related agents: risk factors of biological products. An informal consultation, 1990, Annecy, France.
3. International meeting. Virological aspects of the safety of biological products, 1990, London. In: International Association of Biological Standardization, ed. Developments in biological standards, vol 75. Basel: Karger, 1991.
4. BSE Seminar, 1990, Brussels (Belgium). In: Bradley R, Savey M, Marchant B, eds. Sub-acute spongiform encephalopathies. Dordrecht, The Netherlands: Kluwer Academic Publishers, 1991.
5. Mad cow disease (BSE)—danger by food and drugs? 1991, Bonn, Germany.
6. Roundtable/workshop on public health issues associated with bovine spongiform encephalopathy. Bethesda, MD: National Institutes of Health, 1991.
7. Transmissible spongiform encephalopathies: their impact on the safety of medicinal products. Lagen, Germany: Paul-Ehrlich-Institut, 1991.
8. Committee for Proprietary Medicinal Products. Ad hoc working party on biotechnology/pharmacy on safety of medicines: EEC regulatory document. Note for guidance. Guidelines for minimizing the risk of transmission of agents causing spongiform encephalopathies via medicinal products, 1991, Brussels, Belgium. Biologicals 1992;20:155-8.
9. WHO. Consultation on public health issues, related to animal and human spongiform encephalopathies, 1991, Geneva: WHO Bull 1992;70:183-90.
10. International Meeting on Transmissible Spongiform Encephalopathies. Impact on animal and human health, 1992, Heidelberg, Germany. In: International Association of Biological Standardization, ed. Developments in biological standards, vol 80. Basel: Karger, 1993.
11. Bovine spongiform encephalopathy—where are we now? European Commission Meeting of the Scientific Veterinary Committee with Scientific Experts, 1993, Brussels, Belgium. In: Bradley R, Marchant B, eds. Transmissible spongiform encephalopathies. Brussels: Working Document for the European Commission, Ref.: F.II.3-JC/0003, 1994.
12. Symposium, "Transmissible spongiform encephalopathies," 1993, Berlin.
13. VI International Workshop on Bovine Spongiform Encephalopathy "The BSE Dilemma," 1995, Williamsburg, VA.
14. Savey M, Coudert M, Jobert JL, Fontaine JJ, Belli P. Bovine spongiform encephalopathy (BSE) risk assessment and surveillance in France. In: Bradley R, Marchant B, eds. Transmissible spongiform encephalopathies. Brussels: Working Document for the European Commission, Ref.: F.II.3-JC/0003, 1994:57-67.
15. Guarda F, Castiglione F, Agrimi U, Cardone F, Caracappa S, Pocchiari M. Bovine spongiform encephalopathy in Italy. Eur J Vet Pathol 1995;1:71-2.
16. Fraser H, Bruce ME, Chree A, McConnell I, Wells GAH. Transmission of bovine spongiform and scrapie to mice. J Gen Virol 1992;73:1891-7.
17. Fraser H, Foster JD. Transmission to mice, sheep and goats and bioassay of bovine tissues. In: Bradley R, Marchant B, eds. Transmissible spongiform encephalopathies. Brussels: Working Document for the European Commission, Ref.: F.II.3-JC/0003, 1994:145-59.

18. Hadlow WJ, Kennedy RC, Race RE, Eklund CM. Virologic and neuro-histologic findings in dairy goats affected with natural scrapie. Vet Pathol 1980; 17:187–99.

19. Hadlow WJ, Kennedy RC, Race RE. Natural infection of Suffolk sheep with scrapie virus. J Infect Dis 1982;146:657–64.

20. Taylor DM, Dickinson AG, Fraser H, Robertson PA, Salcinski PR, Lowry PJ. Preparation of growth hormone free from contamination with unconventional viruses. Lancet 1985;2:260–2.

21. Pocchiari M, Macchi G, Peano S, Conz A. Can potential hazard of Creutzfeldt-Jakob disease infectivity be reduced in the production of human growth hormone? Inactivation experiments with the 263K strain of scrapie. Arch Virol 1988;98:131–5.

22. Diringer H, Braig HR. Infectivity of unconventional viruses in dura mater. Lancet 1989;1:439–40.

23. Dormont D, Deslys JP, Bouissin F, et al. Inactivation of unconventional slow viruses and HIV in human extractive biological products used in human therapy. Biotechnol Plasma Proteins 1989;175:355–64.

24. Pocchiari M, Peano S, Conz A, et al. Combination ultrafiltration and 6 M urea treatment of human growth hormone effectively minimizes risk from potential Creutzfeldt-Jakob disease virus contamination. Horm Res 1991;35:161–6.

25. Di Martino A, Safar J, Ceroni M, Gibbs CJ Jr. Purification of non-infectious ganglioside preparations from scrapie-infected brain tissue. Arch Virol 1992;124:111–21.

26. Di Martino A, Safar J, Gibbs CJ Jr. The consistent use of organic solvents for purification of phospholipids from brain tissue effectively removes scrapie infectivity. Biologicals 1994;22:221–5.

27. Pocchiari M. Methodological aspects of the validation of purification procedures of human/animal derived products to remove unconventional slow viruses. Dev Biol Stand 1991;75:87–95.

28. Fraser H. The pathology of natural and experimental scrapie. In: Kimberlin RH, ed. Slow virus disease of animals and man. Amsterdam: North-Holland, 1976:267–305.

29. Brown P. Iatrogenic Creutzfeldt-Jakob disease. Aust NZ J Med 1990;20:633–5.

30. Reed LJ, Muench H. A simple method of estimating 50% endpoints. Am J Hyg 1938;27:493–7.

31. Rohwer RG. Virus-like sensitivity of the scrapie agent to heat inactivation. Science 1984;223:600–2.

32. Rohwer RG. Scrapie infectious agent is virus-like in size and susceptibility to inactivation. Nature 1984;308:658–62.

33. Kimberlin RH, Walker CA. Pathogenesis of experimental scrapie. In: Bock G, Marsh J, eds. Novel infectious agents and the central nervous system. Ciba Foundation Symposium 135. Chichester, UK: John Wiley & Sons, 1988:37–54.

34. Pocchiari M, Salvatore M, Ladogana A, et al. Experimental drug treatment of scrapie: a pathogenetic basis for rationale therapeutics. Eur J Epidemiol 1991;7:556–61.

35. Prusiner SB, Cochran SP, Alpers MP. Transmission of scrapie in hamsters. J Infect Dis 1985;152:971–8.

36. Kimberlin RH, Walker CA. Pathogenesis of scrapie in mice after intragastric infection. Virus Res 1989;12:213–20.

37. Barlow RM, Middleton DJ. Dietary transmission of bovine spongiform encephalopathy to mice. Vet Rec 1990;126:111–2.
38. Taylor DM, Fraser H, McConnell I, et al. Decontamination studies with the agents of bovine spongiform encephalopathy and scrapie. Arch Virol 1994; 139:313–26.
39. Brown P, Cathala F, Rauberats RF, Gajdusek DC, Castaigne P. The epidemiology of Creutzfeldt-Jakob disease: conclusion of a 15-year investigation in France and review of the world literature. Neurology 1987;37:895–904.
40. Alperovitch A, Brown P, Weber T, Pocchiari M, Hofman A, Will R. Incidence of Creutzfeldt-Jakob disease in Europe in 1993. Lancet 1994;343:918.

29

Evaluation of BSE Risk Factors Among European Countries

BEAT HORNLIMANN, DAGMAR HEIM, AND CHRISTIAN GRIOT

Bovine spongiform encephalopathy (BSE), a subacute degenerative disease affecting the central nervous system of cattle, belongs to a group of related diseases known as the transmissible spongiform encephalopathies (TSEs). They are caused by poorly characterized scrapie-like agents (prions) that produce spongiform changes in the brain (1). Sheep scrapie is thought to be the origin of BSE, which was first diagnosed in the United Kingdom in 1986. The disease was confirmed a few years later in domestic cattle in France, Ireland, Portugal, and Switzerland (2), as well as in cattle exported from the United Kingdom to Denmark, Canada, the Falkland Islands, Germany, Italy, Oman, Portugal, and Ireland. Surveillance of BSE and/or scrapie is carried out in some countries. However, there is no test to detect the disease in live animals and the screening for the disease is rather difficult because it is based on the detection of clinically suspect cases (3). An additional approach to ascertain if BSE risk factors are likely to be present in a country is to conduct a qualitative or quantitative risk assessment. This chapter evaluates qualitatively some of the BSE risk factors in Europe.

Material and Methods

Questions

For all European countries, the following questions were asked:

1. What was the cattle feeding practice concerning the use of animal waste-derived by-products such as meat and bone meal (MBM) and meat meal (MM)? Was it permitted and until when?
2. What was the quantity of MBM and MM imported from the United Kingdom through direct routes from 1985 to 1989?

3. What was the number of live cattle imported from the United Kingdom through direct routes to other European countries beginning in 1980?
4. What was the ratio of the sheep population versus the cattle population from 1985 to 1993?
5. What was the scrapie status since 1980: (a) endemic, (b) sporadic, (c) rare, or (d) no occurrence?

The main data sources were Food and Agriculture Organization of the United Nations (FAO) (4), Office International des Epizooties (OIE) (5, 6) and FAO-OIE–World Health Organization (WHO) (7) publications. Additional data were obtained from proceedings of the European Union's scientific BSE committee (8) and from national veterinary authorities (written communications) such as the British export statistics (9).

Using the OIE classification, countries were grouped into high BSE incidence countries, low BSE incidence countries, and countries with no BSE cases. Countries with a low BSE incidence were subdivided into (a) countries with a low BSE incidence in domestic cattle, and (b) countries with a low BSE incidence only due to the importation of live cattle from the United Kingdom. Furthermore, countries with officially no BSE cases were subdivided into (a) countries with BSE notification, and (b) countries without BSE notification. Concerning the occurrence of scrapie, countries were, similarly to the BSE status, distinguished between countries with a high scrapie incidence (disease endemic), a low scrapie incidence (disease sporadic), and very low scrapie incidence (disease rare). Countries in which scrapie had never been recorded were subdivided into (a) countries with or (b) without scrapie notification.

Results

To date, BSE was diagnosed in domestic animals in five out of 26 European countries—United Kingdom, Ireland, Portugal, Switzerland, and France. Furthermore, Denmark, Germany, and Italy reported BSE cases in imported United Kingdom cattle, as did Portugal and Ireland beside the occurrence in domestic cattle (Table 29.1). Data currently available on the following BSE risk factors are summarized in Tables 29.1, 29.2, and 29.3, and in Figure 29.1.

Risk Factors

Feeding of Meat and Bone Meal (MBM) and/or Meat Meal (MM)
to Cattle

The potential BSE risk of feeding MBM and MM from insufficiently heated offal contaminated by scrapie-like agents was first identified in 1988 (10, 11). Furthermore, in 1994 it was suggested that only a relative low dose

TABLE 29.1. The occurrence of BSE in European countries (Feb. 1995) and risk factors related to cattle feeding, and importation of animal waste-derived protein and live cattle.

Country	BSE occurrence[a]		Feeding cattle with animal waster–derived protein[b]	Animal waste– derived United Kingdom protein imported (tons)[c]	Live United Kingdom cattle imported annually[d]	
Albania	(−)		NA	0	NA	
Austria	−		+ 1994*	0	NA	
Belgium	−		+ 1994*	10,797 incl. Lux.	NA	
Bulgaria	(−)		NA	0	NA	
Czech/Slovak	−		NA	0	NA	
Denmark	(+)	1992	+ 1990	81	150	1990
Finland	−		+ 1989[e]/1994*	3	24	1985–1988
France	+	1991	+ 1991	42,196	NA	
Germany	(+)	1993	+ 1994*	1,899	NA	
Greece	−		+ 1994*	0	NA	
Hungary	(−)		NA	0	NA	
Iceland	−		+ 1973	0	0	1980–1994
Ireland	+/(+)	1989	+ 1990	8,367	1,785	1986–1989
Italy	(+)	1994	+ 1994*	130	NA	
Luxembourg	−		NA/1994*	See Belgium	NA	
Malta	(−)		NA	0	NA	
The Netherlands	−		NA/1994*	12,945	NA	
Norway	−		+ 1991	78	NA	
Poland	(−)		NA	0	NA	
Portugal	+/(+)	1990	+ 1994*	590	1,333	1981–1989
Romania	−		NA	0	NA	
Spain	−		+ 1994*	1,029	395	1980–1989
Sweden	−		+ 1991	87	27	1987–1988
Switzerland	+	1990	+ 1990	20	0	1987–1991
United Kingdom	+ +	1986	+ 1988			
Yugoslavia	−		NA	0	NA	

[a] + +, high incidence; +, low incidence; (+), low incidence due to imported cases; −, no cases, notification is mandatory; (−), no cases recorded but notification is not mandatory. Year of the first occurrence is indicated.

[b] +, feeding of animal waste–derived protein to cattle was permitted. The year of the feed ban for animal waste–derived protein to cattle is indicated. NA, data not available.

[c] Direct importations of United Kindgom animal waste–derived protein from 1985 to 1989. Source: UK Exports General Trade (9).

[d] Average number of cattle imported annually. Year or period of which data were available on cattle importation is indicated.

[e] In Finland, imported MBM was not allowed to use in feed for ruminants since 1989. Sources (a–b and d–e): EU and national veterinary offices, written communication; *EU Directive (13).

exposure of MBM or MM is necessary for oral infection of cattle (12). Since it was difficult to separate abattoir waste from different species slaughtered, MBM and MM from the United Kingdom was generally considered as potentially contaminated in this study.

TABLE 29.2. BSE risk factors related to total sheep vs. cattle population of European countries in 1985/1993.

Country	Sheep/Cattle (1985)[a]	Sheep/Cattle (1993)[b]	Change of ratio (%)[c] (1985/1993)	Population of sheep increased
Albania	2.016	2.667	32.25	Yes
Austria	0.082	0.122	47.94	Yes
Belgium-Lux.	0.042	0.050	20.49	Yes
Bulgaria	5.997	5.448	−9.15	No
Czech/Slovak	0.207	0.209	0.89	No
Denmark	0.027	0.055	105.35	Yes
Finland	0.043	0.043	1.48	No
France	0.469	0.521	11.13	Yes
Germany	0.178	0.166	−6.49	No
Greece	11.390	13.88	21.84	Yes
Hungary	1.485	1.187	−20.06	No
Iceland	10.548	6.551	−37.89	No
Ireland	0.461	0.961	108.49	Yes
Italy	1.017	1.333	31.08	Yes
Luxembourg[d]	—	—	—	—
Malta	0.357	0.273	−23.64	Yes
The Netherlands	0.155	0.372	139.94	Yes
Norway	2.490	2.270	−8.82	No
Poland	0.438	0.366	−16.43	No
Portugal	4.305	4.126	−4.17	Yes
Romania	2.744	2.613	−4.78	No
Spain	3.499	4.864	39.00	Yes
Sweden	0.232	0.245	5.85	No
Switzerland	0.185	0.224	20.64	Yes
United Kingdom	1.844	2.486	34.83	Yes
Yugoslavia	1.477			

[a] Ratio of total sheep versus total cattle population in 1985. Source: FAO (4).
[b] Ratio of total sheep versus total cattle population in 1993. Source: FAO-OIE-WHO (7).
[c] The formula $[(b − a)/a] \times 100$ was used to calculate whether the ratio between the sheep and cattle population increased or decreased from 1985 to 1993; the year 1985 was taken as the baseline index.
[d] See Belgium-Lux.; the figures indicated are for both countries together.

Sixteen out of 26 countries, including those countries that recorded BSE cases in domestic cattle, permitted the use of MBM or MM in cattle feed concentrates at some time. No data on the former or current use of MBM and MM was available for ten or eight countries, respectively (Table 29.1).

Iceland (in 1973); the United Kingdom (in 1988); Ireland, Switzerland, and Denmark (in 1990); France, Norway, and Sweden (in 1991); and the European Union (EU) (in 1994) prohibited the use of MBM and MM in cattle feeding (13).

TABLE 29.3. Summary of the scrapie status in European countries from 1980 to 1993.

Country[a]	Scrapie status[b]				
	1993	1991	1990	1985	1980
Albania			0	−	NA
Austria	−	−	−	−	−
Belgium*	1992	−	−	0	(−)
Bulgaria	0	0	0	NA	−
Czech/Slovak*	(+)	(+)	−	−	−
Denmark	0	0	0	0	0
Finland	0	0	0	0	−
France		+	+	+	(+)
Germany*	1990		0	0	−
Greece	0	0	0	0	0
Hungary	0	0	0	0	(−)
Iceland	(+)		+ ( )	+ ( )	+ ( )
Ireland	+	+	+	+	+
Italy	−	NA	NA	NA	(+)
Luxembourg	0	0	0	0	−
Malta	−	−	−	−	−
The Netherlands	?	?	?	+	(+)
Norway*	(+)	(+)	−	+	
Poland	0	0	0	0	−
Portugal	0		−	−	−
Romania	0	0	0	0	−
Spain (*)	0 (*)	NA	−	0 (1984)	0
Sweden*	1986	1986	1986	0	−
Switzerland*	(+)	(+)	?	?	−
United Kingdom	++	+	+	+	+
Yugoslavia*		NA	NA	+	−

[a] *Countries that recorded new scrapie outbreaks since 1980. (*): In contrast to OIE records scrapie was diagnosed in Spain in 1984. Source: Perote and Wilesmith (19).
[b] ++, disease endemic; +, disease present; (+), exceptional occurrence; −, disease not reported; (−), disease probably not present; 0, disease never reported; year, time of last occurrence; ?, disease suspected but not confirmed; ( ), disease confined to certain regions; NA, no data available.
Sources: OIE (5) and Bradley and Matthews (6).

Animal Waste–Derived Protein Imported from the United Kingdom

The importation of United Kingdom animal waste–derived protein, i.e., MBM and MM, is considered as a BSE risk factor (14). Assuming a homogeneous distribution of scrapie-like particles in British MBM or MM between 1985 and 1989, this potential BSE risk has increased proportionally per weight or volume unit imported. Furthermore, taking into consideration the long BSE incubation period (15), cattle exposed to British MBM imported from 1985 to 1989 must still be regarded as a part of the population at risk.

Out of 25 countries, 14 had imported MBM and/or MM from the United Kingdom. Belgium and Luxembourg, the Netherlands, France, and Ireland imported large quantities of MBM and MM. Out of these five countries

FIGURE 29.1. Scrapie occurrence in Europe from 1980 to 1993: endemic (■); sporadic (■); rare (■); not prevalent, no notification, and no control program existing (■); not prevalent, notification and control program existing (□); unshaded countries were not compared.

only France and Ireland recorded BSE cases in domestic cattle, together with Switzerland and Portugal. The latter two countries, however, imported only minor quantities of MBM and MM (Table 29.1).

Live Cattle Importation from the United Kingdom

Because five European countries experienced BSE cases in imported animals from the United Kingdom, importation of cattle from this country is considered a BSE risk factor. In Ireland, for example, out of 78 BSE cases (as of May 1994) 10 involved animals imported from the United Kingdom (16). It was previously suggested that British cattle were infected with the scrapie-like agent starting in 1981 (11), and ending not before 1991, when some cattle in the United Kingdom seem to have been exposed to BSE-contaminated MBM, despite the feed ban of 1988 (17).

At least nine European countries imported live animals from the United Kingdom within the indicated periods (Table 29.1). The list includes only four of five countries where BSE occurred in imported United Kingdom animals because no data was available on Italy. The largest numbers were imported from the United Kingdom by Ireland (7,138 United Kingdom cattle from 1986 to 1989) (16) and Portugal (12,000 United Kingdom cattle during the period from 1981 to 1989) (18). These two countries recorded ten and six BSE cases, respectively, in imported animals. Iceland and Switzerland imported no United Kingdom live cattle imports in the periods indicated (Table 29.1).

Sheep to Cattle Ratio

The larger a scrapie-affected sheep population is in relation to the size of a cattle population, the more likely it is that offal from domestic sheep could become the source of an indigenous BSE infection (15). In 1985 and in 1993 the sheep population was larger relative to the cattle population in 11 out of 26 countries. Among BSE-affected countries, this ratio was >1 in the United Kingdom, Portugal, and Italy in 1985 and 1993, respectively (Table 29.2).

In countries where the sheep population was outgrowing the cattle population, it could become even more likely for the latter species to be exposed to offal-derived meal originating from domestic sheep than in the other countries. From 1985 to 1993, in 12 out of 26 countries this ratio was clearly increasing from approximately 10% to 140%; the United Kingdom, France, Ireland, and Switzerland, but not Portugal, i.e., countries with BSE cases occurred in domestic animals, are included in the 12 countries (Table 29.2). Moreover, movement of live sheep is more likely to increase due to market activities if a country's sheep population is growing. From 1985 to 1993 this was the case in 15 out of 26 countries, including seven of eight countries where BSE occurred (excluding Germany) (Table 29.2).

Scrapie

Scrapie is thought to be the origin disease of BSE (11, 15). Out of 26 countries, 15 reported in 1980 and in 1985 that they did not diagnose scrapie. Seventeen countries stated this in 1990, whereas in 1991 and 1993 only 11 and 12 countries, respectively, stated that scrapie has not been recorded (Table 29.3).

Over the examined 14-year period, in one country (the United Kingdom) scrapie was endemic; in six countries there were sporadic outbreaks (in France, Iceland, Ireland, Norway, the Netherlands, and the former Yugoslavia); and in seven countries scrapie occurred rarely (in Belgium, the Czech/Slovak Republic, Germany, Italy, Spain, Sweden, and Switzerland). In all other countries the disease was not diagnosed; however, in Albania, Bulgaria, Malta, and Poland scrapie is not notifiable (Fig. 29.1).

Spain (in 1984) (19), the former Yugoslavia and Norway (in 1985), Sweden (in 1986), Germany (in 1990), Switzerland and the Czech/Slovak Republic (in 1991), and Belgium (in 1992) began to report scrapie cases. Some of these countries, including Switzerland, confirmed the disease for the first time (Table 29.3).

Of all eight European countries that experienced BSE outbreaks, only Portugal and Denmark never recorded scrapie cases during the 14-year period, whereas the other countries had at least a single scrapie outbreak from 1980 to 1993 (Table 29.3).

Minimum Offal Rendering Requirements

Ruminant-derived offal contaminated by scrapie-like agents that was rendered under insufficient conditions in regard to temperature and duration of the heating process was identified as a BSE risk factor in 1988 and confirmed in 1991 (10, 11). The recommended autoclaving conditions for inactivation of Creutzfeldt-Jakob disease (CJD) is a porous-load cycle at a minimum of 134°C for 18 minutes. The condition of 133°C for 20 minutes, approximately the equivalent of 134°C for 18 minutes, has now been largely adopted as an appropriate decontamination standard for scrapie-like agents generally (20).

Discussion

In this study, data on risk factors in BSE-affected and BSE-free European countries were analyzed. Reliable data were not available for all 26 countries or on all risk factors. However, of the 10 countries on whom data on all parameters were available (Denmark, Finland, France, Germany, Iceland, Ireland, the Netherlands, Spain, Sweden, and Switzerland), none had the same combination of risk factors found in the United Kingdom, i.e., (a) feeding cattle with MBM or MM, (b) a large sheep-to-cattle population ratio, (c) a high scrapie prevalence, and (d) insufficient heating of high-risk abattoir waste (15).

Importation and Cattle Feeding

For other countries than the United Kingdom the importation of British live cattle or the risk factor combination of (a) importation of ruminant-derived protein from the United Kingdom and (b) feeding of ruminants with such material seemed to be most important. For France, Ireland, and Switzerland this is a confirmation of previous studies (14, 16). The data obtained on importation of animal waste-derived protein from the United Kingdom were limited to direct importations only. However, the scrapie-like agent might also have entered a country via indirect routes (14). Any country with considerable importations of MBM and MM from the United Kingdom might have been an intermediate trading station reexporting the material to further countries. MBM could have been passed on undiluted or diluted with domestic animal waste-derived protein originating from the intermediate country. This additional risk factor could have been of major importance for some countries and should be examined in more details in the future.

The importation-related BSE risk factors are currently decreasing due to control measures enforced by risk management of many importing countries. In addition, live cattle importation-related risks are markedly decreasing due to declining infection rate of United Kingdom cattle since 1988 (17). Implementation of the feed ban of ruminant-derived proteins seems to be an efficient disease control element because interruption of the BSE infection chain leads to a decrease of the epidemic curve as it has been shown in the United Kingdom (21). Therefore, veterinary officials of the EU (13) and of some nonmember countries have enforced a BSE control and prevention program that prohibits the use of ruminant-derived proteins in ruminant feeds. Consequently, the feeding-related risk factor becomes also less important.

Scrapie and Sheep Population

Sheep populations are growing in many countries and movement of live sheep is likely to increase, particularly within the EU (free trade). Consequently, and in contrast to the risk factors discussed above, the potential risk of disease spread from a scrapie-affected to an as yet scrapie-free country is potentially increasing. In the examined period, eight countries reported new or even the first scrapie outbreaks in history.

In this context, a large and increasing sheep population may remain a risk factor, that is, if it occurs in combination with scrapie (15). However, and in contrast to the United Kingdom where scrapie prevalence is high (22) and the incidence seems to be increasing (23), scrapie prevalence appears to be low in other European countries, e.g., in the Netherlands (24), Spain (25), and Italy or Switzerland (2). Nonetheless, we support previous suggestions (26) that scrapie eradication should be pursued as a long-term goal by the international community to prevent new scrapie-related transmissible spongiform encephalopathies in other species. So far, scrapie has become a notifiable disease in the EU, as of January 1993 and in some nonmember states.

Acknowledgments. The authors are grateful to Laurent Audigé and Andrea Vicari for their critical comments on the manuscript.

References

1. Wells GAH, McGill IS. Recently described scrapie-like encephalopathies of animals: case definitions. In: Bradley R, Savey M, Marchant B, eds. Subacute spongiform encephalopathies. Proceedings of a seminar ECSC, EEC, EAEC, Brussels and Luxembourg, 1990. Dordrecht, The Netherlands: Kluwer Academic Publishers, 1991:11-24.
2. Hornlimann B, Guidon D. Bovine spongiform encephalopathy (BSE): epidemiology in Switzerland. In: Bradley R, Marchant B, eds. Transmissible spongiform encephalopathies. Proceedings of a consultation on BSE with the scientific Committee of the European Communities, 14–15 September 1993, Brussels, 1994:13-24.
3. Hornlimann B, Braun U. Bovine spongiform encephalopathy (BSE): Clinical signs in Swiss BSE cases. In: Bradley R, Marchant B, eds. Transmissible spongiform encephalopathies. Proceedings of a consultation on BSE with the scientific Committee of the European Communities, 14–15 September 1993, Brussels, 1994:289-99.
4. FAO. Yearbook Production Vol. 41, 1987.
5. OIE. Santé Animal Mondiale en 1993, 1993.
6. Bradley R, Matthews D. Sub-acute, transmissible spongiform encephalopathies: current concepts and future needs. In: Bradley R, Matthews D, eds. Transmissible spongiform encephalopathies of animals. Rev Sci Tech OIE 1992; 11(2):605-34.
7. FAO-OIE-WHO. Yearbook Production, Vol. 41, 1993.
8. Bradley R, Marchant B, eds. Transmissible spongiform encephalopathies. Proceedings of a consultation on BSE with the scientific Committee of the European Communities, 14–15 September 1993, Brussels, 1994.
9. UK Exports General Trade (COD). British Export Statistics, Vol. 1985-1989.
10. Wilesmith JW, Wells GAH, Cranwell MP, Ryan JBM. Bovine spongiform encephalopathy: epidemiological studies. Vet Rec 1988;123:638-44.
11. Wilesmith JW, Ryan JBM, Atkinson MJ. Bovine spongiform encephalopathy: epidemiological studies in the origin. Vet Rec 1991;128:199-203.
12. Kimberlin RH, Wilesmith JW. Bovine spongiform encephalopathy: epidemiology, low dose exposure and risks. Ann NY Acad Sci 1994;724:210-20.
13. EU Directive 94/381, 1994.
14. Hornlimann B. Risk assessment on the importation of BSE. Dtsch Tieraerztl Wschr 1994;101:295-8.
15. Kimberlin RH. Bovine spongiform encephalopathy. In: Bradley R, Matthews D, eds. Transmissible spongiform encephalopathies of animals. Rev Sci Tech OIE 1992;11:347-90.
16. Costelloe JA. Update on the epidemiology, surveillance and risk assessment of BSE in Ireland. In: Bradley R, Marchant B, eds. Transmissible spongiform encephalopathies. Proceedings of a consultation on BSE with the scientific committee of the European Communities, 14–15 September 1993, Brussels, 1994:49-55.

17. Hoinville LJ, Wilesmith JW, Richards MS. An investigation of risk factors for cases of bovine spongiform encephalopathy born after the introduction of the "feed ban." Vet Rec 1995;136(13):312–8.
18. Telo JMR. Bovine spongiform encephalopathy: ten cases discovered in Portugal. Report of the meeting of the OIE ad hoc group on BSE, September 1994:27–8.
19. Perote M, Wilesmith JW. BSE risk assessment and surveillance in Spain. In: Bradley R, Marchant B, eds. Transmissible spongiform encephalopathies. Proceedings of a consultation on BSE with the scientific committee of the European Communities, 14–15 September 1993, Brussels, 1994:69–76.
20. Taylor DM. Deactivation of BSE and scrapie agents: rendering and other UK studies. In: Bradley R, Marchant B, eds. Transmissible spongiform encephalopathies. Proceedings of a consultation on BSE with the scientific committee of the European Communities, 14–15 September 1993, Brussels, 1994:205–223.
21. Bovine spongiform encephalopathy in Great Britain. A progress report from MAFF, November 1994.
22. Morgan KL, Nicholas K, Glover MJ, Hall AP. A questionnaire survey of the prevalence of scrapie in sheep in Britain. Vet Rec 1990;27:373–6.
23. Anonymous. Animal Pharm, March 25, 1994;297:2.
24. Schreuder BEC, De Jong MCM, Pekelder JJ, Vellema P, Bröker AJM, Betcke H. BSE risk assessment in the Netherlands: prevalence and incidence of scrapie. In: Bradley R, Marchant B, eds. Transmissible spongiform encephalopathies. Proceedings of a consultation on BSE with the scientific committee of the European Communities, 14–15 September 1993, Brussels, 1994:69–76.
25. Garcia del Jalon JA, De las Heras M, Ferrer L. Evolution de la enfermedad del prurigo lumbar (scrapie) en Aragon. Med Vet 1991;8:241–4.
26. Detwiler LA. Scrapie. In: Bradley R, Matthews D, eds. Transmissible spongiform encephalopathies of animals. Rev Sci Tech OIE 1992;11(2):491–537.

Author Index

Subject Index

PrP gene alleles associated with, 294–303
replication versus PrP-res accumulation, in hamster brain, 277–278
in sheep, from vaccine, 204
sheep and goats as natural hosts for, 13
and sheep population, 392
spongiform changes in, 122–123
  precallosal frontal cortex, sheep example, 132
in transgenic and nontransgenic mice, incubation times, 206–207
transmission of
  from sheep to cattle, United States, 159
  vertical and horizontal, 59–80
Scrapie amyloid. *See* Prion protein
Scrapie-associated fibrils (SAFs), 63
Scrapie Control Program, lack of participation in, 117
Scrapie Eradication Program, 59
Scrapie incubation-period (SIP) gene, 73
Scrapie-like diseases, species differences in pathogenesis, 39
Scrapie strains, 214–215
  87A, for investigation of glycation and amyloidogenesis, 241–243
  139A
    for investigating glycation and amyloidogenesis, 234
    for investigating infectivity after glycation, 244
    responses to SDS, 2–ME, heat, and PK, 238–239
  139H
    effect of amphotericin B in, 278
    properties in species of hamster, 273
  263K
    for investigating glycation and amyloidogenesis, 234, 241–243
    for investigating infectivity after glycation, 244
    properties in species of hamster, 273
    responses to SDS, 2–ME, heat,

and PK, 238–239
    response to amphotericin B, 272
  C506, response to amphotericin B, 272
  ME7, for investigation of glycation and amyloidogenesis, 242–243
  susceptibility to amphotericin B, 272
  susceptibility to inactivation, and to denaturation, 232
  See also Mutations
Selective breeding, of sheep from scrapie affected flock members, 68–69
Sensitivity, of bioassay versus immunoblot, for detecting scrapie agent, 322
Sequencing, of the coding region of PrP allelic variants, 296
Serial passage, of scrapie infectious agent, in hamsters and mice, 87–89
Servicio Nacional de Sanidad Animal (SENASA), surveillance for bovine spongiform encephalopathy, 145
Sex ratio, in Creutzfeldt-Jakob disease, 370
Sheep, ratio to cattle, and risk of bovine spongiform encephalopathy, 390
Sinc gene, 160–161, 251
  influence on incubation time, prion diseases, 215
  influence on mouse scrapie incubation time, 193
  linkage to PrP, 216
Sip gene, control of scrapie in sheep by, 160–161
Slaughter, practices in Argentina, 141, 143
Slow virus diseases, 202
Sodium dodecyl sulfate, treatment of PrP^{Sc} with, 234–235, 236–237
Solid state aggregates, amyloids as, 182–183
Solvent memory effect, in proteins, 187
Solvents, effect on conformation of prion protein, 187

PROCEEDINGS IN THE SERONO SYMPOSIA USA SERIES

Continued from page ii